13th International Workshop on Health Text Mining and Information Analysis (LOUHI 2022)

Abu Dhabi, United Arab Emirates and Online
7 December 2022

ISBN: 978-1-7138-9112-3

Printed from e-media with permission by:

Curran Associates, Inc.
57 Morehouse Lane
Red Hook, NY 12571

proceedings.com

Some format issues inherent in the e-media version may also appear in this print version.

Association for Computational Linguistics
209 N. Eighth Street
Stroudsburg, Pennsylvania 18360

Phone: 1-570-476-8006
Fax: 1-570-476-0860

acl@aclweb.org

Additional copies of this publication are available from:

Curran Associates, Inc.
57 Morehouse Lane
Red Hook, NY 12571 USA
Phone: 845-758-0400
Fax: 845-758-2633
Email: curran@proceedings.com
Web: www.proceedings.com

LOUHI 2022

13th International Workshop on Health Text Mining and Information Analysis

Proceedings of the Workshop

December 7, 2022

Introduction

The International Workshop on Health Text Mining and Information Analysis (LOUHI) provides an interdisciplinary forum for researchers interested in automated processing of health documents. Health documents encompass electronic health records, clinical guidelines, spontaneous reports for pharmacovigilance, biomedical literature, health forums/blogs or any other type of health-related documents. The LOUHI workshop series fosters interactions between the Computational Linguistics, Medical Informatics and Artificial Intelligence communities. The 12 previous editions of the workshop were co-located with SMBM 2008 in Turku, Finland, with NAACL 2010 in Los Angeles, California, with Artificial Intelligence in Medicine (AIME 2011) in Bled, Slovenia, during NICTA Techfest 2013 in Sydney, Australia, co-located with EACL 2014 in Gothenburg, Sweden, with EMNLP 2015 in Lisbon, Portugal, with EMNLP 2016 in Austin, Texas; in 2017 was held in Sydney, Australia; in 2018 was co-located with EMNLP 2018 in Brussels, Belgium; in 2019 was co-located with EMNLP 2019 in Hong Kong; in 2020 was co-located with EMNLP 2020 and took place online due to the COVID-19 pandemics; and in 2021 was co-located with EACL 2021 and took place online due to the persistence of the COVID-19 pandemics. This year the workshop is co-located with EMNLP 2022 and takes place with a hybrid modality.

The aim of the LOUHI 2022 workshop is to bring together research work on topics related to health documents, particularly emphasizing multidisciplinary aspects of health documentation and the interplay between nursing and medical sciences, information systems, computational linguistics and computer science. The topics include, but are not limited to, the following Natural Language Processing techniques and related areas:

- Techniques supporting information extraction, e.g. named entity recognition, negation and uncertainty detection
- Classification and text mining applications (e.g. diagnostic classifications such as ICD-10 and nursing intensity scores) and problems (e.g. handling of unbalanced data sets)
- Text representation, including dealing with data sparsity and dimensionality issues
- Domain adaptation, e.g. adaptation of standard NLP tools (incl. tokenizers, PoS-taggers, etc) to the medical domain
- Information fusion, i.e. integrating data from various sources, e.g. structured and narrative documentation
- Unsupervised methods, including distributional semantics
- Evaluation, gold/reference standard construction and annotation
- Syntactic, semantic and pragmatic analysis of health documents
- Anonymization/de-identification of health records and ethics
- Supporting the development of medical terminologies and ontologies
- Individualization of content, consumer health vocabularies, summarization and simplification of text
- NLP for supporting documentation and decision making practices
- Predictive modeling of adverse events, e.g. adverse drug events and hospital acquired infections
- Terminology and information model standards (SNOMED CT, FHIR) for health text mining

- Bridging gaps between formal ontology and biomedical NLP

The call for papers encouraged authors to submit papers describing substantial and completed work but also focus on a contribution, a negative result, a software package or work in progress. We also encouraged to report work on low-resourced languages, addressing the challenges of data sparsity and language characteristic diversity.
This year we received 56 submissions. Each submission went through a double-blind review process which involved three program committee members. Based on comments and rankings supplied by the reviewers, we accepted 25 papers. The selection was entirely based on the scores provided by the reviewers. The overall acceptance rate is 45%.
Our special thanks go to Tim Baldwin for accepting to give an invited talk.
Finally, we would like to thank the members of the program committee for providing balanced reviews in a very short period of time, and the authors for their submissions and the quality of their work.

Organizing Committee

Organizers

Alberto Lavelli, FBK, Trento, Italy
Eben Holderness, Brandeis University, USA
Antonio Jimeno Yepes, RMIT University, Australia
Anne-Lyse Minard, LLL, CNRS, University of Orléans, France
James Pustejovsky, Brandeis University, USA
Fabio Rinaldi, IDSIA, University of Zurich, Switzerland, and FBK, Trento, Italy

Program Committee

Reviewers

Rafael Berlanga Llavori

Leonardo Campillos Llanos, Francisco M. Couto

Hercules Dalianis

Natalia Grabar, Cyril Grouin

Thierry Hamon, Eben Holderness

Antonio Jimeno Yepes

Yoshinobu Kano

Alberto Lavelli, Analia Lourenco

David Martinez, Sérgio Matos, Timothy Miller, Anne-Lyse Minard, Hans Moen, Diego Molla, Roser Morante, Danielle L Mowery

Aakanksha Naik, Mariana Lara Neves, Aurélie Névéol

Jong C. Park, Laura Plaza, James Pustejovsky

Fabio Rinaldi, Thomas Brox Røst

Tapio Salakoski, Maria Skeppstedt, Amber Stubbs, Hanna Suominen

Suzanne Tamang

Pierre Zweigenbaum

Keynote Talk: Deep Phonology: Analysing Antimicrobial Stewardship in Veterinary Clinics through NLP

Tim Baldwin
Mohamed bin Zayed University of Artificial Intelligence, UAE

Abstract: Antimicrobial stewarship refers to guidelines on the appropriate use of antimicrobials to optimise patient health and minimise microbial resistance. In this talk, I will present work on the large-scale analysis of veterinary clinical records to perform fine-grained analysis to aid in the implementation and monitoring of antimicrobial stewardship programmes in Australia.

Bio: Tim Baldwin is Associate Provost (Academic and Student Affairs) and Head of the Department of Natural Language Processing, Mohamed bin Zayed University of Artificial Intelligence in addition to being a Melbourne Laureate Professor in the School of Computing and Information Systems, The University of Melbourne. His primary research focus is on natural language processing (NLP), including social media analytics, deep learning, and computational social science.
Tim completed a BSc(CS/Maths) and BA(Linguistics/Japanese) at The University of Melbourne in 1995, and an MEng(CS) and PhD(CS) at the Tokyo Institute of Technology in 1998 and 2001, respectively. Prior to joining The University of Melbourne in 2004, he was a Senior Research Engineer at the Center for the Study of Language and Information, Stanford University (2001-2004). His research has been funded by organisations including the Australia Research Council, Google, Microsoft, Xerox, ByteDance, SEEK, NTT, and Fujitsu, and has been featured in MIT Tech Review, IEEE Spectrum, The Times, ABC News, The Age/Sydney Morning Herald, Australian Financial Review, and The Australian. He is the author of well over 400 peer-reviewed publications across diverse topics in natural language processing and AI, with around 20,000 citations and an h-index of 66 (Google Scholar), in addition to being an ARC Future Fellow, and the recipient of a number of awards at top conferences.

Table of Contents

Program

Wednesday, December 7, 2022

09:00 - 09:10 *Opening Remarks*

09:10 - 10:00 *Invited Talk*

10:00 - 10:30 *TBD*

Can Current Explainability Help Provide References in Clinical Notes to Support Humans Annotate Medical Codes?
Byung-Hak Kim, Zhongfen Deng, Philip Yu and Varun Ganapathi

10:30 - 11:00 *Coffee Break*

11:00 - 12:30 *Session 2*

Assessing the Limits of Straightforward Models for Nested Named Entity Recognition in Spanish Clinical Narratives
Matias Rojas, Casimiro Pio Carrino, Aitor Gonzalez-Agirre, Jocelyn Dunstan and Marta Villegas

A Quantitative and Qualitative Analysis of Schizophrenia Language
Amal Alqahtani, Efsun Sarioglu Kayi, Sardar Hamidian, Michael Compton and Mona Diab

Enriching Deep Learning with Frame Semantics for Empathy Classification in Medical Narrative Essays
Priyanka Dey and Roxana Girju

12:30 - 14:00 *Lunch Break*

14:00 - 15:30 *Session 3*

Exploring the Influence of Dialog Input Format for Unsupervised Clinical Questionnaire Filling
Farnaz Ghassemi Toudeshki, Anna Liednikova, Philippe Jolivet and Claire Gardent

DDI-MuG: Multi-aspect Graphs for Drug-Drug Interaction Extraction
Jie Yang, Yihao Ding, Siqu Long, Josiah Poon and Soyeon Caren Han

Wednesday, December 7, 2022 (continued)

Divide and Conquer: An Extreme Multi-Label Classification Approach for Coding Diseases and Procedures in Spanish
Jose Barros, Matias Rojas, Jocelyn Dunstan and Andres Abeliuk

15:30 - 16:00 *Coffee Break*

16:00 - 17:15 *Session 4 (Poster Session)*

17:15 - 17:30 *Mini Break*

17:30 - 19:00 *Session 5*

Integration of Heterogeneous Knowledge Sources for Biomedical Text Processing
Parsa Bagherzadeh and Sabine Bergler

How Long Is Enough? Exploring the Optimal Intervals of Long-Range Clinical Note Language Modeling
Samuel Cahyawijaya, Bryan Wilie, Holy Lovenia, Huan Zhong, MingQian Zhong, Yuk-Yu Nancy Ip and Pascale Fung

Proxy-based Zero-Shot Entity Linking by Effective Candidate Retrieval
Maciej Wiatrak, Eirini Arvaniti, Angus Brayne, Jonas Vetterle and Aaron Sim

Exploring the Influence of Dialog Input Format for Unsupervised Clinical Questionnaire Filling

Farnaz Ghassemi Toudeshki[&,#], **Anna Liednikova**[&,†], **Philippe Jolivet**[&], **Claire Gardent**[γ]
[&] ALIAE, [#] IDMC, Université de Lorraine
[†] Loria, Université de Lorraine, [γ] CNRS
{farnaz.ghassemi, anna.liednikova, philippe.jolivet}@aliae.io
claire.gardent@loria.fr

Abstract

In the medical field, we have seen the emergence of health-bots that interact with patients to gather data and track their state. One of the downstream application is automatic questionnaire filling, where the content of the dialog is used to automatically fill a pre-defined medical questionnaire. Previous work has shown that answering questions from the dialog context can successfully be cast as a Natural Language Inference (NLI) task and therefore benefit from current pre-trained NLI models. However, NLI models have mostly been trained on text rather than dialogs, which may have an influence on their performance. In this paper, we study the influence of content transformation and content selection on the questionnaire filling task. Our results demonstrate that dialog pre-processing can significantly improve the performance of zero-shot questionnaire filling models which take health-bots dialogs as input.

1 Introduction

Work on Question Answering (QA) and Machine Reading Comprehension (MRC) mostly focuses on wh-questions of arbitrary types (who, what, where etc.) whose answer can be found in text. The answer can be extractive where a short span of the text is identified as the answer (Pearce et al., 2021) or it can be abstractive where a free-form answer is generated from the question and some support document (Bauer et al., 2018).

Here, we focus instead on a QA setting where questions are restricted to polar (yes/no) and Agreement Likert Scale (ALS) questions and where answers are contained in a dialog rather than a paragraph text. As illustrated in Figure 1, this setting is useful for automatic questionnaire filling (AQF) in the medical field. Given a dialog between a patient and a health bot, the goal of automatic questionnaire filling is to answer a set of predefined questions from a medical questionnaire (here the Pain

Dialog
bot: What is the most difficult for you about your sleep ? **patient:** I have back pain that prevents me from sleeping. **bot:** I'm sorry to hear that. How long have you had back pain? **patient:** Since I've been working out, I've had constant back pain at night. **bot:** Do you think pain can last for long? **patient:** I think it will stop once I stop playing sports. **bot:** Should we let time fix the pain? **patient:** My doctor thinks that I need to get used to doing sports and that the pain will disappear after a while.
Questionnaire
(1) My pain is a temporary problem in my life. CQ: ☐No ■**Yes** ☐NA ALS: ☐Totally disagree ☐Rather disagree ☐Agree ■**Totally agree** ☐NA

Figure 1: An example of a dialog and a question from the PBPI Questionnaire, answered in CQ and ALS format

Beliefs and Perceptions Inventory (PBPI) questionnaire (Williams and Thorn, 1989)) based on the dialog content.

In previous work, Toudeshki et al. (2021) compared three ways of deriving answers to questions from dialogs: Natural Language Inference, Question Answering and Text Classification. For polar and ALS questions, they found that Natural Language Inference (NLI) performs best. One possible limitation of their approach however is that they apply NLI models to dialogs while NLI models are trained on non-dialogic text.

In this paper, we propose different ways of transforming and selecting dialog content before applying NLI to answer questions, and we analyse the impact of these operations on NLI-based questionnaire filling. Our hypothesis is that transforming the input dialog into a format closer to the text format on which NLI models are trained, should help these models perform better. Our experimental results confirm this hypothesis: it demonstrates that, in a zero-shot setting, transforming and select-

Proceedings of the 13th International Workshop on Health Text Mining and Information Analysis (LOUHI), pages 1 - 13
December 7, 2022 ©2022 Association for Computational Linguistics

ing dialog content yields significant improvements over a baseline which takes the full dialog content as input.

2 Related work

We briefly situate our work with respect to three tasks which have similarities with Automatic Questionnaire Filling namely, Machine Reading Comprehension, Question Answering and Aspect-Based Sentiment Analysis (ABSA).

MRC/QA. Given a text and a question, MRC and QA models aim to derive the answer to that question from some input document (Zeng et al., 2020).

Similar to our approach, Ren et al. (2020) focus on filling in medical questionnaires consisting of polar questions about medical terms. However, in their case, the input to the model is a text (patient records) rather than a dialog. Furthermore, QA is modeled as a classification task which restricts the approach to a limited set of possible questions and answers. Finally, the questions are restricted to polar questions about terms whereas we consider polar and ALS questions about full sentences.

Recently, some work has focused on answering questions from dialogs rather than text. A simple approach for modeling a multi-turn dialog is to concatenate all turns (Zhang et al., 2019; Adiwardana et al., 2020). However, for retrieval-based response selection, Zhang et al. (2018); Yuan et al. (2019) showed that turns-aware aggregation methods can achieve a better understanding of dialogs compared to considering all turns equally . Similarly for MRC on dialogs, turns-aware approach have been proposed which select turns in the conversation that are related to the input question: Zhang et al. (2021) uses embedding-based similarity to select such turns while Li et al. (2020) uses a pre-trained language model fine-tuned on NLI tasks. Their results showed that eliminating irrelevant turns effectively improves results. Our work extends on this work showing that both content selection and content transformation help improve MRC on dialogs.

Aspect-Based Sentiment Analysis. Aspect based sentiment analysis (ABSA) is the process of determining sentiment polarity for a specific aspect in a given context. An aspect term is generally a word or a phrase which describes an aspect of an entity (Jiang et al., 2019). For instance, (Jang et al., 2021; Sun, 2022) investigate aspect-based sentiment analysis on user tweets related to COVID-19. While AQF could be viewed as an ABSA task where each item should be labelled with one of three (polar question) or five (ALS question) sentiment value (agree, disagree, etc.), two key differences between ABSA and AQF is that (i) labels apply to sentences rather than aspect terms and (ii) contrary to these terms, the questions used in medical questionnaire can be very similar semantically (e.g., "Is your pain constant?" "Is your pain a temporary problem?") making it harder to extract the correct answer from the input dialog.

Closest to our work, Toudeshki et al. (2021) showed that pre-trained NLI models can be used to fill in questionnaires from dialogs in a zero-shot setting. We depart from their work in that we propose different ways of transforming and selecting dialog content and investigate how this impact zero-shot, dialog-based, automatic questionnaire filling.

3 Automatic Questionnaire Filling (AQF)

Task. Given a dialog D and a questionnaire Q, the Automatic Questionnaire Filling task consists in providing an answer a_i for each question $q_i \in Q$.

We address the task in a zero-shot setting (no training data). For evaluation, we provide a test set consisting of 100 dialogs and their associated questions and answers.

Questionnaire. We consider two types of questions: Closed Questions (CQ) and Agreement Likert Scale (ALS) questions. CQ have three possible answers (yes, no or Not Applicable, i.e. the dialog does not address the question) and ALS has five (totally disagree, rather disagree, agree, totally agree, NA). As illustrated in Figure 1, questions are reformulated as declarative statements with multiple choice answers. With the emergence of health-bots, AQF can help transform human-bot dialogs into structured data which can be used by physicians to track patients condition. In particular, it can be used to fill in questionnaires such as the Pain Beliefs and Perceptions Inventory (PBPI, (Williams and Thorn, 1989)) questionnaire which includes 16 questions and is standardly used in the context of clinical studies.

Collecting dialogs that include information for all of these questions is a difficult task however. To facilitate data collection for the creation of the test set, we therefore decrease the number of questions

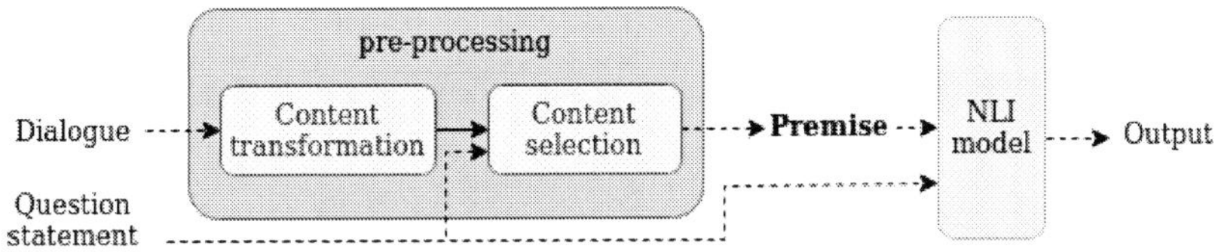

Figure 2: Dialog pre-processing schema

by selecting five questions out of sixteen. Because the questions in the PBPI are often very similar, and knowing the answer to one of them allows deriving the answer to others, we chose questions that are semantically distinct from one another. The list of all PBPI questions is given in Appendix C and the five selected questions are indicated in bold.

Test Data. To evaluate our approach, we create a test set of 100 dialogs and their associated question/answer pairs.

The creation of the test data involves first, collecting human-bot dialogs and second, extracting answers to the PBPI questions from the collected dialogs.

Collecting Dialogs. We collect the dialogs using the Amazon Mechanical Turk platform and asking Turkers to interact with the ComBot health bot (Liednikova et al., 2021) while behaving as if they had chronic pain issues. To avoid Turkers introducing the PBPI questions verbatim in the dialog, they were given a list of topics to be mentioned rather than the questions themselves (See details in Appendix D). In this way, we ensure that the collected dialogs address the questions to be answered while encouraging their diversified paraphrasing during the conversation. Turkers received bonuses each time they mention a topic. Turkers were also given the ability to modify the bot utterance in order to redirect the conversation more easily: they could reject the current candidate in which case, the turn with the next highest confidences score would be displayed by the bot. More information about Turkers payments is provided in the Ethic section (Sec. A). Details of the instructions given to the Turkers and a screenshot of the annotation interface are given in the Appendix.

Identifying Question Answers. Two annotators with good English proficiency were asked to select the correct answer for each of the five selected questions based on each of the 100 collected dialogs. We computed agreement between the two annotators on all Q/A pairs and all 100 dialogs. The Kappa score is 0.94 for CQ and 0.86 for ALS question type. Thereafter, we used adjudication to decide on the final answer for all cases where the two annotators disagreed. The annotators were the first two authors of this paper.

The final test corpus consists of 100 dialogs, each associated with 10 questions (5 yes/no questions and 5 ALS questions) and their answers. Dialog length varies from 4 to 70 turns and from 47 to 593 tokens, with 17.1 turns and 218.7 tokens on average.

4 Approach

Following Toudeshki et al. (2021), we model question answering as an NLI task where the premise is derived from the dialog, the hypothesis from the question and the answer from the NLI result. Given a question and a dialog, our model, illustrated in Figure 2, answers the question in three steps as follows.

Deriving an NLI Premise from the dialog. The NLI premise is derived from the input dialog using first, Content Transformation and second, Content Selection. As detailed in Section 5, we experiment with different ways of transforming and selecting content.

Deriving an NLI hypothesis from a question. To derive an NLI hypothesis from a question, we simply represent questions as statements (E.g., "I have pain regularly" instead of "Do you have pain regularly?"). Since the PBPI questionnaire questions are already in the form of a statement, we did not make any changes to them and used them as they are.

Deriving the answer. We use RoBERTa large (Liu et al., 2019) [1] fine-tuned on the MNLI dataset (Williams et al., 2018) to determine the entailment relation. We then derive the answer from the entailment relation between dialog and question as

[1] https://huggingface.co/roberta-large-mnli

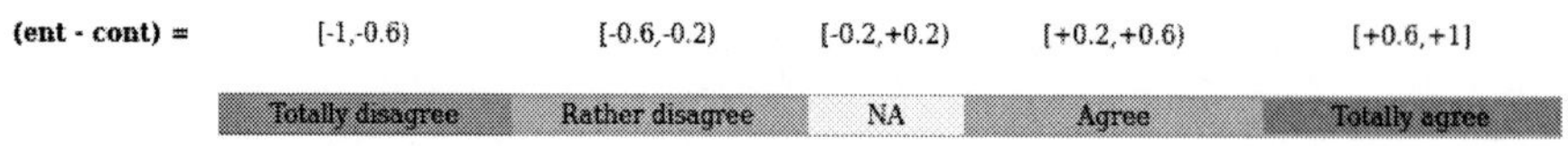

Figure 3: Map NLI scores to ALS answer types

follows.

For Close Questions, we set the answer to "Yes" if NLI returns an entailment, "No" if it returns a contradiction and "NA" if it returns "neutral".

For ALS questions, we map the NLI result to agreement choices as follows. If "neutral" has the highest score, the answer is "NA". Else, the contradiction score is subtracted from the entailment score. The subtraction result lies in a range of (-1,1) which is uniformly divided into 5 segments corresponding to the 5 ALS answer types, as shown in figure 3.

5 NLI-oriented Dialog Pre-processing

We consider different ways of transforming and selecting dialog content.

We also study the impact of the NLI model used, comparing DeBERTa, the model used in Toudeshki et al. (2021), with RoBERTa (Liu et al., 2019), the model used in our approach.

The DeBERTa model (He et al., 2020)[2] extends the BERT architecture with two innovative techniques: disentangled attention mechanism and an enhanced mask decoder. We compare AQF models with and without pre-processing and based on RoBERTa vs. DeBERTa, and find that whereas, when no pre-processing is applied, a DeBERTa model generally outperforms a RoBERTa-based model, the reverse is true when pre-processing is applied. This shows that while the improved DeBERTa-based, NLI model helps bridge the gap between dialog and text, explicit pre-processing still yields better results.

5.1 Content transformation

Null Transformation (CT_{null}) A null transformation baseline where we simply concatenate the turns of the input dialog. To encode the speaker information in each turn, the utterance is accompanied by the speaker role (patient/bot) at the beginning.

Summary (CT_{sum}) Pairs of adjacent turns are summarized, and the resulting summaries are concatenated. In this way, the input dialog is transformed into a sequence of two-turn summaries. We also tried summarizing the whole dialog in one go but found that applying summarization on each two turns rather than on the whole dialog gives better results.We use the **BART-large** model[3] (Lewis et al., 2020) fine-tuned on the News summarization corpus XSUM (Narayan et al., 2018) and on the dialog summarization corpus SAMSum (Gliwa et al., 2019). The model achieves ROGUE-L score of 0.44 on SAMSum test set [4].

Long Answers (CT_{answer}) In information seeking dialog, adjacent turns often are question-answer pairs. Based on this observation, we map each pair of adjacent turns in the dialog into a single declarative sentence assuming that the first turn is a question (e.g., "Which drug did you take?"), the second is a short answer to that question (e.g., "Doliprane") and the sentence derived from the mapping is a long answer to the question (e.g., "I took Doliprane"). To learn this mapping, we fine-tune T5 (Raffel et al., 2019), a pre-trained encoder-decoder model, on two datasets of (question, incomplete answer, full answer) triples, one for wh- and one for yes-no (YN) questions. For wh-questions, we use 3,300 entries of the dataset consisting of (question, answer, declarative answer sentence) triples gathered by Demszky et al. (2018) using Amazon Mechanical Turk workers. For YN questions, we used the SAMSum corpus, (Gliwa et al., 2019) which contains short dialogs in chit-chat format. We created 1,100 (question, answer, full answer) triples by automatically extracting YN (question, answer) pairs from this corpus and manually associating them with the corresponding declarative answer. Data was splitted into train and test (9:1) and the fine-tuned model achieved 0.90 ROUGE-L score on the test set.

This fine-tuned model was applied to each two subsequent turns of the input dialogs, and the resulting declarative sentences were then concatenated to

[2] `https://github.com/microsoft/DeBERTa`

[3] `https://huggingface.co/Salesforce/bart-large-xsum-samsum`

[4] `https://paperswithcode.com/sota/abstractive-text-summarization-on-samsum`

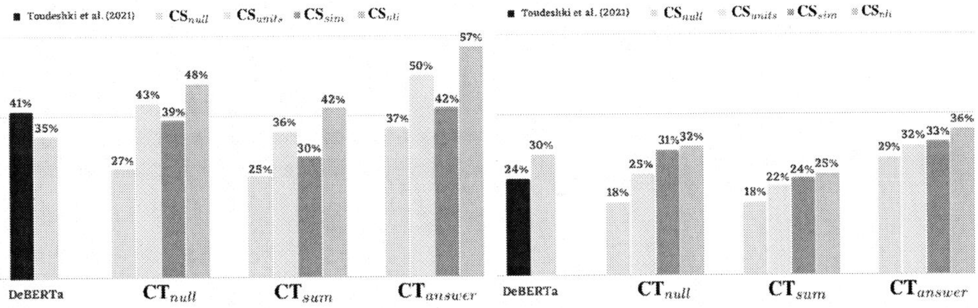

Figure 4: F1 macro average for Close Questions (on the left) and ALS questions (on the right) for the RoBERTa variant of our model. The two most left columns indicate the performance of (Toudeshki et al., 2021)'s model on their (dark blue) and our (light green) test set. The best results are obtained by the CT_{answer}, CS_{nli} model.

form the declarative transform of the whole dialog.

5.2 Content selection

The transformation operations described in the previous section yield sequences of dialog turns, two-turn summaries or full answers. We call these "input units" and consider three ways of pre-selecting the input units that will be used as premise when testing for entailment.

Null Content Selection (CS_{null}) A null content selection baseline where the premise is the concatenation of all the input units produced by the content transformation operations (dialog turns, sequence of two turn summaries, sequence of full form answers).

Unit-Based (CS_{units})**.** Each question is assessed against each input item. Given an input sequence I_n of length n, the answer a_i to a question q is then determined by aggregating the resulting entailment probabilities as follows:

- $a_i = NA$ if for all input items $i \in I_n$, the NA probability is highest.
- $a_i = Yes$ (resp. $a_i = No$) if for at least one item $i \in I_n$, the Yes (resp. No) probability is highest and the highest Yes (resp. No) probability is higher than the highest No (resp. Yes) probability.

Similarity (CS_{sim})**.** For each question q, we select a subset of input units that are semantically similar to q. We encode question and input units using SBERT[5] (Reimers and Gurevych, 2019) and compute *cosine similarity* for each (q, input unit) pair. We then select items whose similarity score is higher than 0.5, concatenate them and use the result as the NLI premise.

NLI (CS_{nli})**.** For each question q in the questionnaire, we select the input units that are related to q using the NLI model (RoBERTa-Large). Specifically, we select sentences which have an entailment or contradiction score higher than 0.5. All selected sentences are then concatenated to form the NLI premise.

5.3 Baseline and Comparison

Our baseline is the null method (CT_{null}+CS_{null}) i.e., the approach where question answering applies to the untransformed, unfiltered dialog. To compare our approach with Toudeshki et al. (2021), we also report the performance of their model on both their test set (10 dialogs) and on ours (100 dialogs).

6 Results

We evaluate our approach using macro and weighted F1 score.

6.1 How much does pre-processing help improve performance ?

Figure 4 shows the results for all combinations of our content transformation and selection methods[6].

Improvement over the baseline. Comparing our best model (CT_{answer}, CS_{nli}) with the no-preprocessing CT_{null}, CS_{null} baseline, we see (Figure 4) that pre-processing can multiply the

[5] `https://huggingface.co/sentence-transformers/paraphrase-distilroberta-base-v2`

[6] We first focus on the results of our RoBERTa based model and delay the comparison with DeBERTa based models to Section 6.4.

Two turns
bot: do you feel anxiety or stress during nights awakenings ? **patient:** I feel stressed during night awakenings although I am not feeling guilty about being in pain.
Generated summary Patient feels stressed during night awakenings although **he's not in pain**.

Table 1: An example of the summarization model performance on two subsequent turns, showing missing and **inconsistent** information in the output summary

macro and weighted F1 scores by two. The best pre-processing method combines a question+answer to sentence transformation (CT_{answer}) with the entailment-based content selection method (CS_{nli}).

Content transformation The CT_{answer} question+answer transform, which merges pairs of adjacent dialog turns into declarative statements, consistently yields the best results. A possible explanation is that this transform yields an input, a declarative sentence, which is consistent with the format of the training data used for NLI models.

Conversely, summarization (CT_{sum}) has the lowest performance. This could be due to errors such as hallucinations or omissions known to be produced by summarization systems (Zhao et al., 2020). Table 1 shows an example of such errors when applying the CS_{sum} transformation.

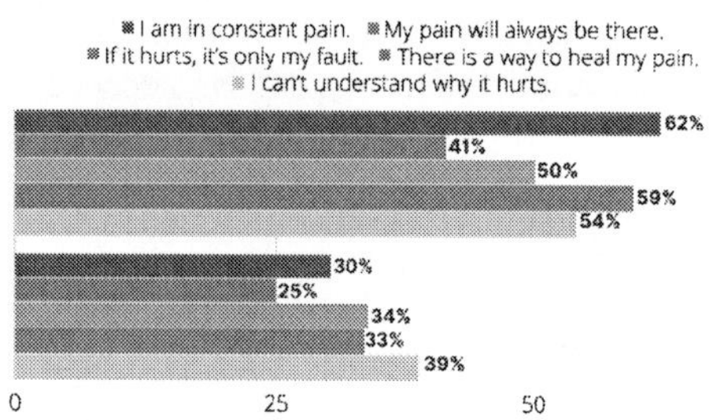

Figure 5: Break down of F1 macro average scores for each question based on out-performed model ($CT_{answer} + CS_{nli}$) results

Content selection The NLI-based content selection method (CS_{nli}) consistently outperforms other content selection approaches. This is consistent with Toudeshki et al. (2021)'s findings that for automatic questionnaire filling in a medical setting, NLI models performed better on average on polar and ALS question types.

We also see that the second best performing content selection method varies depending on the question type. As CS_{unit} first filters question/item pairs with highest probability, the method works well on CQ questions but struggles to handle more nuanced ALS questions which leads to an overall drop in performance on ALS questions.

6.2 Impact of pre-processing on different question/answer types

Table 2 shows the results for all combinations of pre-processing steps for each question/answer type.

Agreement answers (Yes, Totally agree) have the highest accuracy (about 70% in the best case) in both CQ and ALS questions, which suggests that the NLI model is better at confirming rather than rejecting a statement.

On *CQ questions*, various content selection methods have different impacts on each answer type. CS_{sim} shows much lower (3-4 times lower) performance on 'No' class than on 'NA' or 'Yes', CS_{null} has higher accuracy for the 'NA' class than for 'Yes' or 'No' classes and CS_{nli} performs better on 'Yes' and 'No' answers than on 'NA'. Both CS_{nli} and CS_{units} gives the most balanced F1 distribution across classes.

For *ALS questions*, CS_{nli} and CS_{sim} show the best results. While the CS_{nli} model is best at identifying 'Totally agree' and 'Totally disagree' classes, CS_{sim} distinguishes well whether the answer is absent ('NA') or whether it belongs to the 'Totally agree' class.

Performance on ALS questions is always lower. This can be explained by choice of threshold that distinguishes classes 'Totally agree' and 'Agree' as well as 'Totally disagree' and 'Rather disagree'. As mentioned above, CS_{units} favorizes the extreme classes which leads to a higher performance drop in comparison with CS_{sim} on ALS.

6.3 Break down of results for each question

Figure 5 presents the results of our best model (CT_{answer}+CS_{nli}) for each PBPI question separately.

The question "I am in constant pain." obtains highest score in CQ, while it performs poorly in ALS, demonstrating that the model is effective at detecting the presence of consistent pain but bad at predicting the level of agreement. The same behavior can be seen for the question "There is a way to heal my pain". On the other hand, for question "My pain will always be there" gets lowest score

	CQ					ALS						
	NA	YES	NO	macro	weighted	NA	TD	RD	A	TA	macro	weighted
support	142	228	130			142	54	79	115	110		
CT_{null}												
CS_{null}	0.39	0.15	0.27	0.27	0.25	0.28	0.11	0.26	0.07	0.16	0.18	0.18
CS_{units}	0.33	0.48	0.46	0.43	0.43	0.33	0.25	0.02	0.07	0.58	0.25	0.25
CS_{sim}	0.52	0.55	0.10	0.39	0.42	0.54	0.07	0.09	**0.23**	0.60	0.31	0.36
CS_{nli}	0.34	0.60	0.48	0.48	0.50	0.34	0.29	0.08	0.21	**0.67**	0.32	0.34
CT_{sum}												
CS_{null}	0.41	0.11	0.23	0.25	0.23	0.36	0.12	0.21	0.11	0.10	0.18	0.20
CS_{units}	0.32	0.33	0.43	0.36	0.35	0.32	0.23	0.06	0.02	0.44	0.22	0.23
CS_{sim}	0.49	0.40	0.02	0.30	0.32	0.51	0.00	0.05	0.21	0.46	0.24	0.30
CS_{nli}	0.37	0.43	0.46	0.42	0.42	0.31	0.28	0.10	0.09	0.48	0.25	0.26
CT_{answer}												
CS_{null}	0.45	0.28	0.37	0.37	0.35	0.41	0.27	**0.27**	0.17	0.33	0.29	0.30
CS_{units}	0.40	0.59	0.51	0.50	0.52	0.41	0.29	0.17	0.16	0.57	0.32	0.33
CS_{sim}	**0.53**	0.60	0.13	0.42	0.46	**0.55**	0.10	0.20	**0.23**	0.59	0.33	**0.38**
CS_{nli}	0.45	**0.70**	**0.57**	**0.57**	**0.59**	0.42	**0.35**	0.16	**0.23**	0.65	**0.36**	**0.38**

Table 2: F1-Scores for RoBERTa for closed (CQ) and agreement Likert scale (ALS) question types; TD - totally disagree, RD - rather disagree, A - agree, TA - totally agree. CT: content transformation, CS: content selection.

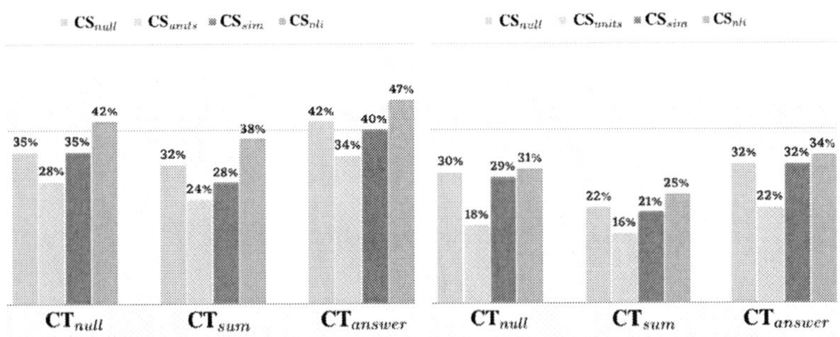

Figure 6: F1 macro average for the DeBERTa variant of our model on Closed Questions (CQ) on the left and Agreement Likert Scale (ALS) on the right. Test set of 100 dialogs with 10 questions each (5 yes/no questions and 5 ALS questions).

for both question types. The presence of the term "always" in the question turns it into a strong statement and consequently the model mostly rejects the statement unless it has been explicitly mentioned in the dialog.

6.4 Comparison with previous work and a different classifier (RoBERTa vs. DeBERTa)

Our model differs from previous work by Toudeshki et al. (2021) in two ways: it includes a pre-processing phase and uses the RoBERTa classifier whereas Toudeshki et al. (2021) applies DeBERTa to the whole input dialog. We compare our model with (i) the same model using DeBERTa and (ii) Toudeshki et al. (2021)'s model both on their and our test set.

Comparison with previous work In Figure 4, the two columns on the far left show the performance of Toudeshki et al. (2021)'s model on two test sets: the test set they used (10 instances and 16 questions) and our test set (100 instances and 5 questions).

Unsurprisingly, Toudeshki et al. (2021)'s results vary with the test set: while they report F1 score of 41 for CQ and 24 for ALS questions on their test set, these change to 35 and 30 on ours.

We also see that Toudeshki et al. (2021)'s DeBERTa-based, no pre-processing model outperforms our RoBERTa-based, null-preprocessing model (CT_{null}, CS_{null}) on both test sets. We conjecture that this difference can be explained by DeBERTa's improved attention mechanism, which se-

	CQ					ALS						
	NA	YES	NO	macro	weighted	NA	TD	RD	A	TA	macro	weighted
support	142	228	130			142	54	79	115	110		
CT_{null}												
CS_{null}	0.43	0.33	0.31	0.35	0.35	0.41	0.23	0.19	0.22	0.47	0.30	0.32
CS_{units}	0.15	0.29	0.40	0.28	0.28	0.15	0.21	0.07	0.00	0.45	0.18	0.17
CS_{sim}	0.51	0.45	0.09	0.35	0.37	**0.54**	0.03	0.05	**0.24**	0.60	0.29	0.35
CS_{nli}	0.34	0.51	0.40	0.42	0.43	0.29	0.23	0.17	0.21	**0.63**	0.31	0.32
CT_{sum}												
CS_{null}	0.40	0.29	0.26	0.32	0.31	0.37	0.16	0.11	0.16	0.31	0.22	0.25
CS_{units}	0.20	0.18	0.33	0.24	0.23	0.20	0.17	0.11	0.05	0.26	0.16	0.16
CS_{sim}	0.48	0.34	0.01	0.28	0.29	0.49	0.00	0.00	0.17	0.40	0.21	0.27
CS_{nli}	0.39	0.39	0.35	0.38	0.38	0.37	0.20	0.07	0.15	0.45	0.25	0.27
CT_{answer}												
CS_{null}	0.44	0.48	0.35	0.42	0.43	0.43	0.26	0.14	0.19	0.57	0.32	0.34
CS_{units}	0.19	0.45	0.39	0.34	0.36	0.19	0.20	0.07	0.10	0.55	0.22	0.23
CS_{sim}	**0.52**	0.51	0.16	0.40	0.42	0.53	0.09	0.15	0.20	0.61	0.32	**0.37**
CS_{nli}	0.40	**0.60**	**0.42**	**0.47**	**0.50**	0.41	**0.30**	**0.21**	0.16	**0.63**	**0.34**	0.36

Table 3: F1-Scores for DeBERTa for closed (CQ) and agreement Likert scale (ALS) question types; TD - totally disagree, RD - rather disagree, A - agree, TA - totally agree. CT: content transformation, CS: content selection.

lects relevant information in the input dialog with respect to the hypothesis.

However, our best model outperforms Toudeshki et al. (2021)'s approach by 22 points F1 for CQ questions and 6 points for ALS questions which indicates that pre-processing better helps bridge the gap between dialog and NLI-based QA.

DeBERTa vs. RoBERTa figure 6 and Table 3 show the result of our model when using DeBERTa instead of RoBERTa.

When using pre-processing, we see that the best RoBERTa model (CT_{answer}, CS_{nli}) outperforms the best DeBERTa model by 10 points F1 for CQ questions and 2 points for ALS questions.

Conversely, when no pre-processing is applied, the DeBERTa variant of our model outperforms the RoBERTa variant which is consistent with the results discussed in the previous paragraph. For the DeBERTa variant, we observe that the CS_{null} baseline is no longer the lowest performing content selection approach, while the performance of CS_{units} and CS_{sim} becomes lower than the baseline (CS_{null}). This highlights the fact that the DeBERTa model performs better without weak content selection approaches. On the other hand, it can be seen that the impact of content selection and transformation approaches is significant in RoBERTa, although using a weaker classifier, and our model outperforms previous work. This shows that the proposed select-and-transform pre-processing approach improves results in both RoBERTa and DeBERTa, though this improvement is more significant in RoBERTa, suggesting that this latter model is more sensitive to the form and size of the input content.

7 Conclusion

In this paper, we studied how dialog pre-processing can impact the task of filling medical questionnaires based on patient-bot interactions. Our experimental results show that converting pairs of adjacent turns to declarative sentences and selecting input units based on their entailment relation with the question can significantly enhance performance.

References

Daniel Adiwardana, Minh-Thang Luong, David R So, Jamie Hall, Noah Fiedel, Romal Thoppilan, Zi Yang, Apoorv Kulshreshtha, Gaurav Nemade, Yifeng Lu, et al. 2020. Towards a human-like open-domain chatbot. *arXiv preprint arXiv:2001.09977.*

Lisa Bauer, Yicheng Wang, and Mohit Bansal. 2018. Commonsense for generative multi-hop question answering tasks. *arXiv preprint arXiv:1809.06309.*

Dorottya Demszky, Kelvin Guu, and Percy Liang. 2018. Transforming question answering datasets into natural language inference datasets. *arXiv preprint arXiv:1809.02922.*

Bogdan Gliwa, Iwona Mochol, Maciej Biesek, and Aleksander Wawer. 2019. Samsum corpus: A human-annotated dialogue dataset for abstractive summarization. *arXiv preprint arXiv:1911.12237*.

Pengcheng He, Xiaodong Liu, Jianfeng Gao, and Weizhu Chen. 2020. DeBERTa: Decoding-enhanced BERT with disentangled attention. *arXiv preprint arXiv:2006.03654*.

H Jang, E Rempel, D Roth, G Carenini, and NZ Janjua. 2021. Tracking COVID-19 discourse on twitter in North America: Topic modeling and aspect-based sentiment analysis. *Journal of Medical Internet Research*.

Qingnan Jiang, Lei Chen, Ruifeng Xu, Xiang Ao, and Min Yang. 2019. A challenge dataset and effective models for aspect-based sentiment analysis. In *Proceedings of the 2019 Conference on Empirical Methods in Natural Language Processing and the 9th International Joint Conference on Natural Language Processing (EMNLP-IJCNLP)*, pages 6280–6285.

Mike Lewis, Yinhan Liu, Naman Goyal, Marjan Ghazvininejad, Abdelrahman Mohamed, Omer Levy, Veselin Stoyanov, and Luke Zettlemoyer. 2020. BART: Denoising sequence-to-sequence pre-training for natural language generation, translation, and comprehension. In *Proceedings of the 58th Annual Meeting of the Association for Computational Linguistics*, pages 7871–7880, Online. Association for Computational Linguistics.

Junlong Li, Zhuosheng Zhang, and Hai Zhao. 2020. Knowledgeable dialogue reading comprehension on key turns. *arXiv preprint arXiv:2004.13988*.

Anna Liednikova, Philippe Jolivet, Alexandre Durand-Salmon, and Claire Gardent. 2021. Gathering information and engaging the user combot: A task-based, serendipitous dialog model for patient-doctor interactions. In *Proceedings of the Second Workshop on Natural Language Processing for Medical Conversations*, pages 21–29.

Yinhan Liu, Myle Ott, Naman Goyal, Jingfei Du, Mandar Joshi, Danqi Chen, Omer Levy, Mike Lewis, Luke Zettlemoyer, and Veselin Stoyanov. 2019. Roberta: A robustly optimized bert pretraining approach. *arXiv preprint arXiv:1907.11692*.

Shashi Narayan, Shay B Cohen, and Mirella Lapata. 2018. Don't give me the details, just the summary! topic-aware convolutional neural networks for extreme summarization. *arXiv preprint arXiv:1808.08745*.

Kate Pearce, Tiffany Zhan, Aneesh Komanduri, and Justin Zhan. 2021. A comparative study of transformer-based language models on extractive question answering. *arXiv preprint arXiv:2110.03142*.

Colin Raffel, Noam Shazeer, Adam Roberts, Katherine Lee, Sharan Narang, Michael Matena, Yanqi Zhou, Wei Li, and Peter J Liu. 2019. Exploring the limits of transfer learning with a unified text-to-text transformer. *arXiv preprint arXiv:1910.10683*.

Nils Reimers and Iryna Gurevych. 2019. Sentence-bert: Sentence embeddings using siamese bert-networks. In *Proceedings of the 2019 Conference on Empirical Methods in Natural Language Processing*. Association for Computational Linguistics.

Jiangtao Ren, Naiyin Liu, and Xiaojing Wu. 2020. Clinical questionnaire filling based on question answering framework. *International Journal of Medical Informatics*, 141:104225.

Mary Sun. 2022. *Natural Language Processing for Health System Messages: Deep Transfer Learning Approach to Aspect-Based Sentiment Analysis of COVID-19 Content*. Ph.D. thesis, Harvard University.

Farnaz Ghassemi Toudeshki, Philippe Jolivet, Alexandre Durand-Salmon, and Anna Liednikova. 2021. Zero-shot clinical questionnaire filling from human-machine interactions. In *Proceedings of the 3rd Workshop on Machine Reading for Question Answering*, pages 51–62.

Adina Williams, Nikita Nangia, and Samuel Bowman. 2018. A broad-coverage challenge corpus for sentence understanding through inference. In *Proceedings of the 2018 Conference of the North American Chapter of the Association for Computational Linguistics: Human Language Technologies, Volume 1 (Long Papers)*, pages 1112–1122. Association for Computational Linguistics.

David A. Williams and Beverly E. Thorn. 1989. An empirical assessment of pain beliefs. *Pain*, 36(3):351–358.

Chunyuan Yuan, Wei Zhou, Mingming Li, Shangwen Lv, Fuqing Zhu, Jizhong Han, and Songlin Hu. 2019. Multi-hop selector network for multi-turn response selection in retrieval-based chatbots. In *Proceedings of the 2019 conference on empirical methods in natural language processing and the 9th international joint conference on natural language processing (EMNLP-IJCNLP)*, pages 111–120.

Changchang Zeng, Shaobo Li, Qin Li, Jie Hu, and Jianjun Hu. 2020. A survey on machine reading comprehension—tasks, evaluation metrics and benchmark datasets. *Applied Sciences*, 10(21):7640.

Yizhe Zhang, Siqi Sun, Michel Galley, Yen-Chun Chen, Chris Brockett, Xiang Gao, Jianfeng Gao, Jingjing Liu, and Bill Dolan. 2019. DialoGPT: Large-scale generative pre-training for conversational response generation. *arXiv preprint arXiv:1911.00536*.

Zhuosheng Zhang, Jiangtong Li, Pengfei Zhu, Hai Zhao, and Gongshen Liu. 2018. Modeling multi-turn conversation with deep utterance aggregation. *arXiv preprint arXiv:1806.09102*.

Zhuosheng Zhang, Junlong Li, and Hai Zhao. 2021. Multi-turn dialogue reading comprehension with pivot turns and knowledge. *IEEE/ACM Transactions on Audio, Speech, and Language Processing*, 29:1161–1173.

Zheng Zhao, Shay B. Cohen, and Bonnie Webber. 2020. Reducing quantity hallucinations in abstractive summarization. In *Findings of the Association for Computational Linguistics: EMNLP 2020*, pages 2237–2249, Online. Association for Computational Linguistics.

A Ethics

Regarding Regulation (EU) 2017/745, described software is intended for general uses, even when used in a healthcare environment, it is intended for uses relating to lifestyle or well-being that do not constitute any a medical prediction and medical prognosis function without doctors validation or correction.

We gathered dialogs for experiments using Amazon Mechanical Turk. Because of the task's difficulty and estimated completion time, we set the initial reward at 1$. We assigned 0.5$ bonus for each key point mentioned by the user during the dialogue. If the user was successful in mentioning all five key points, he was awarded a bonus of 2.5$ in total.

B Experiment time estimation

The experiments were conducted with a laptop having Intel® Core™ i7-10610U CPU @ 1.80GHz * 8 and NVIDIA Quadro P520.

C Questionnaire

PBPI questionnaire statements are provided in table 4.

D Data Collection

Instructions used for data collection in Amazon Mechanical Turk and the interface are shown in figures 7, 8 and 9.

We requested the Turkers to converse with the heath-bot for at least 10 turns in total.

Id	Question
1	No one is able to tell me why it hurts.
2	I thought my pain could be healed, but now I'm not so sure.
3	There are times when it doesn't hurt.
4	My pain is difficult for me to understand.
5	**My pain will always be there.**
6	**I am in constant pain.**
7	**If it hurts, it's only my fault.**
8	I don't have enough information about my pain.
9	My pain is a temporary problem in my life.
10	I feel like I wake up with pain and fall asleep with it.
11	I am the cause of my pain.
12	**There is a way to heal my pain.**
13	I blame myself when it hurts.
14	**I can't understand why it hurts.**
15	One day, again, I won't have any pain at all.
16	My pain varies in intensity but it is always present with me.

Table 4: List of questions in PBPI questionnaire

Task Description

In this task, you are going to talk to a chatbot about health and quality of your life.

You are supposed to **play the role of a chronic pain patient**, and **share your pain with the bot**.
What is chronic pain? Doctors often define chronic pain as any pain that lasts for 3 to 6 months or more. Chronic pain can have real effects on day-to-day life and mental health.

It is very important that you **mention about all follwing key points during your conversation** (in a seamless way):

1. (Constantly/Temporarly) in pain
2. (Having/Losing) hope for getting healed
3. (Feeling/Not feeling) guiltiness that the pain is your fault
4. (Possibility/Impossibility) of healing
5. (Understanding/Not understanding) the reason of having pain

Try to **give an implicit and seamless reference to these keypoints** in the dialogue (with considering the flow of conversation). **Prevent using the same wording** in your messages.

** BONUS **

Playing the role of a chronic pain patient and mentioning each keypoint will get 0.5 $ bonus. By metioning all keypoints you will get 2.5 $ bonus (do not use the same wording).

note1: It is an information seeking conversation and **you are not expected to ask questions from the bot**.
note2: Wait for the bot message to be appeared completely, and then reply.
note3: When it is your time to reply, **reply only once**.

Figure 7: Instructions (part 1)

Lead the conversation

The chatbot is not developed to ask you explicitly about these key points. Therefore, you have to mention them creatively during the dialog flow.
To make it easier for you, we have given you the authority of **controling chatbot messages**. You can direct the conversation by **changing chatbot reply**. To do that, you can **click on the "next" button** (below the bot message) to change the chatbot utterance and if you found it good enough you can just continue the conversation.

Annotating each user reply

After you entered your answer, you will notice **5 checkpoints apear below your answer**. Each check point refers to each of the keypoints. **If one or mutliple keypoints have been mentioned in your answer (implicitly or explicitly) choose the related checkpoints**.
Please keep in my mind that you have to **fill check points before entering your next response.** They would be disabeled afterwards.

End the conversation

To end the conversation, you can click on **green "Submit" button**. But before that, **wiat for the bot message to be appeared completely, and then press the submit button**.
If you click on the button before reaching to the minimum number of turns (5 messages each user), you will receive an alert error message and be taken back to the conversation to complete the task.

Figure 8: Instructions (part 2)

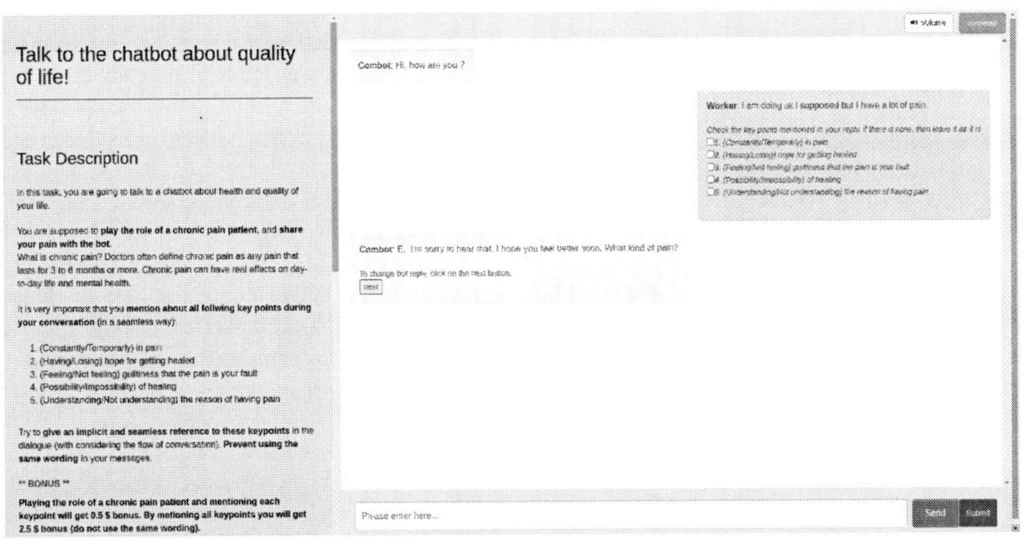

Figure 9: Interface

Assessing the Limits of Straightforward Models for Nested Named Entity Recognition in Spanish Clinical Narratives

Matías Rojas[1], **Casimiro Pio Carrino**[2], **Aitor Gonzalez-Agirre**[2]
Jocelyn Dunstan[1], **and Marta Villegas**[2]
[1]Center for Mathematical Modeling, University of Chile
[2]Text Mining Unit, Barcelona Supercomputing Center

Abstract

Nested Named Entity Recognition (NER) is an information extraction task that aims to identify entities that may be nested within other entity mentions. Despite the availability of several corpora with nested entities in the Spanish clinical domain, most previous work has overlooked them due to the lack of models and a clear annotation scheme for dealing with the task. To fill this gap, this paper provides an empirical study of straightforward methods for tackling the nested NER task on two Spanish clinical datasets, Clinical Trials, and the Chilean Waiting List. We assess the advantages and limitations of two sequence labeling approaches; one based on Multiple LSTM-CRF architectures and another on Joint labeling models. To better understand the differences between these models, we compute task-specific metrics that adequately measure the ability of models to detect nested entities and perform a fine-grained comparison across models. Our experimental results show that employing domain-specific language models trained from scratch significantly improves the performance obtained with strong domain-specific and general-domain baselines, achieving state-of-the-art results in both datasets. Specifically, we obtained F_1 scores of 89.21 and 83.16 in Clinical Trials and the Chilean Waiting List, respectively. Interestingly enough, we observe that the task-specific metrics and analysis properly reflect the limitations of the models when recognizing nested entities. Finally, we perform a case study on an aggregated NER dataset created from several clinical corpora in Spanish. We highlight how entity length and the simultaneous recognition of inner and outer entities are the most critical variables for the nested NER task.

1 Introduction

Named Entity Recognition (NER) is a widely studied task that seeks to identify text spans associated with predefined categories. Nested Named Entity Recognition is a particular case of NER, where entities can be nested within each other (Finkel and Manning, 2009), such as the example in Figure 1. Traditional NER models simplify nested entities through predetermined rules, such as keeping the most external entity and ignoring inner ones. This simplified problem is better known as flat NER and allows solving the task using traditional sequence labeling architectures such as the BiLSTM-CRF (Lample et al., 2016) approach or fine-tuning transformer-based models (Vaswani et al., 2017).

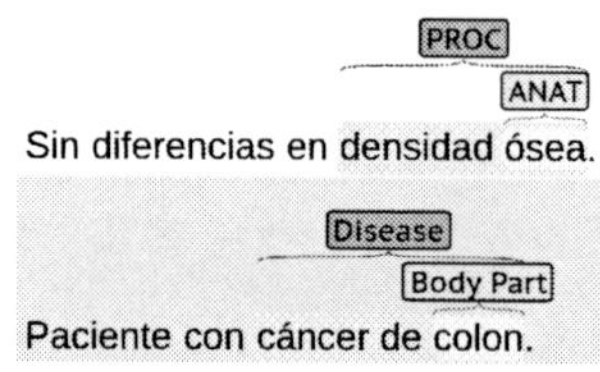

Figure 1: Example of nested entities in the Clinical Trials and Chilean Waiting List datasets.

Regarding the Spanish language, there are several biomedical and clinical datasets containing nested entities, such as the Spanish radiology corpus (Cotik et al., 2017), NUBes (Lima Lopez et al., 2020), the Chilean Waiting List (Báez et al., 2020), Clinical Trials (Campillos-Llanos et al., 2021). However, most previous works transformed the task into a flat NER. As mentioned in Wang et al. (2022), this simplification is due to technological rather than ideological reasons, mainly explained by the difficulty of representing nested entities with the traditional annotation scheme, for example, with the IOB2 sequence labeling format. We argue that treating the nested NER task as flat NER is not optimal since removing part of the entities could result in a loss of information previously annotated by humans, wasting time and resources, and harming the model's performance.

This paper explores simple neural network-based models as a proxy to address the challenging nested

Proceedings of the 13th International Workshop on Health Text Mining and Information Analysis (LOUHI), pages 14 - 25
December 7, 2022 ©2022 Association for Computational Linguistics

NER task. Specifically, we revisited the Multiple LSTM-CRF (MLC) and the Joint Labeling architectures and performed experiments on two Spanish clinical corpora. The former consists of training a flat NER model for each entity type following the IOB2 format, while the latter transforms the nested NER task into a flat NER using an annotation scheme that allows preserving the nested entities. We analyze the impact of using pre-trained language models trained on specific domains compared to general-domain ones.

To evaluate the performance of our models, we provide a detailed analysis of task-specific evaluation metrics that adequately measure the effectiveness of the models in recognizing nested entities, considering variables such as entity length, the nesting depth level, and the different types of nested entities. In addition, to better understand the limitations of these models, we created an aggregated corpus formed from several Spanish clinical NER corpora.

In summary, the main contributions of our work are the following:

- We show that straightforward architectures leveraging domain-specific models can tackle the nested NER task, achieving state-of-the-art performances on two clinical datasets in Spanish.
- We conduct an empirical study that compares the impact of using domain-specific language models against general-domain ones, either by using contextualized embeddings or fine-tuning the model in the task.
- We performed an in-depth analysis of the advantages and limitations of the previous approaches by testing our models on an aggregated clinical corpus in Spanish exhibiting complex annotations.

2 Related Work

The nested nature of named entities has recently gained special attention from the NLP research community. Several models have been proposed to handle the nesting problem, which can be mainly divided into three categories: region-based, hypergraph-based, and sequence labeling-based models.

Region-based models list potential span candidates and then classify them into predefined categories. In Sohrab and Miwa (2018), they used an exhaustive neural model enumerating all possible spans within a limited length and then predicted the entity types of those regions using boundary and average internal token representation. Zheng et al. (2019) used a sequence labeling layer to identify candidate spans and then classified the selected regions into their entity category labels. Another region-based model was proposed by Yu et al. (2020), who used contextual representations models to encode sentences and two separate MLPs to create start and end token representations. They then ranked all possible start-end regions in the sentence using nested constraints to predict the labels. Recently, Shen et al. (2021) used a two-stage identifier, using a filter and a regressor to identify high-quality candidate spans and then classifying them into their entity types.

Hypergraph-based models learn the nested structure of entities in the sentence through hypergraphs. The aim is to capture the relations between inner and outer entities to leverage the extraction of nested entities. In Lu and Roth (2015), they proposed a mention hypergraph representation for both extracting entity boundaries and predicting entity labels. Similarly, Katiyar and Cardie (2018) designed a directed hypergraph using LSTM features to learn the nesting structure. In Luo and Zhao (2020), they used a flat NER module for recognizing the most external entities and a graph module for inner entities.

Sequence labeling-based models formulate the nested NER task as several flat NER models. Early work from Alex et al. (2007) introduced three CRF-based methods to reduce the nested NER as several IOB2 tagging problems. Ju et al. (2018) took advantage of inner entity information to improve outer entity recognition. They dynamically stacked LSTM-CRF layers predicting entities in an inside-to-outside manner. In contrast, Shibuya and Hovy (2020) recognized entities from outermost to inner ones using a recursive method based on separate CRFs. This method was improved in Wang et al. (2021), demonstrating that inner to outermost recognition is best for modeling this task. Finally, Wang et al. (2020) recursively introduced the embedding of tokens and regions into flat NER layers simulating the shape of a pyramid and extracting nested entities from the innermost to the outermost entities. The models used in our experiments fall into this category.

3 Nested NER Models

In recent years, contextual representational models have improved the performance of many neural network-based models, making it possible to achieve state-of-the-art in several NLP tasks. Unlike traditional word embeddings, language models can represent words according to the sentence-level context. Regarding the NER task, using contextual word embeddings or fine-tuning a pre-trained language model to a specific domain has boosted the performance of models in datasets from several domains.

Previous work in clinical NER showed that using domain-specific language models improves results considerably compared to general-domain language models. However, no studies show this behavior occurs when there is a nested structure in the entities, especially in low-resource languages such as Spanish. In this work, we study whether this trend is confirmed in nested NER datasets using two sequence labeling-based architectures, the Joint Labeling, and the Multiple LSTM-CRF models.

3.1 Joint Labeling Model

The Joint Labeling architecture (Agrawal et al., 2022) consists of formulating nested NER as a flat NER task using an appropriate annotation scheme. Since nested entities allow a token to have more than one entity type, all the token labels are merged into a single token label using a delimiter. This scheme allows solving the problem using traditional sequence labeling architectures that treats the problem as a token-level classification.

We decided to use this architecture due to its high performance on the nested NER task in other languages, such as English and German. Therefore, it is interesting to study the performance of this approach on Spanish datasets, which have been less explored. To solve the token-level classification, we followed the classic approach of fine-tuning transformer-based language models on the NER task. In other words, we fine-tuned language models trained on giant text corpora and added a linear layer to perform the token-level classification.

3.2 Multiple LSTM-CRF

The second approach uses the Multiple LSTM-CRF (MLC) architecture (Rojas et al., 2022a), which trains separate flat NER models for each entity type. The predicted labels of the input sentences correspond to the union of the outputs of each model, thus retrieving both nested entities and text spans tagged with multiple labels.

Each flat NER module consists of three main layers: the embedding layer, the encoding layer with a BiLSTM, and the classification layer, where the most likely sequence of labels is obtained using the CRF algorithm. Regarding the embedding layer, we incorporated contextualized word representations retrieved from a language model, replacing traditional representations such as word and character-level embeddings.

As for the previous model, we tested several domain-specific and general-domain transformer-based language models. The vector representation of words was computed by averaging the representations retrieved from all hidden states. Since BERT-based language models use WordPiece tokenization, we calculated word embeddings using the embedding of the first subtoken. In addition, we tested Clinical Flair (Rojas et al., 2022b), a character-level language model trained on Spanish clinical narratives. Being a character-level model, it is particularly effective for handling out-of-vocabulary and misspelled words, which are very common in clinical texts.

4 Experiments

In this section, we present the datasets, settings, and evaluation metrics used in our experiments.

4.1 Datasets

We conducted our experiments with two corpora containing nested entities.

- **Chilean Waiting List**[1] (Báez et al., 2020): clinical corpus annotated from real diagnoses of the Chilean healthcare system. It is composed of 87,024 entity mentions and seven entity types. From a nested NER point of view, it is a good resource since 48.23% of the entities are involved in nesting.

- **Clinical Trials**[2] (Campillos-Llanos et al., 2021): clinical corpus created from 500 abstracts of journal articles about clinical trials and 700 announcements of trial protocols. It consists of 46,518 entity mentions and four

[1] https://zenodo.org/record/3926705

[2] http://www.lllf.uam.es/ESP/nlpmedterm_en

	Chilean Waiting List			Clinical Trials		
	Train	Test	Dev	Train	Test	Dev
tokens	291,561	36,963	34,987	202,541	67,281	67,661
sentences	15,290	1,912	1,911	7,604	2,522	2,550
avg sent len	19.07	19.33	18.31	26.64	26.68	26.53
entities	69,847	8,837	8,340	27,967	8,940	9,611
avg entity len	2.73	2.71	2.74	1.89	1.86	1.88
nested entities	33,667	4,182	4,126	7,373	2,333	2,580
nested entities (%)	48.20	47.32	49.47	26.36	26.10	26.84

Table 1: Statistics of the datasets used in our experiments.

entity types, which belong to a subset of semantic groups from the Unified Medical Language System (UMLS).

Table 1 shows the overall statistics for each corpus. Compared to other well-known nested NER datasets such as GENIA (Kim et al., 2003) and GermEval (Benikova et al., 2014), where the nesting percentage is less than 20%, these two datasets are a valuable resource for the nested NER task. Especially the Chilean Waiting List corpus, which contains more than twice as much nesting compared to the datasets mentioned above.

4.2 Settings

To analyze the impact of domain-specific language models in Spanish, we used the biomedical version of RoBERTa (*bsc-bio-es*[3]) and the clinical version of RoBERTa (*bsc-bio-ehr-es*[4]) (Carrino et al., 2022). We compared these models with a general-domain Spanish model (*BETO*) (Cañete et al., 2020), a multilingual model (*mBERT*) (Devlin et al., 2019), and two domain-specific models based on continuous pre-training: *mBERT-Galén* (based on mBERT) and *BETO-Galén* (based on BETO) (López-García et al., 2021). As previously mentioned, the MLC model uses these models as contextualized embeddings, while for Joint Labeling, we used them to perform a fine-tuning and solve the token-level classification task.

To train the Joint Labeling model, we used the Adam optimizer and searched for an optimal learning rate out of 1e-5, 5e-5, 5e-6, and 1e-6, with linear decay and no warm-up steps. We trained the model up to a maximum of 20 epochs using a batch size of 8 sequences with a maximum length of 512 tokens and a gradient accumulation of 2 steps, resulting in a total batch size of 16. The training took

[3]https://huggingface.co/PlanTL-GOB-ES/bsc-bio-es

[4]https://huggingface.co/PlanTL-GOB-ES/bsc-bio-ehr-es

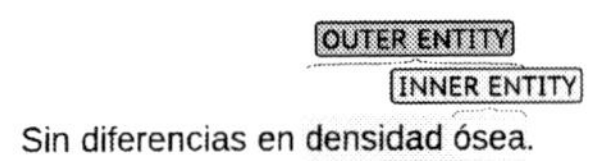

Figure 2: Example of different types of entities.

approximately 45 minutes for each dataset, using 2 AMD MI50 GPUs with 32 GB of VRAM each.

Regarding the MLC architecture, to train the model of each entity type, we used the SGD optimizer to a maximum of 100 epochs, with mini-batches of size 16 and a learning rate of 0.1. We set the number of RNN layers to 1 and the hidden size to 256. To control overfitting, we employed a learning rate scheduler and an early stopping strategy based on the performance of the validation partition. We also applied dropout regularization after the embedding layer and BiLSTM. The training for each entity type took at most 7 hours under the same hardware settings as Joint Labeling. Since the model of each entity type is independent of the others, this allows us to perform parallel training, reducing the computational cost of this approach.

4.3 Metrics

To evaluate the performance of our models, we computed the micro-average precision, recall, and F_1 score over all entities, which is the standard metric used by the research community for evaluating NER systems. In this context, precision is the percentage of entities found by our system that belonged to the test set, while recall is the percentage of entities from the test set found by our system. This metric follows a strict evaluation approach since an entity is considered correct when both entity types and boundaries are predicted correctly. However, one of the main drawbacks of the above metrics is that they do not differentiate nested entities from flat entities. Since flat entities are the most frequent in nested NER datasets, this could overestimate the model's performance on the task.

	Chilean Waiting List			Clinical Trials		
	P	R	F_1	P	R	F_1
Joint Labeling w/ mBERT cased	$74.33_{0.84}$	$78.08_{1.02}$	$76.16_{0.93}$	$83.34_{0.64}$	$85.97_{0.55}$	$84.64_{0.55}$
Joint Labeling w/ mBERT-Galén	$75.16_{0.56}$	$79.24_{0.33}$	$77.15_{0.43}$	$82.53_{0.48}$	$84.83_{0.37}$	$83.67_{0.41}$
Joint Labeling w/ BETO	$75.93_{0.89}$	$79.10_{0.52}$	$77.48_{0.67}$	$84.96_{0.43}$	$87.19_{0.17}$	$86.06_{0.21}$
Joint Labeling w/ BETO-Galén	$74.52_{0.46}$	$78.79_{0.39}$	$76.59_{0.10}$	$82.47_{0.38}$	$85.49_{0.13}$	$83.95_{0.15}$
Joint Labeling w/ Biomedical RoBERTa	$76.55_{0.23}$	$80.32_{0.33}$	$78.39_{0.24}$	$87.92_{0.14}$	$90.20_{0.24}$	$89.04_{0.10}$
Joint Labeling w/ Clinical RoBERTa	$77.31_{0.40}$	$81.27_{0.46}$	$79.24_{0.39}$	$88.03_{0.34}$	$\mathbf{90.43}_{0.12}$	$\mathbf{89.21}_{0.14}$
MLC w/ mBERT cased	$79.41_{0.16}$	$71.31_{0.34}$	$75.14_{0.15}$	$84.67_{0.11}$	$83.90_{0.17}$	$84.28_{0.05}$
MLC w/ mBERT-Galén	$78.94_{0.18}$	$75.56_{0.09}$	$77.21_{0.13}$	$84.99_{0.26}$	$81.67_{0.30}$	$83.29_{0.24}$
MLC w/ BETO	$79.33_{0.58}$	$72.26_{0.25}$	$75.63_{0.40}$	$86.04_{0.81}$	$81.02_{0.73}$	$83.46_{0.76}$
MLC w/ BETO-Galén	$79.14_{0.30}$	$74.67_{0.17}$	$76.84_{0.23}$	$85.91_{0.18}$	$82.21_{0.26}$	$84.02_{0.22}$
MLC w/ Bio RoBERTa	$80.30_{0.19}$	$75.40_{0.35}$	$77.77_{0.27}$	$87.97_{0.06}$	$84.84_{0.43}$	$86.37_{0.20}$
MLC w/ Clinical RoBERTa	$80.71_{0.51}$	$76.13_{1.09}$	$78.35_{0.82}$	$\mathbf{88.80}_{0.23}$	$85.90_{0.07}$	$87.32_{0.13}$
MLC w/ Clinical Flair	$\mathbf{84.31}_{0.37}$	$\mathbf{82.04}_{0.68}$	$\mathbf{83.16}_{0.28}$	$88.38_{0.13}$	$85.21_{0.13}$	$86.76_{0.06}$

Table 2: Overall results on two nested NER datasets. The reported results correspond to the average of three evaluation rounds using different seeds. Subscript numbers indicate the standard deviations.

To address the above issue, we compute task-specific metrics proposed in Rojas et al. (2022a) that allow analyzing the predictions in detail according to the nested NER task. Specifically, we compute a score for entities not involved in nestings (m_{flat}), entities involved in nestings (m_{nested}), inner entities in nestings (m_{inner}), outer entities in nestings (m_{outer}), and complete nestings ($m_{nesting}$). In this context, a nesting is composed of inner and outer entities, and m_{nested} encompasses the m_{inner} and m_{outer} metrics.

These task-specific metrics were calculated using micro-average precision, recall, and F_1 score. Using Figure 2 as an example to better understand the different types of entities, the inner entity is *ósea*, while the outer entity is *densidad ósea*. Both inner and outer entities compose a nesting of depth 2, and there are no flat entities to measure. All experiments and models are freely available to ensure reproducibility[5].

5 Overall Results

Table 2 shows the overall results of our experiments. We observe that across all the experiments, the Joint Labeling model obtains a lower precision than the recall, while in the case of the MLC model, the opposite occurs. As expected, in both models and datasets, the incorporation of domain-specific contextual representation models contributes to significant improvements in the performance compared to general-domain models. However, in some cases, it occurred that the BETO-Galén and mBERT-Galén models did not provide improvements over the general-domain base models. One plausible reason may be found in the domain-specific vocabulary since the Galén model was trained with the continuous training technique, unlike the RoBERTa-based models, which were trained from scratch.

Although the MLC and Joint Labeling architectures appear to be simple approaches for solving nested NER, we observe that their results are pretty high. Specifically, the best setting for the Chilean Waiting List corpus is the MLC model with embeddings retrieved from the Clinical Flair model. Using the same data splits, we obtained state-of-the-art results with an improvement of almost three micro F_1 points over the best system to date, as reported in Báez et al. (2022), where they achieved a micro F_1 score of 80.27. This excellent performance could be explained since Clinical Flair is a character-level language model, particularly beneficial in datasets with many misspelled and out-of-vocabulary words, such as diagnoses from public hospitals.

On the other hand, the best setting in Clinical Trials is the Joint Labeling approach with the clinical version of RoBERTa. To date, the only result reported on Campillos-Llanos et al. (2021) achieved a micro F_1 score of 86.74 without considering the nested entities. In contrast, we obtained a micro F_1 score of 89.21, achieving state-of-the-art in the corpus and demonstrating the importance of considering nested entities.

6 Discussion and Analysis

6.1 Nested NER Performance

For a more detailed analysis of the above results, we employ the metrics introduced in Section 4.3 that decompose the model's performances for different types of nested entities. Table 3 shows the

[5]`https://github.com/TeMU-BSC/clinical-nested-ner`

	Chilean Waiting List					Clinical Trials				
	m_{flat}	m_{inner}	m_{outer}	m_{nested}	$m_{nesting}$	m_{flat}	m_{inner}	m_{outer}	m_{nested}	$m_{nesting}$
Joint Labeling w/ mBERT cased	$76.64_{0.89}$	$82.90_{0.40}$	$65.95_{1.89}$	$75.62_{0.97}$	$54.81_{1.60}$	$84.59_{0.38}$	$87.84_{0.89}$	$81.32_{1.72}$	$84.79_{1.24}$	$72.17_{1.70}$
Joint Labeling w/ mBERT-Galén	$77.06_{1.00}$	$83.44_{0.56}$	$69.11_{0.31}$	$77.27_{0.23}$	$57.25_{0.17}$	$83.67_{0.43}$	$87.13_{0.51}$	$79.73_{0.57}$	$83.64_{0.47}$	$70.55_{1.06}$
Joint Labeling w/ BETO	$78.22_{1.15}$	$83.05_{0.29}$	$68.23_{0.10}$	$76.64_{0.13}$	$56.35_{0.21}$	$86.11_{0.13}$	$89.09_{0.61}$	$82.32_{0.36}$	$85.93_{0.48}$	$74.07_{0.43}$
Joint Labeling w/ BETO-Galén	$76.18_{0.41}$	$83.22_{0.74}$	$68.81_{0.32}$	$77.07_{0.42}$	$57.13_{0.71}$	$83.85_{0.12}$	$88.06_{0.29}$	$79.95_{0.32}$	$84.25_{0.24}$	$71.57_{0.18}$
Joint Labeling w/ Bio RoBERTa	$78.40_{0.19}$	$85.05_{0.12}$	$69.42_{0.74}$	$78.37_{0.36}$	$58.80_{0.34}$	$89.09_{0.23}$	$91.54_{0.40}$	$85.95_{0.21}$	$88.91_{0.29}$	$78.62_{0.62}$
Joint Labeling w/ Clinical RoBERTa	$79.50_{0.56}$	$84.76_{0.33}$	$71.22_{0.73}$	$78.94_{0.27}$	$59.89_{0.40}$	$\mathbf{89.16}_{0.15}$	$\mathbf{91.85}_{0.16}$	$\mathbf{86.58}_{0.28}$	$\mathbf{89.36}_{0.19}$	$\mathbf{78.94}_{0.68}$
MLC w/ mBERT cased	$75.57_{0.30}$	$82.45_{0.17}$	$64.32_{0.28}$	$74.66_{0.03}$	$52.30_{0.14}$	$84.63_{0.10}$	$85.89_{0.17}$	$80.41_{0.28}$	$83.30_{0.10}$	$67.93_{0.18}$
MLC w/ mBERT-Galén	$78.13_{0.41}$	$82.63_{0.58}$	$67.75_{0.18}$	$76.20_{0.41}$	$53.82_{0.58}$	$83.60_{0.16}$	$84.49_{0.12}$	$80.15_{0.88}$	$82.43_{0.47}$	$67.71_{0.70}$
MLC w/ BETO	$76.43_{0.29}$	$81.12_{0.52}$	$66.32_{0.65}$	$74.73_{0.58}$	$52.05_{0.59}$	$83.90_{0.82}$	$84.38_{0.92}$	$79.61_{0.96}$	$82.15_{0.72}$	$66.88_{2.36}$
MLC w/ BETO-Galén	$77.46_{0.35}$	$82.47_{0.58}$	$67.81_{0.29}$	$76.15_{0.21}$	$53.80_{0.50}$	$84.51_{0.22}$	$84.19_{0.25}$	$80.89_{0.25}$	$82.62_{0.25}$	$68.42_{0.39}$
MLC w/ Bio RoBERTa	$78.47_{0.45}$	$83.70_{0.28}$	$68.15_{0.05}$	$77.00_{0.15}$	$55.52_{0.28}$	$86.59_{0.20}$	$87.68_{0.16}$	$83.60_{0.33}$	$85.76_{0.24}$	$73.06_{0.61}$
MLC w/ Clinical RoBERTa	$79.34_{0.73}$	$83.73_{0.70}$	$68.71_{1.37}$	$77.26_{0.94}$	$55.72_{1.67}$	$87.47_{0.12}$	$89.46_{0.20}$	$84.04_{0.18}$	$86.90_{0.18}$	$74.72_{0.19}$
MLC w/ Clinical Flair	$\mathbf{84.11}_{0.27}$	$\mathbf{88.62}_{0.19}$	$\mathbf{73.41}_{0.85}$	$\mathbf{82.09}_{0.34}$	$\mathbf{62.82}_{0.86}$	$86.69_{0.03}$	$90.69_{0.29}$	$82.76_{0.16}$	$86.98_{0.23}$	$74.77_{0.48}$

Table 3: Task-specific metrics for nested NER.

results according to task-specific metrics. Interestingly, we note that the nesting metric score, which consists of simultaneously recognizing inner and outer entities, is between 10 and 20 F_1 points lower than the standard F_1 metric across models and datasets. In fact, in all cases, the models fail more in recognizing outermost entities than inner ones, suggesting that straightforward methods for nested NER cannot correctly model existing relations between the components of a nested entity. Presumably, since outermost entities are longer in the number of tokens, it is easier for the model to make mistakes when using a strict evaluation metric. Therefore, despite the high score obtained with the standard F_1 metric (see Table 2), this finding points out the importance of using suitable metrics to test the limitations of nested NER approaches. Finally, we can notice that the best models, according to the standard metric, also get the best results according to the nested metrics, proving that the standard metric is consistent but insufficient according to the above findings.

Another point to analyze is the multilabel entities. These entities correspond to text spans associated with more than one entity type, as in the case of the medical term *HTN*, which is both an Abbreviation and Disease. In the Chilean Waiting List corpus, $1,030$ entities participate in this type of nesting. Considering only the F_1 score of both models on these types of entities, the MLC approach with Clinical Flair obtained 85.1, while Joint Labeling with Clinical RoBERTa obtained 84.21. Therefore, the difference in the standard metric cannot be explained by the performance of these types of nestings. In the following sections, we perform a detailed analysis of the model predictions, looking for information that explains the difference in performance between the Joint Labeling and MLC approaches beyond the domain in which they were trained.

Level removed	MLC [CF]	Joint Labeling [CR]	ΔF_1
None	$\mathbf{83.16}_{0.28}$	$79.24_{0.39}$	3.92
≥ 3 (88)	$\mathbf{82.86}_{0.29}$	$78.96_{0.35}$	3.90
≥ 2 (1,875)	$\mathbf{79.08}_{0.55}$	$75.91_{0.49}$	3.17

Table 4: Overall results of our two best models in the Chilean Waiting List when removing deeper entities. ΔF_1 corresponds to the subtraction in the performance between two models. Here, CF stands for Clinical Flair, while CR is Clinical RoBERTa. The values in parentheses correspond to the support.

6.2 Nesting Depth

An interesting point to analyze between both approaches is the variation in the standard metric when deeper nesting level entities are removed. In Table 4, we show the results in the Chilean Waiting List when entities of depths 2, 3, and 4 entities are removed. Here, depth 1 are the outermost entities, while entities in level 4 are the innermost. First, we notice that by removing nested entities of depths 3 and 4, the ΔF_1 score between both models remains similar. However, when we removed entities of depth 2, the difference was reduced by 1 F_1 point. This might suggest that removing inner entities within a nesting implies a higher decay in MLC performance compared to the Joint Labeling approach. To support this hypothesis, we will analyze the performance of both architectures according to entity length.

6.3 Entities of Different Length

In Figure 3, we separate the results obtained in Table 2 depending on the entity's length. The left side of the figure shows that when the entity length increases, the MLC curve gets closer to the Joint Labeling curve, suggesting that the performance on shorter entities is better for MLC. This finding is confirmed when observing the Clinical Trials

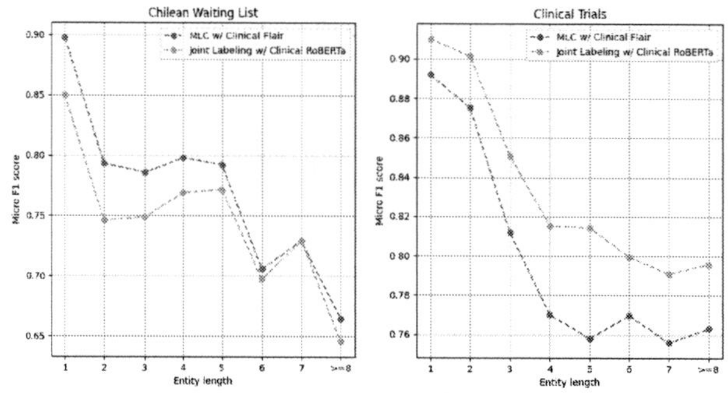

Figure 3: Results of our models according to the entity length.

Length	ΔF_1 Chilean Waiting List	ΔF_1 Clinical Trials
1	$4.76\ (4,198)$	$-1.82\ (5,312)$
2	$4.75\ (1,522)$	$-2.59\ (1,780)$
3	$3.74\ (976)$	$-3.93\ (917)$
4	$2.90\ (667)$	$-4.51\ (442)$
5	$2.08\ (470)$	$-5.62\ (207)$
6	$0.86\ (289)$	$-2.97\ (116)$
7	$-0.01\ (223)$	$-3.49\ (61)$
≥ 8	$1.8\ (492)$	$-3.28\ (105)$

Table 5: ΔF_1 Score between MLC with Clinical Flair and Joint Labeling with Clinical RoBERTa depending on the length of entities. The values in parentheses correspond to the support.

figure, where the curves move further apart as the length increases.

In Table 5, we see this behavior more explicitly using the ΔF_1, which corresponds to the subtraction of the F_1 scores of both models. In the case of MLC, we can see that the most significant difference in the Chilean Waiting List occurs in shorter entities, which could be influencing the standard NER metric. In contrast, in Clinical Trials, although the MLC approach does not outperform Joint Labeling according to the standard metric, the ΔF_1 score decreases as the entities become smaller.

In the following section, we perform a case study on a synthetic dataset created from several clinical corpora in Spanish. The aim is to study if this behavior is repeated in a dataset containing a similar percentage of nested entities compared to the Chilean Waiting List. Note that this dataset was not used in the experiments section since it is not publicly available for privacy reasons; thus, future works could not reproduce the experiments.

7 Case Study

In order corroborate the conclusions presented above, we have created a synthetic nested NER corpus by aggregating the datasets from the PharmaCoNER (Gonzalez-Agirre et al., 2019), CODIESP (Miranda-Escalada et al., 2020), and the recent DisTEMIST (Miranda-Escalada et al., 2022) shared tasks. These datasets are based on the SPACCC corpus[6], a collection of $1,000$ clinical cases from SciELO. Since all the datasets are annotated on the same plain text, merging the annotation of the different tasks is possible. The aggregated dataset is composed of seven entity types, where three are from the PharmaCoNER corpus, two from CODIESP, and one from DisTEMIST.

	Train	Test	Dev
tokens	$240,381$	$29,600$	$31,364$
sentences	$9,482$	$1,120$	$1,230$
avg sent len	25.35	26.43	25.50
entities	$18,912$	$2,283$	$2,597$
avg entity len	2.15	2.21	2.14
nested entities	$8,167$	$1,019$	$1,147$
- entities at level 1	$6,577$	827	938
- entities at level 2	$1,572$	191	209
- entities at level 3	18	1	0

Table 6: Statistics of the SPACCC Aggregated dataset.

To generate the aggregated dataset, some important factors have been considered. First, CODIESP is not a NER task but a clinical coding task. However, the authors annotated not only the ICD-10 codes but also the textual evidence that supports the assigned codes. For this experiment, we used the textual evidence from CODIESP as if they were named entities. Secondly, we have found that some textual evidences are either discontinuous or partially contained within other evidences, better known as crossing entities. Both cases are beyond the scope of this research, so we decided to discard them. Thirdly, DisTEMIST is an ongoing task, and we do not have access to the test set

[6]https://zenodo.org/record/2560316

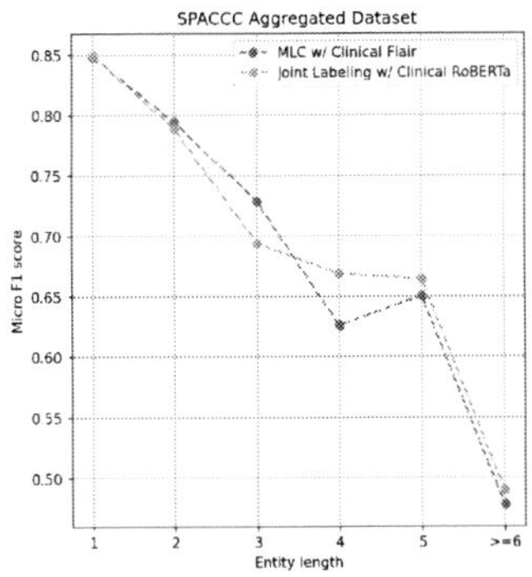

Figure 4: Results of both models on the SPACCC aggregated dataset depending on the entity length.

Metric	MLC	Joint Labeling	Support
standard F_1	**78.53**$_{0.16}$	$78.25_{0.09}$	2,283
m_{flat}	**77.99**$_{0.48}$	$77.48_{0.12}$	1,264
m_{inner}	**79.27**$_{0.10}$	$76.44_{0.41}$	520
m_{outer}	$79.07_{0.60}$	**82.21**$_{0.31}$	499
m_{nested}	$79.18_{0.24}$	**79.23**$_{0.36}$	1,019
$m_{nesting}$	$63.76_{0.60}$	**64.32**$_{0.08}$	499
m_{level_1}	$72.68_{0.03}$	**77.08**$_{0.32}$	827
m_{level_2}	**51.57**$_{2.83}$	$48.71_{0.67}$	191

Table 7: Standard and nested metrics on the SPACCC aggregated dataset.

annotations. For this reason, we only have annotations for 750 clinical cases of the SPACCC corpus. Finally, once the three datasets were aggregated, we found that there were discontinuous annotations between CODIESP and DisTEMIST in 17 of the documents. We removed these documents from the corpus, leaving us with 733 documents. The dataset was divided into 80% for training, 10% for validation, and 10% for testing. The statistics of the corpus are shown in Table 6, and in Appendix A, we show examples of nested entities in this corpus.

According to the standard NER metric, the results for the MLC and Joint Labeling approaches are 78.53 and 78.25, respectively. Although the performance was comparable between both models, analyzing Figure 4, we note that the behavior in the two previous datasets is repeated. The MLC curve is higher than the Joint Labeling curve for the smallest entities, but as the number of tokens increases, the Joint Labeling model obtains slightly better results than MLC.

Considering the m_{nested} and $m_{nesting}$ metrics shown in Table 7, we see that Joint Labeling achieves 79.23 and 64.32 F_1 scores, while MLC obtains 79.18 and 63.76. Therefore, the former architecture handles better the nested entities in this corpus. One possible reason why MLC performs better on the standard evaluation metric is that this model achieves the best results according to the m_{inner} metric by a wide margin, obtaining 79.27 versus 76.44. In contrast, using the m_{outer} metric, MLC achieves 3.14 points less than Joint Labeling. These findings reaffirm our hypothesis that MLC is better at recognizing smaller entities. For example, if we analyze the metrics in each nesting depth level (m_{level_1} and m_{level_2}), we can see how the MLC model obtains better results in recognizing entities in level 2, which are the innermost entities within a nesting. Finally, and as seen in the other corpora, the results according to the $m_{nesting}$ metric are low, and the standard metric cannot reflect this limitation.

8 Conclusions and Future Work

Since most previous works on nested NER have focused on solving the task in English, this paper contributes to the exploration of diverse models for solving the task on two Spanish clinical datasets, resulting in the state-of-the-art in both corpora. Specifically, we explore the advantages and limitations of the Multiple LSTM-CRF approach, which consists of training one model for each entity type, and the Joint Labeling approach, which through an appropriate annotation scheme, allows solving the task by fine-tuning transformer-based models.

To assess the limitations, we studied task-specific metrics for the nested NER task, which consider variables such as the entity position in the nesting, the impact of nesting depth, and entity length. Although our approaches achieve high results according to the standard metric, we found limitations concerning the recognition of nested entities. The main drawbacks of these architectures are the low performance when recognizing complete nestings and the outermost entities of a nesting. In addition, the MLC approach combined with a character-level language model performs less when recognizing entities with many tokens.

We believe this work can contribute to the NLP community to re-think how the nested NER task is being evaluated, considering task-specific metrics beyond the traditional micro F_1 score. Furthermore, our case study on the SPACCC aggregated dataset points out many of the challenges of the nested NER task, especially when complex annotations are allowed due to the aggregation process. Therefore, future work will analyze the performance of other existing architectures beyond the sequence labeling-based approach and compare their performance against our models. We also plan to propose new methods to treat the cases of discontinuous entities and crossing entities, which are entities that overlap others but are not fully contained, to address the nested NER task fully.

Limitations

Although both approaches achieved excellent results across all the datasets in this research, they have clear limitations. The main drawback is that both models cannot handle the case of nested entities of the same type. This is explained since the file format used for training these architectures cannot incorporate this type of nesting. The second major limitation of both models is that they cannot capture the existing relations between inner and outer entities, leading to poor performance in recognizing complete nestings. These limitations could be addressed by using architectures that separate the problem of detecting entity boundaries from classifying the entity type or hypergraph-based models.

Another significant limitation of the MLC architecture is the high computational cost. Although the models of each entity type can be trained in parallel, when scaling to a dataset with many entity types, the training and inference time could increase considerably compared with other models. On the other hand, we have shown that using character-level language models in this architecture obtains low performance when recognizing longer entities.

Finally, despite the Joint labeling approach employing one model for all the entities, its label space increase exponentially with the number of entities involved, resulting in a bigger classification layer and thus requiring more computational resources than standard NER classification layers.

Ethics Statement

We defend that this work meets EMNLP Code of Ethics requirements. Our findings have been corroborated by creating an aggregated dataset and replicating the experiments. Furthermore, the aggregated corpus has been generated using other corpora with a free license (Creative Commons Attribution 4.0 International),[7] and the documents where the annotation quality might have been compromised have been removed from the final corpus.

Acknowledgements

This work was funded by ANID Chile: Basal Funds for Center of Excellence FB210005 (CMM); Millennium Science Initiative Program ICN17_002 (IMFD) and ICN2021_004 (iHealth), and Fondecyt grant 11201250. In addition, it was funded by the Spanish State Secretariat for Digitalization and Artificial Intelligence (SEDIA) within the framework of the Plan-TL[8]. Regarding hardware, the research was partially supported by the supercomputing infrastructure of the NLHPC (ECM-02) and the Patagón supercomputer of Universidad Austral de Chile (FONDEQUIP EQM180042).

References

Ankit Agrawal, Sarsij Tripathi, Manu Vardhan, Vikas Sihag, Gaurav Choudhary, and Nicola Dragoni. 2022. Bert-based transfer-learning approach for nested named-entity recognition using joint labeling. *Applied Sciences*, 12:976;.

Beatrice Alex, Barry Haddow, and Claire Grover. 2007. Recognising nested named entities in biomedical text. In *Biological, translational, and clinical language processing*, pages 65–72, Prague, Czech Republic. Association for Computational Linguistics.

Pablo Báez, Felipe Bravo-Marquez, Jocelyn Dunstan, Matías Rojas, and Fabián Villena. 2022. Automatic extraction of nested entities in clinical referrals in spanish. *ACM Trans. Comput. Healthcare*, 3(3).

Pablo Báez, Fabián Villena, Matías Rojas, Manuel Durán, and Jocelyn Dunstan. 2020. The Chilean waiting list corpus: a new resource for clinical named entity recognition in Spanish. In *Proceedings of the 3rd Clinical Natural Language Processing Workshop*, pages 291–300, Online. Association for Computational Linguistics.

Darina Benikova, Chris Biemann, and Marc Reznicek. 2014. NoSta-D named entity annotation for German: Guidelines and dataset. In *Proceedings of the Ninth International Conference on Language Resources and Evaluation (LREC'14)*, pages 2524–2531, Reykjavik, Iceland. European Language Resources Association (ELRA).

Leonardo Campillos-Llanos, Ana Valverde-Mateos, Adrián Capllonch-Carrión, and Antonio Moreno-Sandoval. 2021. A clinical trials corpus annotated with umls entities to enhance the access to evidence-based medicine. *BMC Medical Informatics and Decision Making*, 21.

[7] https://creativecommons.org/licenses/by/4.0/legalcode

[8] https://plantl.mineco.gob.es/Paginas/index.aspx

Casimiro Pio Carrino, Joan Llop, Marc Pàmies, Asier Gutiérrez-Fandiño, Jordi Armengol-Estapé, Joaquín Silveira-Ocampo, Alfonso Valencia, Aitor Gonzalez-Agirre, and Marta Villegas. 2022. Pretrained biomedical language models for clinical NLP in Spanish. In *Proceedings of the 21st Workshop on Biomedical Language Processing*, pages 193–199, Dublin, Ireland. Association for Computational Linguistics.

José Cañete, Gabriel Chaperon, Rodrigo Fuentes, Jou-Hui Ho, Hojin Kang, and Jorge Pérez. 2020. Spanish pre-trained bert model and evaluation data. In *PML4DC at ICLR 2020*.

Viviana Cotik, Darío Filippo, Roland Roller, Hans Uszkoreit, and Feiyu Xu. 2017. Annotation of entities and relations in Spanish radiology reports. In *Proceedings of the International Conference Recent Advances in Natural Language Processing, RANLP 2017*, pages 177–184, Varna, Bulgaria. INCOMA Ltd.

Jacob Devlin, Ming-Wei Chang, Kenton Lee, and Kristina Toutanova. 2019. BERT: Pre-training of deep bidirectional transformers for language understanding. In *Proceedings of the 2019 Conference of the North American Chapter of the Association for Computational Linguistics: Human Language Technologies, Volume 1 (Long and Short Papers)*, pages 4171–4186, Minneapolis, Minnesota. Association for Computational Linguistics.

Jenny Rose Finkel and Christopher D. Manning. 2009. Nested named entity recognition. In *Proceedings of the 2009 Conference on Empirical Methods in Natural Language Processing*, pages 141–150, Singapore. Association for Computational Linguistics.

Aitor Gonzalez-Agirre, Montserrat Marimon, Ander Intxaurrondo, Obdulia Rabal, Marta Villegas, and Martin Krallinger. 2019. PharmaCoNER: Pharmacological substances, compounds and proteins named entity recognition track. In *Proceedings of The 5th Workshop on BioNLP Open Shared Tasks*, pages 1–10, Hong Kong, China. Association for Computational Linguistics.

Meizhi Ju, Makoto Miwa, and Sophia Ananiadou. 2018. A neural layered model for nested named entity recognition. In *Proceedings of the 2018 Conference of the North American Chapter of the Association for Computational Linguistics: Human Language Technologies, Volume 1 (Long Papers)*, pages 1446–1459, New Orleans, Louisiana. Association for Computational Linguistics.

Arzoo Katiyar and Claire Cardie. 2018. Nested named entity recognition revisited. In *Proceedings of the 2018 Conference of the North American Chapter of the Association for Computational Linguistics: Human Language Technologies, Volume 1 (Long Papers)*, pages 861–871, New Orleans, Louisiana. Association for Computational Linguistics.

Jin-Dong Kim, Tomoko Ohta, Yuka Tateisi, and Jun'ichi Tsujii. 2003. Genia corpus—a semantically annotated corpus for bio-textmining. *Bioinformatics (Oxford, England)*, 19 Suppl 1:i180–2.

Guillaume Lample, Miguel Ballesteros, Sandeep Subramanian, Kazuya Kawakami, and Chris Dyer. 2016. Neural architectures for named entity recognition. In *Proceedings of the 2016 Conference of the North American Chapter of the Association for Computational Linguistics: Human Language Technologies*, pages 260–270, San Diego, California. Association for Computational Linguistics.

Salvador Lima Lopez, Naiara Perez, Montse Cuadros, and German Rigau. 2020. NUBes: A corpus of negation and uncertainty in Spanish clinical texts. In *Proceedings of the 12th Language Resources and Evaluation Conference*, pages 5772–5781, Marseille, France. European Language Resources Association.

Wei Lu and Dan Roth. 2015. Joint mention extraction and classification with mention hypergraphs. In *Proceedings of the 2015 Conference on Empirical Methods in Natural Language Processing*, pages 857–867, Lisbon, Portugal. Association for Computational Linguistics.

Ying Luo and Hai Zhao. 2020. Bipartite flat-graph network for nested named entity recognition. In *Proceedings of the 58th Annual Meeting of the Association for Computational Linguistics*, pages 6408–6418, Online. Association for Computational Linguistics.

Guillermo López-García, José M. Jerez, Nuria Ribelles, Emilio Alba, and Francisco J. Veredas. 2021. Transformers for clinical coding in spanish. *IEEE Access*, 9:72387–72397.

Antonio Miranda-Escalada, Luis Gascó, Salvador Lima-López, Eulàlia Farré-Maduell, Darryl Estrada, Anastasios Nentidis, Anastasia Krithara, Georgios Katsimpras, Georgios Paliouras, , and Martin Krallinger. 2022. Overview of DISTEMIST at BioASQ: Automatic detection and normalization of diseases from clinical texts: results, methods, evaluation and multilingual resources.

Antonio Miranda-Escalada, Aitor Gonzalez-Agirre, Jordi Armengol-Estapé, and Martin Krallinger. 2020. Overview of automatic clinical coding: Annotations, guidelines, and solutions for non-english clinical cases at codiesp track of CLEF ehealth 2020. In *Working Notes of CLEF 2020 - Conference and Labs of the Evaluation Forum, Thessaloniki, Greece, September 22-25, 2020*, volume 2696 of *CEUR Workshop Proceedings*. CEUR-WS.org.

Matias Rojas, Felipe Bravo-Marquez, and Jocelyn Dunstan. 2022a. Simple yet powerful: An overlooked architecture for nested named entity recognition. In *Proceedings of the 29th International Conference on Computational Linguistics*, pages 2108–2117, Gyeongju, Republic of Korea. International Committee on Computational Linguistics.

Matías Rojas, Jocelyn Dunstan, and Fabián Villena. 2022b. Clinical flair: A pre-trained language model for Spanish clinical natural language processing. In *Proceedings of the 4th Clinical Natural Language Processing Workshop*, pages 87–92, Seattle, WA. Association for Computational Linguistics.

Yongliang Shen, Xinyin Ma, Zeqi Tan, Shuai Zhang, Wen Wang, and Weiming Lu. 2021. Locate and label: A two-stage identifier for nested named entity recognition. In *Proceedings of the 59th Annual Meeting of the Association for Computational Linguistics and the 11th International Joint Conference on Natural Language Processing (Volume 1: Long Papers)*, pages 2782–2794, Online. Association for Computational Linguistics.

Takashi Shibuya and Eduard H. Hovy. 2020. Nested named entity recognition via second-best sequence learning and decoding. *Transactions of the Association for Computational Linguistics*, 8:605–620.

Mohammad Golam Sohrab and Makoto Miwa. 2018. Deep exhaustive model for nested named entity recognition. In *Proceedings of the 2018 Conference on Empirical Methods in Natural Language Processing*, pages 2843–2849, Brussels, Belgium. Association for Computational Linguistics.

Ashish Vaswani, Noam Shazeer, Niki Parmar, Jakob Uszkoreit, Llion Jones, Aidan N Gomez, Ł ukasz Kaiser, and Illia Polosukhin. 2017. Attention is all you need. In *Advances in Neural Information Processing Systems*, volume 30. Curran Associates, Inc.

Jue Wang, Lidan Shou, Ke Chen, and Gang Chen. 2020. Pyramid: A layered model for nested named entity recognition. In *Proceedings of the 58th Annual Meeting of the Association for Computational Linguistics*, pages 5918–5928, Online. Association for Computational Linguistics.

Yiran Wang, Hiroyuki Shindo, Yuji Matsumoto, and Taro Watanabe. 2021. Nested named entity recognition via explicitly excluding the influence of the best path. In *Proceedings of the 59th Annual Meeting of the Association for Computational Linguistics and the 11th International Joint Conference on Natural Language Processing (Volume 1: Long Papers)*, pages 3547–3557, Online. Association for Computational Linguistics.

Yu Wang, Hanghang Tong, Ziye Zhu, and Yun Li. 2022. Nested named entity recognition: A survey. *ACM Trans. Knowl. Discov. Data*. Just Accepted.

Juntao Yu, Bernd Bohnet, and Massimo Poesio. 2020. Named entity recognition as dependency parsing. In *Proceedings of the 58th Annual Meeting of the Association for Computational Linguistics*, pages 6470–6476, Online. Association for Computational Linguistics.

Changmeng Zheng, Yi Cai, Jingyun Xu, Ho-fung Leung, and Guandong Xu. 2019. A boundary-aware neural model for nested named entity recognition. In *Proceedings of the 2019 Conference on Empirical Methods in Natural Language Processing and the 9th International Joint Conference on Natural Language Processing (EMNLP-IJCNLP)*, pages 357–366, Hong Kong, China. Association for Computational Linguistics.

A Examples from the SPACCC Aggregated Dataset

As discussed in Section 7, the SPACCC aggregated dataset represents a challenging case study since it may pose severe limitations to straightforward approaches addressing nested NER tasks, mainly due to entity annotations such as discontinuous entities, nested entities, and different entity types. To better visualize such complex entity annotations, we selected some sentences from the SPACCC aggregated dataset before we removed them to perform our experiments. Specifically, Figure 5 shows three different entities, namely, disease entity (DIS_ENFERMEDAD), ICD diagnosis (CIE_DIAGNOSTICO), and protein names (PHA_PROTEINAS), from the PharmaCoNER, CODIESP, and DisTEMIST datasets are presented in different colors to highlight the amount of overlap and crossing between them.

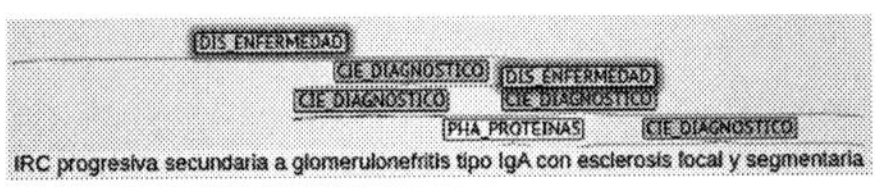

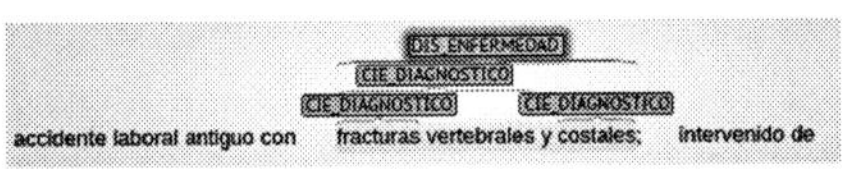

Figure 5: Example of annotations from the SPACCC aggregated dataset with different types of entities belonging to the PharmaCoNER, CODIESP, and DisTEMIST datasets.

Can Current Explainability Help Provide References in Clinical Notes to Support Humans Annotate Medical Codes?

Byung-Hak Kim[†,1] **Zhongfen Deng**[2] **Philip S. Yu**[2] **Varun Ganapathi**[1]

[†]project lead

[1]AKASA, Inc. [2]University of Illinois Chicago

Abstract

The medical codes prediction problem from clinical notes has received substantial interest in the NLP community, and several recent studies have shown the state-of-the-art (SOTA) code prediction results of full-fledged deep learning-based methods. However, most previous SOTA works based on deep learning are still in early stages in terms of providing textual references and explanations of the predicted codes, despite the fact that this level of explainability of the prediction outcomes is critical to gaining trust from professional medical coders. This raises the important question of how well current explainability methods apply to advanced neural network models such as transformers to predict correct codes and present references in clinical notes that support code prediction. First, we present an explainable **R**ead, **A**ttend, and **C**ode (xRAC) framework and assess two approaches, attention score-based xRAC-ATTN and model-agnostic knowledge-distillation-based xRAC-KD, through simplified but thorough human-grounded evaluations with SOTA transformer-based model, RAC. We find that the supporting evidence text highlighted by xRAC-ATTN is of higher quality than xRAC-KD whereas xRAC-KD has potential advantages in production deployment scenarios. More importantly, we show for the first time that, given the current state of explainability methodologies, using the SOTA medical codes prediction system still requires the expertise and competencies of professional coders, even though its prediction accuracy is superior to that of human coders. This, we believe, is a very meaningful step toward developing explainable and accurate machine learning systems for fully autonomous medical code prediction from clinical notes.

1 Introduction

Within current medical systems, the prediction of medical codes from clinical notes is a practical and essential need for every healthcare delivery organization (Dev, 2021). A human coder or health care provider scans medical documentation in electronic health records, identifying important information and annotating codes for that specific treatment or service. With a diverse range of medical services and providers (primary care clinics, specialty clinics, emergency departments, mother-baby units, outpatient and inpatient units, etc.), the complexity of human coders' tasks grows, while productivity standards fall as charts take longer to review. Thus, even partial automation of the annotation workflow will save significant time and effort that human coders currently spend. The biggest challenge, however, is directly identifying appropriate medical codes from thousands of high-dimensional codes from unstructured free-text clinical notes (Dong et al., 2022).

Lately, advanced deep learning-based methods for predicting medical codes based on clinical notes (Kim and Ganapathi, 2021; Sun et al., 2021; Liu et al., 2021; Yuan et al., 2022) have achieved state-of-the-art prediction performance and even reached parity with human coders' performance (Kim and Ganapathi, 2021). However, most current works on medical code prediction based on deep learning models do not provide the end-user with references from the clinical notes to explain why the predicted codes were presented/chosen. There have been some related works that provide the rationales or text highlights from clinical notes to explain why the predictions were made to support humans clinical decision making (Taylor et al., 2021; Cao et al., 2020; Mullenbach et al., 2018; Wood-Doughty et al., 2022). However, to the best of our knowledge, there is still a gap in studies that have thoroughly analyzed explainability to extract supporting text for code prediction, especially made by state-of-the-art (SOTA) transformer-based models such as the RAC model (Kim and Ganapathi, 2021).

Two examples are the attention score-based ap-

Proceedings of the 13th International Workshop on Health Text Mining and Information Analysis (LOUHI), pages 26 - 34
December 7, 2022

proach first introduced in Mullenbach et al. (2018) and the model-independent knowlege-distillation based method recently initiated in Wood-Doughty et al. (2022). The first approach utilizes the per-label attention mechanism to select key sentences for prediction decisions; however, if the model does not have the per-label attention layer, it cannot generate text snippets, and even worse, applying this method to transformer-based architecture and deploying it to production comes with a range of compute and memory challenges (Vaswani et al., 2017). In the second knowledge-distillation approach, a large neural network is distilled into linear student models in a post-hoc manner without sacrificing much accuracy of the teacher model while retaining many advantages of linear models including explainability and smaller model size, which is beneficial for deployment.

This paper makes two primary contributions:

- First, we present a general and explainable xRAC framework that generates evidentiary text snippets for a predicted code, which are oriented towards the needs of a deployment scenario of the RAC model. Then, to better assess the explainability of our xRAC framework, human-grounded evaluations is conducted with two groups of internal annotators, one group with and one group without medical coding expertise. We find that the proposed xRAC framework can benefit professional coders but not lay annotators who lack relevant expertise and competencies.

- Second, we propose code-prior matching and text-prior matching losses to augment the original binary cross-entropy (BCE) loss used to train the RAC model. Because trained models with BCE loss typically tend to focus more on the frequent medical codes and their associated clinical notes portions of the dataset, these new losses are to help distribute the gradient update evenly across all of the codes and note tokens, regardless of code's frequency and token's relevance to codes, so as to improve the xRAC model's prediction as a whole.

2 xRAC Framework

2.1 xRAC-ATTN

The original RAC architecture is built on the code-title guided attention module that considerably improves the per-label attention mechanism first introduced in Mullenbach et al. (2018). This enhanced attention module is to address the extreme sparsity of the large code output space with so called code-title embedding. Because code titles (or descriptions) contain important semantic information and meaning of the codes, the RAC model obtains its embedding from its textual description as shown in the Table 2 examples. Specifically, the code description is fed into an embedding layer, which is then followed by a CNN and Global Max Pooling layer to learn the code embedding.

Therefore, the first xRAC-ATTN directly leverages the attention scores learned in the RAC model to generate the evidence text for each code i. In particular, the attention scores $\mathbf{w}_i^{\text{ATTN}} = (w_{i,1}^{\text{ATTN}}, ..., w_{i,n_x}^{\text{ATTN}})$ on the input tokens for code-i is computed as follows:

$$\mathbf{w}_i^{\text{ATTN}} = \text{Softmax}\left(\frac{\mathbf{e_i}\mathbf{U}_\mathbf{x}^\mathbf{T}}{\sqrt{d}}\right), \quad (1)$$

where $\mathbf{e_i} \in \mathbb{R}^{1\times d}$ is one row of $\mathbf{E_t} \in \mathbb{R}^{n_y \times d}$ which is the code embeddings from the code descriptions, $\mathbf{U_x} \in \mathbb{R}^{n_x \times d}$ is the text representation outputted by the Reader, d is the dimension of code embedding, n_x is the number of tokens in the input document, and n_y is the number of codes in the dataset.

2.2 xRAC-KD

The application of the original idea of knowledge distillation (Hinton et al., 2014) requires specific adjustments to the problem setting of medical codes prediction. Knowledge distillation is typically used to train a compact neural network from a large or ensemble of neural network models. Unlike those standard approaches, xRAC-KD transfers the large RAC-based "teacher" model into a set of reliable and explainable "student" linear models by distilling the predictions made by the large teacher model.

Assume that we have a trained "teacher" neural network $f_{\text{teacher}}(\mathbf{x}_t)$ and training data.[1] xRAC-KD approximates $f_{\text{teacher}}(\mathbf{x}_t)$ with a collection of student linear models $f_{\text{student}}(\mathbf{x}_s) = (f_{s,0}(\mathbf{x}_s), ..., f_{s,n_y}(\mathbf{x}_s))$ defined as

$$f_{s,i}(\mathbf{x}_s) = \mathbf{w}_i^{\text{KD}}\mathbf{x}_s, \quad (2)$$

[1]Note that because there is flexibility in using different representations for the same clinical note, we use different notations $\mathbf{x}_t$ and $\mathbf{x}_s$ to denote a tokenized clinical note. We use Word2Vec for $\mathbf{x}_t$ and bag of words for $\mathbf{x}_s$.

where $\mathbf{w}_i^{\text{KD}} = (w_{i,1}^{\text{KD}}, ..., w_{i,n_x}^{\text{KD}})$. $f_{\text{teacher}}(\mathbf{x}_t)$ produces predicted probability vector $\hat{\mathbf{y}}_t = (\hat{y}_{t,1}, ..., \hat{y}_{t,n_y})$. First xRAC-KD converts $\hat{\mathbf{y}}_t$ to $\mathbf{q}_t = (q_{t,1}, ..., q_{t,n_y})$ which is defined as

$$q_{t,i} = T\text{logit}(\hat{y}_{t,i}) = T\log\left(\frac{\hat{y}_{t,i}}{1-\hat{y}_{t,i}}\right), \quad (3)$$

where a temperature parameter T is to adjust the logit values and set it to 1 for convenience.

Then, as a distillation loss to train the student models $f_{\text{student}}(\mathbf{x}_s)$, xRAC-KD uses the L1 regularized regression loss between $\mathbf{q}_t$ and the student's predicted output vectors $\mathbf{q}_s$ written as follows

$$||\mathbf{q}_t - \mathbf{q}_s||_2 + \lambda||\mathbf{w}_i^{\text{KD}}||_1, \quad (4)$$

with λ parameter. xRAC-KD does not use any additional loss term with respect to the training data's hard labels (either 0 or 1). Once the distilled student models $f_{\text{student}}(\mathbf{x}_s)$ are ready, xRAC-KD finally transforms the output vector $\mathbf{q}_s$ back to the prediction vector $\hat{\mathbf{y}}_s = (\hat{y}_{s,0}, ..., \hat{y}_{s,n_y})$ easily as follows:

$$\hat{y}_{s,i} = \text{expit}\left(\frac{q_{s,i}}{T}\right) = \frac{1}{1+\exp(-q_{s,i}/T)}. \quad (5)$$

The logit and expit transforms defined in Eq. (3) and (5) pairs that are inverse to each other are a fundamental improvement over the initial method presented in Wood-Doughty et al. (2022). Previously, the distilled models showed consistently low precision scores and it was hypothesized for the independence of the distilled linear models. However, by comparing the first and last rows of Table 1, it turns out that this new pair has resulted in a clear outperformance across the board over the logistic regression baseline unlike the initial approach.

2.3 Supporting Text Extraction

Lastly, the evidence text of the xRAC-ATTN and xRAC-KD models is constructed by first locating the *n*-gram with the highest average weight score for each code i calculated as

$$\arg\max_j \sum_{n-gram} w_{i,j}, \quad (6)$$

then *m* tokens on either side of the *n* gram are included to obtain the final subsequence of evidence with length of $n + 2m$. We set n to 4 and m to 5.

2.4 xRAC with Augmented Losses

The RAC model utilizes a transformer encoder and an attention-based architecture to attain SOTA performance. It also makes use of code descriptions to obtain code embeddings. Although the code embeddings obtained from the code description capture the semantic meaning of each code, due to the natural characteristics of medical coding, most of the codes appear just a few times compared to other common codes associated with common diseases.

Similarly, not all tokens in a given piece of text can be learned sufficiently and equally during the training process; therefore, frequent code embedding (as well as token embeddings) will receive more updates than infrequent codes (and tokens). In other words, trained models with BCE loss tend to focus more on the frequent codes and their associated clinical notes portions in the dataset; therefore, we propose code-prior matching and text-prior matching losses to supplement the BCE loss to encourage the models better handle imbalance issues and improve the model's overall prediction.

Code Prior Matching (CPM): To alleviate the issue of frequent codes receiving more updates than infrequent codes during training, CPM is applied to the second to the last output of the Coder, $\mathbf{V_x} \in \mathbb{R}^{n_y \times d}$ defined as

$$\mathbf{V_x} = \text{Softmax}\left(\frac{\mathbf{E_t U_x^T}}{\sqrt{d}}\right)\mathbf{U_x}. \quad (7)$$

The CPM can help the model learn evenly across all codes, regardless of frequency, by imposing constraints on the learned $\mathbf{V_x}$. This prior matching module is implemented by a discriminator D_{cpm}, which shares the same structure as D_{lpm} in Deng et al. (2021) and introduces a regularization loss for each code as

$$\begin{aligned} l_c^i = -(&\mathbb{E}_{\mathbf{c_p}\sim\mathbb{Q}}[\log D_{\text{cpm}}(\mathbf{c_p})] + \\ &\mathbb{E}_{\mathbf{v_i}\sim\mathbb{P}}[\log(1 - D_{\text{cpm}}(\mathbf{v_i}))]), \end{aligned} \quad (8)$$

where $\mathbf{v_i} \in \mathbb{R}^{1\times d}$ is one row of $\mathbf{V_x}$ which is the vector for one code in the dataset, $\mathbb{P}$ is the code embedding distribution learned by the model, $\mathbf{c_p}$ is a prior vector of the same size as $\mathbf{v_i}$ for the given code generated by a uniform distribution $\mathbb{Q}$ in the interval of $[0, 1)$, and l_c^i is the prior matching loss for code-i.[2] We take the average of l_c^i losses

[2] We chose a compact uniform distribution on $[0, 1)$ as the

from all codes to obtain the final CPM loss L_C as follows:

$$L_C = \frac{1}{n_y}\sum_{i=1}^{n_y} l_c^i. \tag{9}$$

Text Prior Matching (TPM): In the RAC model, not all tokens in a particular clinic text note can be learned equally, as the Reader focuses more on tokens related to frequent codes in the data set. To help the model's gradient be updated equally for all tokens in the input, a TPM loss is applied on $\mathbf{U_x}$ output of the Reader. The TPM is also implemented by a discriminator D_{tpm} similar to D_{cpm}, where it introduces another prior matching loss L_T shown as

$$\begin{aligned} l_t^i = &-(\mathbb{E}_{\mathbf{t_p}\sim\mathbb{Q}}[\log D_{tpm}(\mathbf{t_p})] + \\ &\mathbb{E}_{\mathbf{u_i}\sim\mathbb{P}}[\log(1 - D_{tpm}(\mathbf{u_i}))]), \end{aligned} \tag{10}$$

$$L_T = \frac{1}{n_x}\sum_{i=1}^{n_x} l_t^i, \tag{11}$$

where $\mathbf{u_i}$ is one row of U_x that is the embedding of a token in the input document, $\mathbb{P}$ is the distribution of text embedding learned by the model, $\mathbf{t_p}$ is the prior embedding vector for the given token in the input document also generated by a uniform distribution $\mathbb{Q}$ in the interval of $[0, 1)$, and l_t^i is the TPM for a token in the input; we then use the average loss for all tokens in the input document as the final TPM loss L_T, similar to Eq. (9). This loss can make the model evenly learn the embeddings for all tokens in the input, which will be fed to the Coder for code prediction.

Overall Training Loss: Finally, the total augmented loss is written as

$$L_{\text{total}} = L_{\text{BCE}} + \alpha * L_C + \beta * L_T, \tag{12}$$

where α and β are parameters to balance L_C and L_T respectively. The updated RAC model trained with L_{total} instead of BCE loss, is first used for xRAC-ATTN and its performance is shown in the third row of Table 1. Although we used the original RAC model as a teacher model to distill from in xRAC-KD, this updated RAC model can also be used. Comparing the second (RAC model trained with BCE loss) and third rows (updated RAC model trained with L_{total}) in Table 1 shows modest improvements in both standard and hierarchical micro F1 scores, indicating that prior matching modules modestly help to address the imbalanced issues.

prior, which worked better in practice than other priors, such as Gaussian, unit ball, or unit sphere as shown in previous works (Deng et al., 2021; Hjelm et al., 2019).

3 Experimental Results

3.1 MIMIC-III Dataset

The MIMIC-III Dataset (MIMIC v1.4 Johnson et al. (2016)) is a freely accessible medical database that contains de-identified medical data from over 40,000 patients who visited the Beth Israel Deaconess Medical Center between 2001 and 2012.[3] We extract the discharge summaries and the corresponding medical codes, for this study. For a direct comparison with previous works, we use the same data processing, and data split described in (Mullenbach et al., 2018). This processing results in 47,724 samples for training, 1,632 and 3,373 samples for validation and testing, respectively, with an average number of 16 codes assigned to each discharge summary. More dataset statistics, can be found in Table 2 of (Mullenbach et al., 2018).

3.2 Training Details

The xRAC models follow the same training details as the RAC model, which can be found in the original RAC paper (Kim and Ganapathi, 2021). The xRAC-ATTN model is also trained with the same hyperparameters as the RAC model.[4] The xRAC-ATTN model's extra hyperparameters include α and β in Eq. (12), with values of 0.5 and 0.8 respectively. The temperature for the xRAC-KD model is set to 1, λ to 1e-3, and the maximum iteration for the training is set to 800.[5]

3.3 xRAC Model Performance

In addition to the same standard flat metrics used in previous RAC model evaluations, recently introduced hierarchical metrics (e.g. CoPHE (Falis et al., 2021), set-based metrics (Kosmopoulos et al., 2015)) are used. These two metrics take the hierarchical structure of the ICD codes tree into consideration for evaluating codes prediction. The CoPHE

[3] One reason for using the MIMIC-III dataset for this study is that it has been used as standard benchmark in previous studies (Kim and Ganapathi, 2021; Mullenbach et al., 2018), allowing meaningful head-to-head comparisons with our work. We believe that the proposed xRAC model is not limited to the MIMIC-III dataset and will also work well with a MIMIC-IV dataset, but MIMIC-IV-Note is currently not available to the public.

[4] The maximum sequence length is 4096, and there are four stacks of attention layers with single attention head. The code and text embedding dimensions are 300 and the batch size is 16.

[5] For the choices of hyper-parameters, we fine-tuned the model by running a linear search of these hyper-parameters to find the best value at which the model's performance peaks.

Table 1: Medical codes prediction results (in %) by ML systems on the MIMIC-III-full-label testing set as described in Kim and Ganapathi (2021). The bold value shows the best (and highest) value for each column metric. The logistic regression results are taken from Mullenbach et al. (2018), and the RAC results come from Kim and Ganapathi (2021). All numbers are the results of a single run with fixed random seeds, as practiced in the previous literature (Kim and Ganapathi, 2021; Mullenbach et al., 2018) for apples-to-apples comparisons. Note that our baseline is the most recent SOTA model RAC, and our xRAC-ATTN outperforms RAC in most metrics.

Model	**AUC**		**Standard F1**		**Precision@n**			**Hierarchical F1**	
	Macro	Micro	Macro	Micro	5	8	15	CoPHE	Set-Based
Logistic Regression	56.1	93.7	1.1	27.2		54.2	41.1		
RAC	**94.8**	**99.2**	**12.7**	58.6	**82.9**	75.4	**60.1**	62.7	64.0
xRAC-ATTN (ours)	**94.8**	99.1	12.6	**58.8**	**82.9**	**75.6**	**60.1**	**62.9**	**64.3**
xRAC-KD (ours)	93.6	98.7	7.4	46.0	69.4	61.6	48.6	51.8	54.5

metric further utilizes depth-based hierarchical representation and the count of codes at different ancestral levels of the tree to evaluate model's prediction, providing more meaningful evaluation in this context.

The results of the xRAC-ATTN and the xRAC-KD are shown in the last two rows of Table 1 respectively. First, when compared to the prior RAC model trained with BCE loss, the xRAC-ATTN model improves both standard and hierarchical micro F1 scores, as noted by comparing the second and third rows of Table 1, suggesting that the prior matching modules modestly help and effectively improve the SOTA scores. Second, while the xRAC-KD student model (shown in the last row) performs slightly worse than that of the RAC-based teacher model (shown in the second row), it still significantly outperforms the logistic regression baseline (shown in the first row, which was trained from scratch and has the same level of model complexity) across the board, which was not the case in Wood-Doughty et al. (2022).

3.4 Human-Grounded Evaluation

Human Evaluation Design: Human-grounded evaluation is important for evaluating the explainability. Because medical code annotation involves domain knowledge specific to medical coding, human evaluation is challenging; thus, we conducted a human evaluation with two groups of internal annotators. **Group A** had two annotators without medical coding experience and **Group B** had six certified professional coders. While both groups followed the same annotation instructions and guidelines, Group A was supervised by one manager and Group B was supervised by two managers with professional coder management experience to ensure annotation consistency (i.e., inter-annotator agreement) within each group. Group A worked full-time for two weeks to finish all the annotation, while Group B worked part-time for three weeks. Because the two groups of annotators involved in the human evaluation process are well aware that the task involves the medical notes of anonymized patients, the study does not require IRB approval and does not raise any ethical concerns.

Annotation Task Design: We select the overlap of codes predicted between the xRAC-ATTN and xRAC-KD models on the MIMIC-III-full-label testing set and combine the code descriptions and the corresponding textual explanations generated by each model together in a question sheet [6]. We then provide the sheet to Groups A and B for evaluation. Specifically, the question sheet contains six columns which are Question ID, Code and Description, Explanation Text Snippet, Highly Informative, Informative, and Irrelevant (see Table 2 for sample questions). Each code has two different text snippets extracted by two models, respectively. The annotators need to assign one of the three choices, which are highly informative, informative, and irrelevant to every explanation text snippet extracted to support the appearance of the predicted code.

Highly informative is defined as if the text snippet provides an accurate explanation for the pre-

[6]The MIMIC-III dataset's entire test set is used for human evaluation. Specifically, both the xRAC-ATTN and xRAC-KD models take clinical note from each example in the test set as input and predict multiple codes associated with this note. Because each model can predict differently for each example in the test set, we select all the test examples from the two models that are predicted with the same codes to compare their explainability. As a result, there are a total of 3,813 test examples predicted with the same codes by the xRAC-ATTN and xRAC-KD models.

Table 2: Two example questions provided for human evaluation: The codes in these two questions are the same, "521.00, Dental caries, unspecified", however, the two explanation text snippets classified as A) and B) are extracted by two different models, xRAC-ATTN and xRAC-KD. The information about the models is hidden from human annotators, and the order of text snippets for the same code is permuted to prevent the annotators from guessing the models based on the order. Note that **HI**, **I**, and **IR** stand for Highly Informative, Informative, and Irrelevant, respectively.

Question ID	Code and Description	Explanation Text Snippet	HI	I	IR
1	521.00, Dental caries, unspecified	A) surgical or invasive procedure left **and right heart catheterization** coronary angiogram multiple dental extractions			
1	521.00, Dental caries, unspecified	B) balloon s p dental extractions **s p exploratory laparotomy** and cholecystectomy fungal sepsis discharge			

Table 3: The overall informativeness of xRAC-ATTN and xRAC-KD retrieved explanatory text snippets. The left half represents the outcome of Group A's annotation, while the right half represents the outcome of Group B's evaluation. **HI**, **I**, and **IR** stand for Highly Informative, Informative, and Irrelevant, respectively. Percent denotes the ratio of informative text snippets (HI and I) to the total extracted snippets, which is 3,813 (in %).

Model	Group A (Lay Annotators)				Group B (Professional Coders)			
	HI	I	IR	Percent	HI	I	IR	Percent
xRAC-ATTN	1652	1389	772	79.75	1283	1094	1436	62.34
xRAC-KD	865	1318	1630	57.25	145	212	3456	9.36

Table 4: The evaluation agreements on Highly Informative and Informative text snippets between Groups A and B as measured by Jaccard Similarity (in %). Note that we evaluated the annotation consistency between two groups as described in Section 3.4, and the annotation consistency (or correctness) of lay annotators (Group A) is lower than 40% even provided with the same textual references as for professional coders (Group B).

Model	Jaccard Similarty	
	HI	I
xRAC-ATTN	39.2	18.5
xRAC-KD	7.0	5.0

dicted code. Otherwise, it is informative as long as the annotators believe that the text snippet adequately explains the presence of the given code, is related to the code's description, or has a close meaning to the code's description. Because the medical note contains domain knowledge, it is difficult for annotators to assign a finer-grained scale to the textual evidence when deciding between highly informative, informative, and irrelevant.

The final question sheet has a total of 3,813 codes predicted with different supporting text snippets. Unlike all previous studies, which typically collect less than 100 samples from clinicians (e.g., Mullenbach et al. (2018)), the task design of our study is quite unique, as is the volume of questions to our knowledge.

Human Evaluation Results: The results of human evaluation for the explainability of xRAC framework are shown in Tables 3 and 4. Table 3 shows the overall result of the informativeness of the text snippets extracted by xRAC-ATTN and xRAC-KD. The percentage column in Table 3 represents the percentage of explanations annotated as highly informative or informative, excluding irrelevant explanations. Thus, the irrelevant explanations generated by our model are about 20-40% as shown in Table 3.

One can see that there is a much larger gap in xRAC-KD between Group A and Group B than between xRAC-ATTN. Each group of annotators adhered to use the same standard to evaluate the textual explanation and was monitored by managers with professional coder management experience to ensure that there was no annotation variation among annotators in the same group. However, the large deviation between the two groups (Groups A and B) is understandable due to the domain knowledge gap between professional coders and lay annotators. Because of their limited medical knowledge and understanding, lay annotators tend to assign more highly informative and informative to the extracted textual explanation. Whereas, professional coders are much stricter on the informativeness of textual explanations.

In other words, this implies that xRAC-ATTN is a more viable choice than xRAC-KD to extract a text snippet from clinical notes to support code prediction. However, Table 4 shows that the consistency score measured by Jaccard Similarity between two groups is lower than 40% even with xRAC-ATTN. This suggests that the automated extraction system must continue to rely on professional coders' feedback and domain experience, and that text snippets alone are insufficient to replace them. In other words, there is still room to improve explainability for a lay person without expertise to appropriately code.

4 Conclusion

In this paper, a xRAC framework is presented to obtain supporting evidence text from clinical notes that justify the predicted medical codes from medical code prediction systems. We have demonstrated that the proposed xRAC framework may help even complex transformer-based models (e.g., RAC model) to attain high accuracy with a decent level of explainability (which is of high value for deployment scenarios) through quantitative experimental studies and qualitative human-grounded evaluations. It was also shown for the first time that, given the current state of explainability methodologies, using the proposed explainable yet accurate medical codes prediction system still requires professional coders' expertise and competencies.

Limitations

The current human-grounded evaluation studies only a simplified scenario: the impact of clinical-text-based explanations provided alongside predictions on explainability as judged by humans with and without professional coding backgrounds. This exercise sheds light on a key element that is necessary for these AI coding-based models to be useful in real-world deployment scenarios, but does not definitively ascertain that these coding predictions provided alongside explanations of the prediction would enable a transition to AI-driven coding autonomously. First, we have not studied how to incorporate the proposed xRAC framework into a human-in-the-loop situation with human coder feedback, which may be a very common scenario of deployment in practice. Second, we have not compared a full AI-driven coding model with humans-in-the-loop to a human-only process, in terms of speed, manpower needed, and accuracy. Limitations of the prediction model may become relevant in these situations, as human coders must occasionally combine disparate pieces of information together (Dong et al., 2022). Third, while the MIMIC-III dataset provides a useful benchmark for evaluating approaches, it is not representative of the wide range of clinical notes, so it would be beneficial to expand to other data sets with a wider range of codes.

Ethics Statement

First and foremost, an automated and explainable machine learning system for medical code prediction aims to streamline the medical coding workflow, reduce the backlog of human coders by increasing productivity, and assist human coders quickly navigating complex and extended charts while reducing coding errors (Crawford, 2013). Second, an automated and explainable system is designed to lessen the administrative burden on providers, allowing them to focus on providing care rather than mastering the complexities of coding. Furthermore, better automated and explainable software can improve clinical documentation, enhance the overall picture of its quality, and eventually redirect lost healthcare dollars to more meaningful purposes (Shrank et al., 2019).

Acknowledgements

The authors would like to thank everyone who helped with the human-grounded evaluation, especially Juliann Chaparro, Tracy Salyers, Peg Leland, Jasmine Porter, Jenn Arlas, Amber Taylor, Stetson Bauman, Colin Wilde, Beth Cable, Dana Hunter, and Amy Raymond. Angela Kilby and Jiaming Zeng's suggestions to improve the manuscripts are also gratefully acknowledged.

References

Pengfei Cao, Chenwei Yan, Xiangling Fu, Yubo Chen, Kang Liu, Jun Zhao, Shengping Liu, and Weifeng Chong. 2020. Clinical-coder: Assigning interpretable ICD-10 codes to Chinese clinical notes. In *Proceedings of the 58th Annual Meeting of the Association for Computational Linguistics (ACL 2020)*, volume System Demonstrations, page 294–301, Association for Computational Linguistics.

Mark Crawford. 2013. Truth about computer-assisted coding: A consultant, HIM professional, and vendor weigh in on the real CAC impact. *Journal of American Health Information Management Association*, 84(7):24–27.

Zhongfen Deng, Hao Peng, Dongxiao He, Jianxin Li, and Philip Yu. 2021. HTCInfoMax: A global model for hierarchical text classification via information maximization. In *Proceedings of the 2021 Conference of the North American Chapter of the Association for Computational Linguistics: Human Language Technologies*, pages 3259–3265, Online. Association for Computational Linguistics.

Sachin Dev. 2021. The healthcare delivery organization cios' guide to computer-assisted coding. *Gartner Research*.

Hang Dong, Matúš Falis, William Whiteley, Beatrice Alex, Shaoxiong Ji, Jiaoyan Chen, and Honghan Wu. 2022. Automated clinical coding: what, why, and where we are? *npj Digital Medicine*, 5(159).

Matúš Falis, Hang Dong, Alexandra Birch, and Beatrice Alex. 2021. CoPHE: A count-preserving hierarchical evaluation metric in large-scale multi-label text classification. In *Proceedings of the 2021 Conference on Empirical Methods in Natural Language Processing*, pages 907–912, Online and Punta Cana, Dominican Republic. Association for Computational Linguistics.

Geoffrey Hinton, Oriol Vinyals, and Jeff Dean. 2014. Distilling the knowledge in a neural network. In *NeurIPS 2014 Deep Learning and Representation Learning Workshop*.

R Devon Hjelm, Alex Fedorov, Samuel Lavoie-Marchildon, Karan Grewal, Phil Bachman, Adam Trischler, and Yoshua Bengio. 2019. Learning deep representations by mutual information estimation and maximization. In *International Conference on Learning Representations*.

Justin Johnson, Alexandre Alahi, and Li Fei-Fei. 2016. Perceptual losses for real-time style transfer and super-resolution. In *Proceedings of European Conference on Computer Vision 2016 (ECCV 2016)*.

Byung-Hak Kim and Varun Ganapathi. 2021. Read, Attend, and Code: Pushing the limits of medical codes prediction from clinical notes by machines. In *Proceedings of the 6th Machine Learning for Healthcare Conference (MLHC 2021)*, volume 149 of *Proceedings of Machine Learning Research*, pages 196–208. PMLR.

Aris Kosmopoulos, Ioannis Partalas, Eric Gaussier, Georgios Paliouras, and Ion Androutsopoulos. 2015. Evaluation measures for hierarchical classification: a unified view and novel approaches. *Data Mining and Knowledge Discovery*, 29(3):820–865.

Yang Liu, Hua Cheng, Russell Klopfer, Matthew R. Gormley, and Thomas Schaaf. 2021. Effective convolutional attention network for multi-label clinical document classification. In *Proceedings of the 2021 Conference on Empirical Methods in Natural Language Processing (EMNLP 2021)*, pages 5941–5953, Online and Punta Cana, Dominican Republic. Association for Computational Linguistics.

James Mullenbach, Sarah Wiegreffe, Jon Duke, Jimeng Sun, and Jacob Eisenstein. 2018. Explainable prediction of medical codes from clinical text. In *Proceedings of the 2018 Conference of the North American Chapter of the Association for Computational Linguistics: Human Language Technologies, Volume 1 (Long Papers)*, page 1101–1111, New Orleans, Louisiana. Association for Computational Linguistics.

William H. Shrank, Teresa L. Rogstad, and Natasha Parekh. 2019. Waste in the US health care system: Estimated costs and potential for savings. *JAMA Network*, 322(15):1501–1509.

Wei Sun, Shaoxiong Ji, Erik Cambria, and Pekka Marttinen. 2021. Multitask balanced and recalibrated network for medical code prediction. In *European Conference on Machine Learning and Principles and Practice of Knowledge Discovery in Databases (ECML PKDD 2021)*.

Niall Taylor, Lei Sha, Dan W Joyce, Thomas Lukasiewicz, Alejo Nevado-Holgado, and Andrey Kormilitzin. 2021. Rationale production to support clinical decision-making. In *Machine Learning for Health (ML4H 2021)*, volume Extended Abstract.

Ashish Vaswani, Noam Shazeer, Niki Parmar, Jakob Uszkoreit, Llion Jones, Aidan N. Gomez, Lukasz Kaiser, and Illia Polosukhin. 2017. Attention is all you need. In *Advances in Neural Information Processing Systems 30 (NeurIPS 2017)*.

Zach Wood-Doughty, Isabel Cachola, and Mark Dredze. 2022. Model distillation for faithful explanations of medical code predictions. In *Proceedings of the 21st Workshop on Biomedical Language Processing*, page 412–425, Dublin, Ireland. Association for Computational Linguistics.

Zheng Yuan, Chuanqi Tan, and Songfang Huang. 2022. Code Synonyms Do Matter: Multiple synonyms matching network for automatic icd coding. In *Proceedings of the 60th Annual Meeting of the Association for Computational Linguistics (ACL 2022)*.

Distinguishing between focus and background entities in biomedical corpora using discourse structure and transformers

Antonio Jimeno Yepes[1,2] and Karin Verspoor[1,2]

[1]School of Computing Technologies, RMIT University
[2]School of Computing and Information systems, The University of Melbourne
Melbourne, VIC, Australia
[1]{antonio.jose.jimeno.yepes,karin.verspoor}@rmit.edu.au

Abstract

Scientific documents typically contain numerous entity mentions, while only a subset are directly relevant to the key contributions of the paper. Distinguishing these focus entities from background ones effectively could improve the recovery of relevant documents and the extraction of information from documents. To study the identification of focus entities, we developed two large datasets of disease-causing biological pathogens using MEDLINE, the largest collection of biomedical citations, and PubMed Central, a collection of full text articles. The focus entities were identified using human-curated indexing on these collections. Experiments with machine learning methods to identify focus entities show that transformer methods achieve high precision and recall and that document discourse information is relevant. The work lays the foundation for more targeted retrieval/summarisation of entity-relevant documents.

1 Introduction

Scientific documents typically discuss one or more topics linked to key entities of interest. However, entities may also be mentioned incidentally to support argumentation, in discussing related work, or be used in comparison with focus entities of direct interest. Distinguishing between these focus and background entities might improve the selection of information most relevant to a user.

The automatic identification of entities in text is typically achieved using named entity recognition or entity linking methods based on dictionary, rule-based and/or machine learning methods, and aims to identify *all* mentions of entities of the target type(s). However, not all entities correctly identified in a text may be entities relevant for further processing or important to the main conclusions of a document. For example, it has been suggested that only ~10% of chemical mentions play a major role within a chemical patent (Akhondi et al., 2019). Strategies for identifying entities that are in focus in a document enable honing in on critical document information, and can support filtering out entities that are ancillary to the main objectives of the work, e.g. for literature-based discovery applications (Henry and McInnes, 2017).

In this work, we introduce two large datasets annotated with focus and background entities that support experimentation with methods for distinguishing these two types of entity mentions[1]. We evaluated several machine learning algorithms on these dataset, setting baseline results for future work to be done on this task, and laying the foundation for more nuanced treatment of document entities in document retrieval or in summarisation.

2 Related work

Entity salience, relevant to identifying focus entities, has been discussed in previous work. Use of discourse structure has been suggested in previous work on entity salience (Boguraev and Kennedy, 1999; Walker and Walker, 1998). The work of Dunietz and Gillick (2014) evaluates a comprehensive set of features, showing that the discourse structure and centrality may support predicting entity salience. One hypothesis is that the focus and background entities are distributed in specific argumentative sections of a document (Ruch et al., 2007; Jimeno Yepes et al., 2021).

The identification of focus entities has multiple relevant applications. In information retrieval (IR), the objective is to recover documents that are relevant to the user information needs, which is challenging for long documents (Webber et al., 2012) as a larger number of entities are being mentioned. In information extraction (IE), we find the task of named entity recognition (NER), in which the objective is to identify entities of interest, from people and locations to proteins and genes, depending on the domain. In NER, all entities of a certain type

[1]https://zenodo.org/record/5866759

Proceedings of the 13th International Workshop on Health Text Mining and Information Analysis (LOUHI), pages 35 - 40
December 7, 2022 ©2022 Association for Computational Linguistics

are identified, even the ones that are not the main focus (Dunietz and Gillick, 2014).

Our study relates specifically to identification of biological pathogen entities in scientific literature. Pathogen NER has been studied in the Bacteria Biotope shared task (Bossy et al., 2019). The GeoBoost tool (Tahsin et al., 2018) addresses the identification of entities from the gene database GenBank (Benson et al., 2012) and largely includes information about viruses and bacteria.

3 Datasets

Development of large corpora is costly since human annotation is slow and expensive. There are biomedical datasets that have been manually annotated and could be considered as proxy for manual annotation. For this work, we have developed two large corpora automatically using existing resources from the National Center for Biotechnology Information (NCBI) at the NLM. The corpora are targeted to microbial pathogens, some of the most relevant entities for infectious diseases (Balloux and van Dorp, 2017), such as COVID-19.

3.1 MEDLINE citation dataset

MEDLINE[2] is the largest biomedical citation database with over 30 million citations from more than 5,000 journals. MEDLINE is indexed semi-manually (Mork et al., 2013) with the MeSH (Medical Subject Headings) controlled vocabulary[3], providing a resource to identify focus entities in biomedical articles. To identify the pathogens in MEDLINE, we created a dictionary of pathogens and collected MEDLINE citations that indexed these pathogens. MeSH contains 360 of the 2.8k pathogens of interest in our work, which constitutes our focus entities. We applied a dictionary-based approach using ConceptMapper (Tanenblatt et al., 2010; Funk et al., 2014) with evaluation available from Jimeno Yepes and Verspoor (2022).

With the list of PubMed identifiers (PMIDs) obtained using MeSH indexing, we recovered their citations from MEDLINE and annotated the text with the dictionary. Overlapping mentions of the same entity were removed and removed pathogen mentions that could not be identified in MeSH. From the set of selected pathogens identified in the citations, the ones that appeared in the MeSH indexing of the citation were considered focus entities, while the pathogens not mentioned in the indexing were consider background entities. We considered both major and minor MeSH headings. For each pathogen identified in a citation, all of its mentions in text were changed to the string @PATHOGEN$. Table 1 presents the corpus statistics, divided into $2/3$ for training $1/3$ for testing.

3.2 PubMed Central full text dataset

In addition to MEDLINE citations, we also consider full text articles from PubMed Central (Roberts, 2001), a collection of full text articles made available from the NLM. To collect the full text articles from PubMed Central, we used the PMIDs obtained using MeSH indexing and mapped these identifiers to PubMed Central identifiers (PMCIDs). We applied the same methodology to highlight the mentions of a specific pathogen as with the MEDLINE citations. Statistics of the full text collection are available in table 1.

MEDLINE dataset	Training	Testing
Unique citations	622,447	320,318
With more than one pathogen	136,546	70,670
Focus entities	661,470	340,991
Background entities	160,540	82,470
Document avg entities	1.3206	1.3220
Document avg focus entities	1.1250	1.1268
Full text dataset	Training	Testing
Unique articles	79,352	39,677
With more than one pathogen	53,003	26,551
Focus entities	82,922	41,602
Background entities	157,072	78,148
Document avg entities	3.0244	3.0181
Document avg focus entities	1.0450	1.0485

Table 1: Frequency of example documents and statistics on focus and background pathogen entities in MEDLINE and full text datasets.

Full text articles are already divided into discourse sections. We process these sections in two ways, first by concatenating the text in the article following the order in the PMC XML file, in which each section is prefixed by the name of the section starting with the character "@ "and ending with ":", e.g. "@title:". Second, we keep each information in a separate section, which allows only considering text in a specific section and can be used with learning algorithms that leverage this organization. Table 2 shows entities distribution in full text.

[2] https://www.nlm.nih.gov/medline/medline_overview.html

[3] https://www.ncbi.nlm.nih.gov/mesh

Background	Count	Focus	Count
introduction	53,574	abstract	68,098
discussion	53,486	title	46,971
results	37,768	introduction	44,466
abstract	19,860	results	27,177
methods	18,674	discussion	21,313
background	11,483	methods	11,813
title	5,789	background	11,637
conclusions	3,526	conclusions	6,155
the study	969	the study	785
case layout	745	abbreviations	705
all	157,072	all	82,922

Table 2: Frequency of background and focus entities in training full text sections

4 Methods

4.1 Baseline methods

We consider two baselines. The first baseline selects a single focus entity per document on the basis of frequency. We utilised the inverted document frequency of entity mentions to evaluate if frequent entities in the collection should be discounted. The second baseline annotates all entities mentioned as focus entities.

4.2 Bag-of-words entity categorization

In our work, focus entities are identified at the document level. In a sense, we would be categorising the mentions of the entity within a citation as focus or background. In our datasets, the entity of interest has been renamed to @PATHOGEN$.

We trained a linear Support Vector Machine (SVM) (Vapnik, 2013) with modified Huber loss (Zhang, 2004) suited for imbalanced data and AdaBoostM1 (Freund and Schapire, 1997), (both from the MTIMLExtension package[4] optimised for large datasets and using uni-grams and bi-grams) and FastText (Joulin et al., 2017)[5], using default parameters as well for classification.

4.3 Transformer based methods

Focus entities might appear in specific contexts in comparison to background entities. Bag-of-words methods have a limited coverage of the context in which these entities might appear. Recent advances in deep learning have delivered self-attention methods that have led to the Transformer architecture (Vaswani et al., 2017).

BERT (Devlin et al., 2019) is a transformer based method that encodes the input tokens into contextualised embeddings trained on large corpora. Classification is achieved using the output from BERT, pooled on the [CLS] character, and a fully connected layer to predict if an entity is a focus or background one.

BERT supports a maximum size of 512 tokens, while other methods developed using the BERT architecture, such as the Longformer (Beltagy et al., 2020), allow for longer documents. Longformer achieves this by using a sliding window instead of attending to all tokens and by using a global attention mask which we set to the [CLS] token used in text categorisation settings.

Our MEDLINE corpus has an average of 308 tokens per document, with just a 6% of the citations with length above 512 tokens. We have used the SciBERT (Beltagy et al., 2019) pre-trained model[6], truncating documents at 512 tokens. When using Longformer, we considered a maximum document length of 1,250 tokens due to memory limitations. Transformer methods were trained using 80% of the training set for training purposes and 20% as validation set. We used Adam (Kingma and Ba, 2015) with an initial learning rate of 2e-5 for 30 epochs. The model with best performance on the validation set after each epoch was selected.

4.4 Scientific discourse focus entity selection

Scientific articles follow a discourse structure, with information organised into different rhetorical sections. The mention of an entity in a certain section can indicate the relevance of that entity in the document. Only a small number of MEDLINE citations have an explicit discourse structure (Ripple et al., 2011). Hence, we apply a discourse tagger (Li et al., 2021) to annotate sentences of a citation relevant to a discourse section, except to the *title* which is explicitly marked in the metadata. Table 3 shows the frequency of each of the categories.

5 Results

Table 4 shows the results of using the different methods. We observe that the baseline based on classifying all entities identified by our dictionary method as focus entities has maximum recall and already has a high precision. The most-frequent mentioned entity baseline has better precision, with de-

[4] `https://github.com/READ-BioMed/MTIMLExtension`

[5] `https://fasttext.cc`

[6] We have used Huggingface's (Wolf et al., 2020) implementations of tranformer methods.

Category	Background	Focus
fact	33,044	139,290
goal	9,295	56,100
hypothesis	7,544	21,433
implication	14,077	42,828
method	44,132	203,225
none	1,026	4,452
problem	2,858	11,429
result	61,317	181,691
title	44,884	435,100
all	160,540	661,470

Table 3: Frequency of each discourse category in the training MEDLINE dataset

Average	Prec.	Recall	F1
All-focus entities	0.3474	**1.0000**	0.5157
tf baseline	0.7475	0.7078	0.7271
tf-idf baseline	0.7587	0.7184	0.7380
SVM-tag	0.8110	0.6440	0.7179
SVM-all	0.6525	0.7761	0.7090
AdaBoostM1-tag	0.8447	0.8824	0.8631
AdaBoostM1-all	0.7845	0.7580	0.7710
fastText	0.8557	0.7374	0.7922
SciBERT	0.9115	**0.9314**	0.9213
Longformer	**0.9410**	0.9269	**0.9339**

Table 5: Focus entity prediction in PubMed Central. The *All-focus* baseline trivially has perfect Recall.

creased recall. Considering the learning algorithms, SciBERT and Longformer perform better than the bag-of-words algorithms, which is expected since these algorithms do not consider the context of the pathogen mention, even with bigrams. The two deep learning algorithms have similar performance.

Average	Prec.	Recall	F1
All-focus entities	0.8052	**1.0000**	0.8921
tf baseline	0.9047	0.8508	0.8770
tf-idf baseline	0.8838	0.8311	0.8566
SVM	0.8975	0.9450	0.9206
AdaBoostM1	0.8654	**0.9682**	0.9139
fastText	0.8608	0.9572	0.9064
SciBERT	**0.9359**	0.9631	**0.9493**
Longformer	0.9285	0.9679	0.9478

Table 4: Focus entity prediction results on MEDLINE. The *All-focus* baseline trivially has perfect Recall.

Table 5 shows the result of the learning algorithms on the full text dataset. Compared to the MEDLINE corpus, we identify that the baseline methods suffer a substantial drop in performance. This is expected since there are more background entities in the full texts, and the most frequent entity is not always in focus. Bag-of-words methods have a lower performance as well, AdaBoostM1 with tag related words outperforms the other methods, indicating the effectiveness of linking words to article sections. In this set, documents are longer and longformer improves over the SciBERT model, which has a limit of 512 tokens.

6 Discussion

The datasets we have constructed for the identification of focus entities are large, supporting evaluation of a variety of methods and comparison of performance in both short and large documents.

Full text is more challenging compared to citations, consistent with findings on other tasks (Cohen et al., 2010), and mostly due to the higher proportion of focus entities in citations. Machine learning approaches based on bags-of-words tend to improve over simple baseline methods but underperform transformer methods.

The distribution of entities in article sections (table 2) and prediction results in full text (table 5) show that the discourse sections in which entities appear are relevant for the identification of focus entities in scientific articles.

7 Conclusions and future work

We have developed two large datasets of scientific documents for the study of the identification of focus entities. We find that short documents, represented by MEDLINE citations, are easier to process than longer (full-text) documents. Transformer methods showed higher performance.

Future work will address using the proposed methods in scenarios in which focus entities become relevant, and comparing our approach with other existing methods (Lu and Choi, 2021; Dunietz and Gillick, 2014).

8 Acknowledgments

We acknowledge the funding support of the US Army International Pacific Centre, and the support of the US Defence Threat Reduction Agency Biological Materials Information Project team. Experiments were done using the LIEF HPC-GPGPU Facility, supported by LIEF Grant LE170100200, at the University of Melbourne.

References

Saber A Akhondi, Hinnerk Rey, Markus Schwörer, Michael Maier, John Toomey, Heike Nau, Gabriele Ilchmann, Mark Sheehan, Matthias Irmer, Claudia Bobach, Marius Doornenbal, Michelle Gregory, and Jan A Kors. 2019. Automatic identification of relevant chemical compounds from patents. *Database*, 2019. Baz001.

Francois Balloux and Lucy van Dorp. 2017. Q&a: What are pathogens, and what have they done to and for us? *BMC Biology*, 15(1):1–6.

Iz Beltagy, Kyle Lo, and Arman Cohan. 2019. SciBERT: A pretrained language model for scientific text. In *Proceedings of the 2019 Conference on Empirical Methods in Natural Language Processing and the 9th International Joint Conference on Natural Language Processing (EMNLP-IJCNLP)*, pages 3615–3620, Hong Kong, China. Association for Computational Linguistics.

Iz Beltagy, Matthew E Peters, and Arman Cohan. 2020. Longformer: The long-document transformer. *arXiv preprint arXiv:2004.05150*.

Dennis A Benson, Mark Cavanaugh, Karen Clark, Ilene Karsch-Mizrachi, David J Lipman, James Ostell, and Eric W Sayers. 2012. Genbank. *Nucleic acids research*, 41(D1):D36–D42.

Branimir Boguraev and Christopher Kennedy. 1999. Salience-based content characterisation of text documents. *Advances in automatic text summarization*, pages 99–110.

Robert Bossy, Louise Deléger, Estelle Chaix, Mouhamadou Ba, and Claire Nédellec. 2019. Bacteria biotope at BioNLP open shared tasks 2019. In *Proceedings of The 5th Workshop on BioNLP Open Shared Tasks*, pages 121–131, Hong Kong, China. Association for Computational Linguistics.

K Bretonnel Cohen, Helen L Johnson, Karin Verspoor, Christophe Roeder, and Lawrence E Hunter. 2010. The structural and content aspects of abstracts versus bodies of full text journal articles are different. *BMC Bioinformatics*, 11(1):1–10.

Jacob Devlin, Ming-Wei Chang, Kenton Lee, and Kristina Toutanova. 2019. BERT: Pre-training of deep bidirectional transformers for language understanding. In *Proceedings of the 2019 Conference of the North American Chapter of the Association for Computational Linguistics: Human Language Technologies, Volume 1 (Long and Short Papers)*, pages 4171–4186, Minneapolis, Minnesota. Association for Computational Linguistics.

Jesse Dunietz and Daniel Gillick. 2014. A new entity salience task with millions of training examples. In *Proceedings of the 14th Conference of the European Chapter of the Association for Computational Linguistics, volume 2: Short Papers*, pages 205–209, Gothenburg, Sweden. Association for Computational Linguistics.

Yoav Freund and Robert E Schapire. 1997. A decision-theoretic generalization of on-line learning and an application to boosting. *Journal of Computer and System Sciences*, 55(1):119–139.

Christopher Funk, William Baumgartner, Benjamin Garcia, Christophe Roeder, Michael Bada, K Bretonnel Cohen, Lawrence E Hunter, and Karin Verspoor. 2014. Large-scale biomedical concept recognition: an evaluation of current automatic annotators and their parameters. *BMC Bioinformatics*, 15(1):1–29.

Sam Henry and Bridget T. McInnes. 2017. Literature based discovery: Models, methods, and trends. *Journal of Biomedical Informatics*, 74:20–32.

Antonio Jimeno Yepes, Ameer Albahem, and Karin Verspoor. 2021. Using discourse structure of scientific literature to differentiate focusfrom background entities in pathogen characterisation. In *Australasian Language Technology Association*.

Antonio Jimeno Yepes and Karin Verspoor. 2022. Classifying literature mentions of biological pathogens as experimentally studied using natural language processing. *Journal of Biomedical Semantics*.

Armand Joulin, Edouard Grave, Piotr Bojanowski, and Tomas Mikolov. 2017. Bag of tricks for efficient text classification. In *Proceedings of the 15th Conference of the European Chapter of the Association for Computational Linguistics: Volume 2, Short Papers*, pages 427–431, Valencia, Spain. Association for Computational Linguistics.

Diederik P Kingma and Jimmy Ba. 2015. Adam: A method for stochastic optimization. In *International Conference on Learning Representations (Poster volume)*.

Xiangci Li, Gully Burns, and Nanyun Peng. 2021. Scientific Discourse Tagging for Evidence Extraction. In *Proceedings of the 16th Conference of the European Chapter of the Association for Computational Linguistics: Main Volume*, pages 2550–2562, Online. Association for Computational Linguistics.

Jiaying Lu and Jinho D Choi. 2021. Evaluation of unsupervised entity and event salience estimation. In *The International FLAIRS Conference Proceedings*, volume 34.

James G Mork, Antonio Jimeno-Yepes, Alan R Aronson, et al. 2013. The nlm medical text indexer system for indexing biomedical literature. *BioASQ@ CLEF*, 1.

Anna M Ripple, James G Mork, Lou S Knecht, and Betsy L Humphreys. 2011. A retrospective cohort study of structured abstracts in medline, 1992–2006. *Journal of the Medical Library Association: JMLA*, 99(2):160.

Richard J Roberts. 2001. Pubmed central: The genbank of the published literature. *Proceedings of the National Academy of Sciences*, 98(2):381–382.

Patrick Ruch, Celia Boyer, Christine Chichester, Imad Tbahriti, Antoine Geissbühler, Paul Fabry, Julien Gobeill, Violaine Pillet, Dietrich Rebholz-Schuhmann, Christian Lovis, et al. 2007. Using argumentation to extract key sentences from biomedical abstracts. *International journal of medical informatics*, 76(2-3):195–200.

Tasnia Tahsin, Davy Weissenbacher, Karen O'Connor, Arjun Magge, Matthew Scotch, and Graciela Gonzalez-Hernandez. 2018. GeoBoost: Accelerating research involving the geospatial metadata of virus GenBank records. *Bioinformatics*, 34(9):1606–1608.

Michael Tanenblatt, Anni Coden, and Igor Sominsky. 2010. The conceptmapper approach to named entity recognition. In *Proceedings of LREC'10*.

Vladimir Vapnik. 2013. *The nature of statistical learning theory*. Springer Science & Business Media.

Ashish Vaswani, Noam Shazeer, Niki Parmar, Jakob Uszkoreit, Llion Jones, Aidan N Gomez, Łukasz Kaiser, and Illia Polosukhin. 2017. Attention is all you need. In *Advances in Neural Information Processing Systems (NeurIPS) 30*, pages 5998–6008.

Joshi Prince Walker and Marilyn I Walker. 1998. *Centering theory in discourse*. Oxford University Press.

Bonnie Webber, Markus Egg, and Valia Kordoni. 2012. Discourse structure and language technology. *Natural Language Engineering*, 18(4):437–490.

Thomas Wolf, Lysandre Debut, Victor Sanh, Julien Chaumond, Clement Delangue, Anthony Moi, Pierric Cistac, Tim Rault, Remi Louf, Morgan Funtowicz, Joe Davison, Sam Shleifer, Patrick von Platen, Clara Ma, Yacine Jernite, Julien Plu, Canwen Xu, Teven Le Scao, Sylvain Gugger, Mariama Drame, Quentin Lhoest, and Alexander Rush. 2020. Transformers: State-of-the-art natural language processing. In *Proceedings of the 2020 Conference on Empirical Methods in Natural Language Processing: System Demonstrations*, pages 38–45, Online. Association for Computational Linguistics.

Tong Zhang. 2004. Solving large scale linear prediction problems using stochastic gradient descent algorithms. In *Proceedings of the Twenty-first International Conference on Machine Learning*, page 116.

FrenchMedMCQA: A French Multiple-Choice Question Answering Dataset for Medical domain

Yanis Labrak[1] Adrien Bazoge[2] Richard Dufour[2] Béatrice Daille[2]

Pierre-Antoine Gourraud[3] Emmanuel Morin[2] Mickael Rouvier[1]

LIA - Avignon University[1] LS2N - Nantes University[2]

first.lastname@univ-avignon.fr first.lastname@univ-nantes.fr

CHU de Nantes - La clinique des données - Nantes University[3]

Abstract

This paper introduces FrenchMedMCQA, the first publicly available Multiple-Choice Question Answering (MCQA) dataset in French for medical domain. It is composed of 3,105 questions taken from real exams of the French medical specialization diploma in pharmacy, mixing single and multiple answers. Each instance of the dataset contains an identifier, a question, five possible answers and their manual correction(s). We also propose first baseline models to automatically process this MCQA task in order to report on the current performances and to highlight the difficulty of the task. A detailed analysis of the results showed that it is necessary to have representations adapted to the medical domain or to the MCQA task: in our case, English specialized models yielded better results than generic French ones, even though FrenchMedMCQA is in French. Corpus, models and tools are available online.

1 Introduction

Multiple-Choice Question Answering (MCQA) is a natural language processing (NLP) task that consists in correctly answering a set of questions by selecting one (or more) of the given N candidates answers (also called *options*) while minimizing the number of errors. MCQA is one of the most difficult NLP tasks because it requires more advanced reading comprehension skills and external sources of knowledge to reach decent performance.

In MCQA, we can distinguish two types of answers: (1) single and (2) multiple ones. Most datasets focus on single answer questions, such as MCTest (Richardson et al., 2013), ARC-challenge (Clark et al., 2018), OpenBookQA (Mihaylov et al., 2018), QASC (Khot et al., 2019), Social-IQA (Sap et al., 2019), or RACE (Lai et al., 2017). To our knowledge, few studies have been done to construct medical MCQA dataset. We can cite the MedMCQA (Pal et al., 2022) and HEAD-QA (Vilares and Gómez-Rodríguez, 2019) corpora which contain single answer questions in Spanish and English respectively. For the multiple answer questions, MLEC-QA (Li et al., 2021) provides 136k questions in Chinese covering various biomedical sub-fields, such as clinic, public health and traditional Chinese medicine.

The French community has recently greatly increased its efforts to collect and distribute medical corpora. Even if no open language model is currently available, we can cite the named entity recognition (Névéol et al., 2014) and information extraction (Grabar et al., 2018) tasks. However, they remain relatively classic, current approaches already reaching a high level of performance.

In this article, we introduce FrenchMedMCQA, the first publicly available MCQA corpus in French related to the medical field, and more particularly in the pharmacological domain. This dataset contains questions taken from real exams of the French diploma in pharmacy. Among the difficulties related to the task, the questions asked may require a single answer for some and multiple ones for others. We also propose to evaluate state-of-the-art MCQA approaches, including an original evaluation of several word representations across languages.

Main contributions of the paper concern (1) the distribution of an original MCQA dataset in French related to the medical field, (2) a state-of-the-art approach on this task and a first analysis of the results, and (3) an open corpus, including tools and models, all available online on demand.

2 The FrenchMedMCQA Dataset

In this section, we detail the FrenchMedMCQA dataset and discuss data collection and distribution.

2.1 Dataset collection

The questions and their associated candidate answer(s) were collected from real French pharmacy

Proceedings of the 13th International Workshop on Health Text Mining and Information Analysis (LOUHI), pages 41 - 46
December 7, 2022

exams on the remede[1] website. This site was built around a community linked to the medical field (medicine, pharmacy, odontology...), offering multiple information (news, job offers, forums...) both for students and also professionals in these sectors of activity. Questions and answers were manually created by medical experts and used during examinations. The dataset is composed of 2,025 questions with multiple answers and 1,080 with a single one, for a total of 3,105 questions. Each instance of the dataset contains an identifier, a question, five options (labeled from A to E) and correct answer(s). The average question length is 14.17 tokens and the average answer length is 6.44 tokens. The vocabulary size is of 13k words, of which 3.8k are estimated medical domain-specific words (*i.e.* related to the medical field). We find an average of 2.5 medical domain-specific words in each question (17% of words in average of a question) and 2.0 in each answer (36% of words in average of an answer). On average, a targeted medical domain-specific word is present in 2 questions and in 8 answers.

2.2 Dataset distribution

Table 1 presents the proposed FrenchMedMCQA dataset distribution for the train, development (dev) and test sets detailed per number of answers (*i.e.* number of correct responses per question). Globally, 70% of the questions are kept for the train, 10% for validation and last 20% for testing.

# Answers	Training	Validation	Test	Total
1	595	164	321	**1,080**
2	528	45	97	**670**
3	718	71	141	**930**
4	296	30	56	**382**
5	34	2	7	**43**
Total	**2171**	**312**	**622**	**3,105**

Table 1: FrenchMedMCQA dataset distribution.

3 Methods

The use alone of the question to automatically find the right answer(s) is not sufficient in the context of a MCQA task. State-of-the-art approaches then require external knowledge to improve system performances (Izacard and Grave, 2020; Khashabi et al., 2020). In our case, we decide to build a two-step retriever-reader architecture comparable to UnifiedQA (Khashabi et al., 2020), where the retriever job is to extract knowledge from an external corpus and using it by the reader to predict the correct answers for each question. Figure 1 presents the two-step general pipeline, first step being the retriever module, that extracts external context from the question (see Section 3.1), and second step being the reader, called here question-answering module (see Section 3.2), that automatically selects answer(s) to the targeted question.

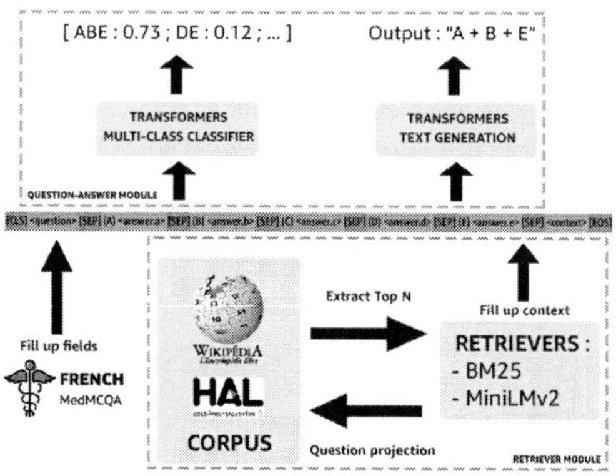

Figure 1: Steps of the pipeline.

3.1 Retriever module

An external medical-related corpus fully composed of French has first been collected from two online sources: Wikipedia life science and HAL, the latter being an open archive run by the French National Centre for Scientific Research (CNRS) where authors can deposit scholarly documents from all academic fields. In our case, we focus on extracting papers and thesis from various specialization, such as Human health and pathology, Cancerology, Public health and epidemiology, Immunology, Pharmaceutical sciences, Psychiatric disorders and Drugs. This results in 1 million of passages (*i.e.* a portion of text that contains at least 100 characters) in HAL and 286k passages in Wikipedia.

This corpus is then used as a context extension for a question. We therefore used a retriever pipeline to automatically assign questions to the most likely passage in the external source. Two retrieval approaches are compared in this article:

- BM25 Okapi (Trotman et al., 2014) for the implementation of the base BM25 algorithm (Robertson and Sparck Jones, 1988).
- SentenceTransformers framework (Reimers

[1] http://www.remede.org/internat/pharmacie/qcm-internat.html

and Gurevych, 2019) is used to perform semantic search using state-of-the-art language representations taken from Huggingface's Transformers library (Wolf et al., 2019).

For both approaches, the goal is to embed each passage of the external corpus into a vector space using one of the two representations. On its side, the question is concatenated with the five options (*i.e.* answers associated to the question) to form a new query embedded in the same vector space. Embeddings from question and passages are finally compared to return the closest passages of a query (here, the cosine similarity is the distance metric). For the SentenceTransformers approach, we used a fast and non domain specific model called `MiniLMv2` (Wang et al., 2020). Note that the 1-best passage is only used in these experiments.

3.2 Question-answer module

A goal of our experiments was to compare baseline approaches regarding two different paradigms. The first one is referred to a discriminative approach and consists in assigning one of N classes to the input based on their projection in a multidimensional space. We also referred to it as a multi-class task. At the opposite, the second method is a generative one which consists of generating a sequence of tokens, also called *free text*, based on a sequence of input tokens identical to the one used for the discriminative approach. The difference with the discriminative approach lies in the fact that we are not outputting a single class, like `ABE` for the question `6234176387997480960`, but a sequence of tokens following the rules of the natural language and referring to a combination of classes like `A + B + E` in the case of our studied generative model (see Section 3.2.2).

3.2.1 Discriminative representations

Four discriminative representations are studied in this paper. We firstly propose to use **CamemBERT** (Martin et al., 2020), a generic French pre-trained language model based on RoBERTa (Liu et al., 2019). Since no language representation adapted to the medical domain are publicly available for French, we propose to evaluate the two pre-trained representations **BioBERT** (Lee et al., 2019) and **PubMedBERT** (Gu et al., 2022), both trained on English medical data and reaching SOTA results on biomedical NLP tasks, including QA (Pal et al., 2022). Finally, we consider a multilingual generic pre-trained model, **XLM-RoBERTa** (Conneau et al., 2020) based on RoBERTa, to evaluate the gap in terms of performance with CamemBERT.

3.2.2 Generative representation

Recently, generative models have demonstrated their interest on several NLP tasks, in particular for text generation and comprehension tasks. Among these approaches, **BART** (Lewis et al., 2019) is a denoising autoencoder built with a sequence-to-sequence model. Due to its bidirectional encoder and left-to-right decoder, it can be considered as generalizing BERT and GPT (Radford et al., 2019), respectively. BART training has two stages: (1) a noising function used to corrupt the input text, and (2) a sequence-to-sequence model learned to reconstruct the original input text. We then propose to evaluate this representation in this paper.

4 Experimental protocol

Each studied discriminative and generative model is fine-tuned on the MCQA task with FrenchMedMCQA training data using an input sequence composed of a question, its associated options (*i.e.* possible answers) and its additional context, all separated with a "[SEP]" token, e.g. [CLS] <question> [SEP] (A) <answer.a> [SEP] (B) <answer.b> [SEP] (C) <answer.c> [SEP] (D) <answer.d> [SEP] (E) <answer.e> [SEP] <context> [EOS].

For each question, the context is the text passage with highest confidence rate and can either be obtained using the BM25 algorithm or semantic search as described in Section 3.1.

Concerning the outputs of the systems, we have for the BART generative model a plain text containing the letter of the answers from A to E separated with plus signs in case of the questions with multiple answers, *e.g.* A + D + E. For the other architectures (*i.e.* discriminative approaches), we simplify the multi-label problem into a multi-class one by classifying the inputs into one of the 31 existing combinations in the corpus. Here, a class may be a combination of multiple labels, *e.g.* if the correct answers are the A and B ones, then we consider the correct class being AB, which explains the number of 31 classes.

4.1 Evaluation metrics

The majority of tasks concentrate either on multi-class or binary classification since they have a single class at a time. However, occasionally, we will

	Without Context		Wiki w/ BM25		HAL w/ BM25		Wiki w/ MiniLMv2		HAL w/ MiniLMv2	
Architecture	Hamming	EMR	Hamming	EMR	Hamming	EMR	Hamming	EMR	Hamming	EMR
BioBERT V1.1	36.19	15.43	**38.72**	16.72	33.33	14.14	35.13	16.23	34.27	13.98
PubMedBERT	33.98	14.14	34.00	13.98	35.66	15.59	33.87	14.79	35.44	14.79
CamemBERT-base	36.24	16.55	34.19	14.46	34.78	15.43	34.66	14.79	34.61	14.95
XLM-RoBERTa-base	37.92	17.20	31.26	11.89	35.84	16.07	32.47	14.63	33.00	14.95
BART-base	31.93	15.91	34.98	**18.64**	33.80	17.68	29.65	12.86	34.65	18.32

Table 2: Performance (in %) on the test set using the Hamming score and EMR metrics.

have a task where each observation has many labels. In this case, we would have different metrics to evaluate the system itself because multi-label prediction has an additional notion of being partially correct. Here, we focused on two metrics called the Hamming score (commonly also multi-label accuracy) and Exact Match Ratio (EMR).

4.1.1 Hamming score

The accuracy for each instance is defined as the proportion of the predicted correct labels to the total number (predicted and actual) of labels for that instance. Overall accuracy is the average across all instances. It is less ambiguously referred to as the Hamming score rather than Multi-label Accuracy.

4.1.2 Exact Match Ratio (EMR)

The Exact Match Ratio (EMR) is the percentage of predictions matching exactly the ground truth answers. To be computed, we sum the number of fully correct questions divided by the total number of questions available in the set. A question is considered *fully correct* when the predictions are exactly equal to the ground truth answers for the question (*e.g.* all multiple answers should be correct to count as a correct question).

5 Results

Table 2 compiled the performance (in terms of Hamming score and EMR) of all the studied architectures and retrievers pipelines. For sake of comparison, the column *Without Context* has been added, considering that no retriever is used (*i.e.* no external passage is present in the QA system).

As we can see, the best performing model is different according to the used metric. **BioBERT V1.1** reaches best performance using the Hamming score and **BART-base** in the case of the EMR. These first observations are quite surprising since both models are trained on English data. While we could expect higher performance with French models (CamemBERT for example), the fact that these models are trained on specialized data for one (BioBERT) and on a model designed for the targeted task (SOTA on question-answering for BART) finally shows that language models trained on generic data are inefficient for the MCQA task on medical domain.

In all considered architectures, context seems to have a small impact on systems performance, with a limited increase or drop depending on the configurations. Clearly, the **RoBERTa** performance is much higher without context (*i.e.* without the use of the retriever part), while models based on **BERT** generally (8 times on 12) outperform their own baseline performances with external context. The fact that we consider the 1-best passage only may explain this impact.

Concerning **XLM-RoBERTa-base** (cross lingual representation), we obtain in the case of the context extracted using BM25 from Wikipedia, the worst Hamming score and EMR out of all the discriminative approaches. This confirms our first observation that a non-specialized model does not allow to achieve the best performance on this task.

Using BM25 promotes better context than semantic search using **MiniLMv2** on both Wikipedia and HAL for most of the runs. Finally, the source depends of the retriever and model used. A majority of the experiments demonstrate that HAL outperforms Wikipedia on BM25 despite the fact that the best model was obtained using Wikipedia.

The scripts to replicate the experiments[2] as well as the pre-trained models[3] are available online.

6 Conclusion

We proposed in this paper FrenchMedMCQA, an original, open and publicly available Multiple-Choice Question Answering (MCQA) dataset in the medical field. This is the first French corpus in this domain, including single and multiple answers to questions. Several state-the art systems have been evaluated to show current performance on the dataset. The analysis of these first results notably

[2]https://github.com/qanastek/FrenchMedMCQA

[3]https://huggingface.co/qanastek/FrenchMedMCQA-BioBERT-V1.1-Wikipedia-BM25/tree/main

highlighted the fact that language models specialized to the medical domain allow us to reach better performance than generic models, even if these have been trained in a different language (here, English biomedical models applied to French).

In future works, we will focus on improving the existing methods for the task of MCQA, considering other strategies for the retriever module (multiple passages, combining contexts...). Likewise, we will also consider the construction of data representation models for French specialized for medical domain.

7 Acknowledgments

This work was financially supported by Zenidoc, the DIETS project financed by the Agence Nationale de la Recherche (ANR) under contract ANR-20-CE23-0005 and the ANR AIBy4 (ANR-20-THIA-0011).

References

Peter Clark, Isaac Cowhey, Oren Etzioni, Tushar Khot, Ashish Sabharwal, Carissa Schoenick, and Oyvind Tafjord. 2018. Think you have solved question answering? try arc, the ai2 reasoning challenge. *ArXiv*, abs/1803.05457.

Alexis Conneau, Kartikay Khandelwal, Naman Goyal, Vishrav Chaudhary, Guillaume Wenzek, Francisco Guzmán, Edouard Grave, Myle Ott, Luke Zettlemoyer, and Veselin Stoyanov. 2020. Unsupervised cross-lingual representation learning at scale. In *Proceedings of the 58th Annual Meeting of the Association for Computational Linguistics*, pages 8440–8451, Online. Association for Computational Linguistics.

Natalia Grabar, Vincent Claveau, and Clément Dalloux. 2018. CAS: French corpus with clinical cases. In *Proceedings of the Ninth International Workshop on Health Text Mining and Information Analysis*, pages 122–128, Brussels, Belgium. Association for Computational Linguistics.

Yu Gu, Robert Tinn, Hao Cheng, Michael Lucas, Naoto Usuyama, Xiaodong Liu, Tristan Naumann, Jianfeng Gao, and Hoifung Poon. 2022. Domain-specific language model pretraining for biomedical natural language processing. *ACM Transactions on Computing for Healthcare*, 3(1):1–23.

Gautier Izacard and Edouard Grave. 2020. Leveraging passage retrieval with generative models for open domain question answering. *arXiv preprint arXiv:2007.01282*.

Daniel Khashabi, Sewon Min, Tushar Khot, Ashish Sabharwal, Oyvind Tafjord, Peter Clark, and Hannaneh Hajishirzi. 2020. Unifiedqa: Crossing format boundaries with a single qa system.

Tushar Khot, Peter Clark, Michal Guerquin, Peter Jansen, and Ashish Sabharwal. 2019. Qasc: A dataset for question answering via sentence composition.

Guokun Lai, Qizhe Xie, Hanxiao Liu, Yiming Yang, and Eduard Hovy. 2017. RACE: Large-scale ReAding comprehension dataset from examinations. In *Proceedings of the 2017 Conference on Empirical Methods in Natural Language Processing*, pages 785–794, Copenhagen, Denmark. Association for Computational Linguistics.

Jinhyuk Lee, Wonjin Yoon, Sungdong Kim, Donghyeon Kim, Sunkyu Kim, Chan Ho So, and Jaewoo Kang. 2019. BioBERT: a pre-trained biomedical language representation model for biomedical text mining. *Bioinformatics*.

Mike Lewis, Yinhan Liu, Naman Goyal, Marjan Ghazvininejad, Abdelrahman Mohamed, Omer Levy, Ves Stoyanov, and Luke Zettlemoyer. 2019. Bart: Denoising sequence-to-sequence pre-training for natural language generation, translation, and comprehension.

Jing Li, Shangping Zhong, and Kaizhi Chen. 2021. MLEC-QA: A Chinese Multi-Choice Biomedical Question Answering Dataset. In *Proceedings of the 2021 Conference on Empirical Methods in Natural Language Processing*, pages 8862–8874, Online and Punta Cana, Dominican Republic. Association for Computational Linguistics.

Yinhan Liu, Myle Ott, Naman Goyal, Jingfei Du, Mandar Joshi, Danqi Chen, Omer Levy, Mike Lewis, Luke Zettlemoyer, and Veselin Stoyanov. 2019. Roberta: A robustly optimized bert pretraining approach.

Louis Martin, Benjamin Muller, Pedro Javier Ortiz Suárez, Yoann Dupont, Laurent Romary, Éric de la Clergerie, Djamé Seddah, and Benoît Sagot. 2020. CamemBERT: a tasty French language model. In *Proceedings of the 58th Annual Meeting of the Association for Computational Linguistics*, pages 7203–7219, Online. Association for Computational Linguistics.

Todor Mihaylov, Peter Clark, Tushar Khot, and Ashish Sabharwal. 2018. Can a suit of armor conduct electricity? a new dataset for open book question answering. In *EMNLP*.

Aurélie Névéol, Cyril Grouin, Jeremy Leixa, Sophie Rosset, and Pierre Zweigenbaum. 2014. The QUAERO French medical corpus: A ressource for medical entity recognition and normalization. In *Proc of BioTextMining Work*, pages 24–30.

Ankit Pal, Logesh Kumar Umapathi, and Malaikannan Sankarasubbu. 2022. Medmcqa: A large-scale multi-subject multi-choice dataset for medical domain question answering. In *Proceedings of the Conference on Health, Inference, and Learning*, volume 174 of *Proceedings of Machine Learning Research*, pages 248–260. PMLR.

Alec Radford, Jeff Wu, Rewon Child, David Luan, Dario Amodei, and Ilya Sutskever. 2019. Language models are unsupervised multitask learners.

Nils Reimers and Iryna Gurevych. 2019. Sentence-bert: Sentence embeddings using siamese bert-networks.

Matthew Richardson, Christopher J.C. Burges, and Erin Renshaw. 2013. MCTest: A challenge dataset for the open-domain machine comprehension of text. In *Proceedings of the 2013 Conference on Empirical Methods in Natural Language Processing*, pages 193–203, Seattle, Washington, USA. Association for Computational Linguistics.

Stephen E. Robertson and Karen Sparck Jones. 1988. *Relevance Weighting of Search Terms*, page 143–160. Taylor Graham Publishing, GBR.

Maarten Sap, Hannah Rashkin, Derek Chen, Ronan LeBras, and Yejin Choi. 2019. Socialiqa: Commonsense reasoning about social interactions.

Andrew Trotman, Antti Puurula, and Blake Burgess. 2014. Improvements to bm25 and language models examined. In *Proceedings of the 2014 Australasian Document Computing Symposium*, ADCS '14, page 58–65, New York, NY, USA. Association for Computing Machinery.

David Vilares and Carlos Gómez-Rodríguez. 2019. HEAD-QA: A healthcare dataset for complex reasoning. In *Proceedings of the 57th Annual Meeting of the Association for Computational Linguistics*, pages 960–966, Florence, Italy. Association for Computational Linguistics.

Wenhui Wang, Hangbo Bao, Shaohan Huang, Li Dong, and Furu Wei. 2020. Minilmv2: Multi-head self-attention relation distillation for compressing pre-trained transformers.

Thomas Wolf, Lysandre Debut, Victor Sanh, Julien Chaumond, Clement Delangue, Anthony Moi, Pierric Cistac, Tim Rault, Rémi Louf, Morgan Funtowicz, Joe Davison, Sam Shleifer, Patrick von Platen, Clara Ma, Yacine Jernite, Julien Plu, Canwen Xu, Teven Le Scao, Sylvain Gugger, Mariama Drame, Quentin Lhoest, and Alexander M. Rush. 2019. Huggingface's transformers: State-of-the-art natural language processing.

A Large-Scale Dataset for Biomedical Keyphrase Generation

Maël Houbre, Florian Boudin and **Béatrice Daille**
Nantes Université, École Centrale Nantes, CNRS, LS2N, UMR 6004, F-44000 Nantes, France
first.last@univ-nantes.fr

Abstract

Keyphrase generation is the task consisting in generating a set of words or phrases that highlight the main topics of a document. There are few datasets for keyphrase generation in the biomedical domain and they do not meet the expectations in terms of size for training generative models. In this paper, we introduce kp-biomed, the first large-scale biomedical keyphrase generation dataset with more than 5M documents collected from PubMed abstracts. We train and release several generative models and conduct a series of experiments showing that using large scale datasets improves significantly the performances for present and absent keyphrase generation. The dataset is available under CC-BY-NC v4.0 license at https://huggingface.co/datasets/taln-ls2n/kpbiomed.

1 Introduction

Keyphrase generation aims at automatically generating a set of keyphrases, that is, words and phrases that summarize a given document. Since they distill the important information from documents, keyphrases have showed to be useful in many applications, most notably in information retrieval (Fagan, 1987; Zhai, 1997; Jones and Staveley, 1999; Song et al., 2006; Boudin et al., 2020) and summarization (Zha, 2002; Wan et al., 2007; Qazvinian et al., 2010).

Current models for generating keyphrases are built upon the sequence-to-sequence architecture (Sutskever et al., 2014) and are able to generate absent keyphrases that is, keyphrases that do not appear in the source text. However, training these models require large amounts of labeled data (Meng et al., 2021). Unfortunately, such data is only available for limited domains and languages which greatly limits the applicability of these models (Ye and Wang, 2018). This work addresses this issue and introduces kp-biomed, the first large-scale dataset for keyphrase generation in the biomedical domain.

Creating labeled data for keyphrase generation is a challenging task, requiring expert annotators and great effort (Kim et al., 2010; Augenstein et al., 2017). A commonly-used approach to cope with this task is to collect scientific abstracts and use keyphrases provided by authors as a proxy for expert annotations. Authors provide keyphrases without any vocabulary constraint to highlight important points of their article; whereas indexers use a specific vocabulary and focus on indexing the article within a collection (Névéol et al., 2010). Therefore, keyphrases may differ from MeSH headings which are another indexing resource in the biomedical domain. Fortunately, author keyphrases are becoming increasingly available in the biomedical domain (Névéol et al., 2010), since they can be incorporated into search strategies in PubMed to improve retrieval effectiveness (Lu and Kipp, 2014). Despite this, the largest keyphrase-labeled biomedical dataset that we know of has about 3k abstracts, all of which are labeled with present-only keyphrases (Gero and Ho, 2019). In this paper, we take advantage of the expansive PubMed database to build a sufficiently large dataset to train biomedical keyphrase generation models[1]. We then compare models trained with different training set sizes to highlight the impact of dataset sizes in keyphrase generation. Our contributions are as follows:

- kp-biomed, a large, publicly available dataset for keyphrase generation in the biomedical domain, available through the Huggingface dataset platform[2];
- Transformer-based models for biomedical keyphrase generation, providing open bench-

[1] KP20k is currently considered as the reference dataset size ($\geq 500k$) to train keyphrase generation models
[2] https://huggingface.co/datasets/taln-ls2n/kpbiomed

Proceedings of the 13th International Workshop on Health Text Mining and Information Analysis (LOUHI), pages 47 - 53
December 7, 2022 ©2022 Association for Computational Linguistics

marks to stimulate further work in the area[3];

- Performance analysis of our models, which provides valuable insights into their generalization ability to other domains.

2 Dataset

We employ the December 2021 baseline set of MEDLINE/PubMed citation records[4] as a resource for collecting abstracts, which contains over 33 million records. We extracted all the records (5.9 million) that include a title, an abstract and some author keyphrases. Records of papers published between 1939 and 2011 only account for a small fraction of these extracted records (3%) and were further filtered out to avoid possible diachronic issues. Last, we went through the remaining records to split the semicolon-separated list of author keyphrases and discard those having keyphrases with punctuation in it. The resulting dataset is composed of 5.6 million abstracts and was randomly and evenly divided by publishing year into training, validation and test splits. To investigate the impact of the amount of training data on the quality of the generated keyphrases, the training split was further divided into increasingly large subsets: small (500k), medium (2M) and large (5.6M). The training splits are also evenly divided by publishing year.

Statistics of the `kp-biomed` dataset are detailed in Table 1 along with other commonly-used datasets for keyphrase generation and extraction. We are aware of only two datasets in the biomedical domain: `NamedKeys` (Gero and Ho, 2019) which is made up of MEDLINE/PubMed abstracts and is therefore mostly included in `kp-biomed`, and Schutz (Schutz, 2008) which is composed of full-text articles from the same source. It is worth noting that these datasets are very limited in size (3k and 1.3k documents respectively) compared to recent keyphrase generation datasets `KP20k` (Meng et al., 2017), `KPTimes` (Gallina et al., 2019) and `LDKP10k` (Mahata et al., 2022). Table 1 shows that thanks to the amount of papers available in MEDLINE/PubMed, `kp-biomed` is the largest of all aforementioned datasets, being more than 10 times larget than `KP20k` which is the current reference dataset for keyphrase generation. The average number of keyphrases per document (#kp) in `kp-biomed` is roughly the same than in `KP20k` and `LDKP10k` which have their keyphrases assigned by authors as well. However, we see that this number is way below the average number of keyphrases assigned by professional indexers like in Inspec (Hulth, 2003) or when authors' keyphrases are combined with readers' as in SemEval-2010 (Kim et al., 2010). The unusually high number of keyphrases per document in `NamedKeys`, despite having author assigned keyphrases, is because of two restrictive criteria. Indeed, each article has at least 5 keyphrases all of which have to occur in the source text. The average number of words per keyphrase (#kp_len) is also comparable for all scientific datasets regardless of the kind of annotators.

Using keyphrases as proxies for indexing or expanding documents with queries composed of words that do not appear in the source text, has been proven more useful to enhance document retrieval than using words occurring in the text (Boudin et al., 2020; Nogueira et al., 2019). In keyphrase generation, we call those keyphrases absent keyphrases, for which several definitions are being used. We refer to the definition from (Meng et al., 2017) "we denote phrases that do not match any contiguous subsequence of source text as absent keyphrases" which was then precised in (Boudin and Gallina, 2021). In (Gero and Ho, 2019) the keyphrase "anesthesia" is considered present if the word "postanesthesia" is in the source text. In our case, it is considered absent which is why `NamedKeys` does not appear with 100% present keyphrases in Table 1. The main difference between `kp-biomed` and `NamedKeys`, despite the number of documents, is the proportion of absent keyphrases. `kp-biomed` contains about 34% of absent keyphrases which is in the same range as scientific datasets `KP20k` and `LDKP10k` that were designed to train neural generative approaches (Meng et al., 2017; Mahata et al., 2022).

3 Experiments

3.1 Models

In keyphrase generation, the architectures are currently mainly based on autoencoders with Recurrent Neural Networks (Meng et al., 2017; Chen et al., 2018, 2019; Chan et al., 2019) or Transformers (Meng et al., 2021; Ahmad et al., 2021).

Following the work of (Meng et al., 2021) that obtained state-of-the-art results with Transform-

[3]https://huggingface.co/datasets/taln-ls2n/kpbiomed-models

[4]https://ftp.ncbi.nlm.nih.gov/pubmed/baseline/

Domain	Dataset	#train	#val	#test	#doc len	#kp	#kp len	P A
Biomedical	**kp-biomed** (ours)	5.6M	20k	20k	271	5.3	1.9	
	NamedKeys	–	–	3k	276	14.3	1.9	
	Schutz	–	–	1.3k	5.4k	5.4	1.9	
General scientific articles	KP20k	530k	20k	20k	175	5.3	2.1	
	SemEval-2010	144	–	100	192	15.4	2.1	
	Inspec	1k	500	500	138	9.8	2.3	
	LDKP10k	1.3M	10k	10k	4.9k	6.9	2.1	
News	KPTimes	260k	10k	20k	921	5.0	1.5	

Table 1: Statistics of the proposed dataset. For comparison purposes, we also report statistics of commonly-used and other biomedical datasets. Columns P and A are respectively the percentage of keyphrases occurring in the source text and absent ones.

ers, we used two different generative BART models (Lewis et al., 2020) and compared their performances on different domains. However, in this article we did not seek to get state-of-the-art results, but rather introduce kp-biomed to the community with results on well known baselines, which is why we employed pre-trained models that we just fine-tuned for keyphrase generation (Chowdhury et al., 2022). The models are BioBART-base (Yuan et al., 2022) which is already pre-trained on PubMed and BART-base (Lewis et al., 2020) which is pre-trained on news, books and webtext. To the best of our knowledge, there is no generic scientific BART model. Therefore, we chose BioBART for fine-tuning on scientific datasets rather than BART. Models are available via the huggingface platform.

For comparison with extractive approaches, we considered MultipartiteRank (Boudin, 2018) as a baseline, which is state-of-the-art in unsupervised graph-based keyphrase extraction. We used the implementation available in the keyphrase extraction toolkit pke[5] with the default settings.

3.2 Experimental settings

We followed the One2Seq paradigm (Meng et al., 2021) for training which consists of generating the keyphrases of an input article as a single sequence. For each article, we concatenated the ground truth keyphrases as a single sequence with a special delimiter. Following (Meng et al., 2021), present keyphrases were ordered by their first occurrence in the source text followed by the absent ones.

We trained each model for 10 epochs with a batch size of 128. We set the input length limit at 512 tokens for the text and 128 tokens for the reference keyphrase sequence. All the parameters and the training were handled with the hugging-face trainer API[6]. Hyperparameters and hardware details are available in appendix A. Training the BioBART-base model on the small training split for 10 epochs took about 9 hours and about 110 hours on the large training split. Once models were trained, we over-generated keyphrase sequences using beam search with a beam width of 20 for evaluation. Inference on test sets took around 50 minutes each.

3.3 Evaluation

We evaluated our models on 3 datasets, kp-biomed for biomedical data, KP20k for generic scientific documents and KPTimes for news articles. We did not use NamedKeys as a test set as we noticed a substantial overlap with our training set. We evaluated present and absent keyphrase generation separately to get better insights of our models' performances. To that end, we only compared each model's output to the present (respectively absent) keyphrases of the ground truth. For present keyphrases we employed F1@M and F1@10. F1@M is the F1 measure applied on the first keyphrase sequence generated by the model whereas F1@10 evaluates the top ten generated keyphrases. We evaluated absent keyphrase generation with R@10 which is the recall on the top 10 generated keyphrases. As F1@10 and R@10 require 10 keyphrases, if we did not have enough unique keyphrases with

[5]https://github.com/boudinfl/pke

[6]Our code is available for reproducibility. https://github.com/MHoubre/kpbiomed

Model	kp-biomed		KP20k		KPTimes	
	F1@10	F1@M	F1@10	F1@M	F1@10	F1@M
MultipartiteRank	15.3	–	12.9	–	16.7	–
BioBART-small	31.4	32.5	25.2	27.1	22.0	24.4
BioBART-medium	32.5†	33.8†	26.2†	28.2†	22.1	24.6
BioBART-large	**33.1**†	**34.7**†	26.9†	28.9†	23.5†	26.2†
BioBART-KP20k	28.2	29.5	**28.6**†	**31.9**†	16.8	19.2
BART-KPTimes	9.1	9.6	3.6	2.7	**29.7**†	**39.4**†

Table 2: Performances of the models on present keyphrase generation. †means significant improvements over BioBART-small. Second best results are underlined.

our over generation, we added the token "<unk>" until we reached 10 keyphrases. The generated keyphrases and the reference were stemmed with the Porter Stemmer to reduce matching errors. To measure statistical significance, we opted for Student's t-test at $p < 0.01$.

3.4 Results

The macro-averaged results of the evaluation are reported in Table 2 and Table 3. BioBART-KP20k (respectively BART-KPTimes) stands for the BioBART (respectively BART) model which has been fine-tuned on KP20k (respectively KPTimes). For BioBART models, we add the size of the kp-biomed training split in the name for clarity.

Model	kp-biomed	KP20k	KPTimes
	R@10	R@10	R@10
BioBART-small	3.3	1.8	2.6
BioBART-medium	3.6†	1.9	2.7
BioBART-large	**4.1**†	1.9	2.1
BioBART-KP20k	2.9	**5.5**†	1.6
BART-KPTimes	1.5	0.8	**39.1**†

Table 3: Performances of the models on absent keyphrase generation. †means significant improvements over BioBART-small. Second best results are underlined.

Transformer based approaches achieve the best results but only on the datasets they were trained on as previously showed for RNN based approaches in (Gallina et al., 2019). For present keyphrase generation, BioBART-large achieves significant improvements compared to its small and medium counterparts in all datasets. This shows that using more data does improve the performances of the generative approaches in predicting present keyphrases in in and out of domain data. The performance drop of BioBART-KP20K on kp-biomed is interestingly much more controlled than BioBART models' on KP20k. Compared to BioBART-small which has been trained on the same amount of data, the drop in F1@M is only of 7.5% relative for BioBART-KP20k when it is of 16.6% relative for BioBART-small. We think that BioBART's pre-training may be beneficial for BioBART-KP20k on kp-biomed. On news articles though, BioBART-KP20k shows a relative drop of 35%, when it is only of 25% relative for BioBART-small. When used on out of domain data, BART-KPTimes performs even worse than MultipartiteRank.

In absent keyphrase generation, models fail in attaining significant improvements outside of their domain. Using more data does not seem to help for out of domain absent keyphrase generation. We can explain the high results of BART-KPTimes on its test set by the fact that many of the absent keyphrases are common to numerous articles.

We also think that the keyphrase order that we chose for training is one reason for the models' poor abstractive results. To verify this hypothesis, we compute the average percentage of the models' predictions appearing in the source text. Results are reported in Table 4. For @10, we removed all the added <unk> tokens before computing. It is clear that the extraction percentage of each model decreases when using top 10 predictions on all datasets. This shows that models prioritize generating present keyphrases which can then lead to low quality absent candidates.

Model	kp-biomed		KP20k		KPTimes	
	@M	@10	@M	@10	@M	@10
BioBART-large	96.3	92.2	94.8	88.5	93.5	84.6
BioBART-KP20k	95.4	84.5	91.8	82.7	83.7	66.6
BART-KPTimes	46.0	31.2	21.4	17.4	65.8	50.7

Table 4: Extraction percentage in top M and top 10 predictions

4 Conclusion

This paper introduces `kp-biomed`, the first large scale dataset for biomedical keyphrase generation. We hope this new dataset will stimulate new research in biomedical keyphrase generation. Several generation models have been trained on this dataset and showed that having more data significantly improves the performances for present and absent keyphrase generation. However, models still perform very poorly on absent keyphrase generation even when using larger amounts of data. In future work, we will focus on how to use `kp-biomed` to improve biomedical absent keyphrase generation.

5 Broader Impact and Ethics

`kp-biomed` contains some abstracts that are part of copyright protected articles. As the "all rights reserved" statement is optional to be copyright protected, removing articles with this statement does not solve the problem (i.e no copyright statement does not mean free of use data). To be able to collect, work with these data and share the dataset to the research community, we complied with the conditions of US fair use and the exceptions from the 2019/79 EU guideline on using copyright content in text and data mining for research purposes. One of those criteria was to not use the data for commercial purposes which is why we opted for the Creative Commons Non Commercial use license CC-BY-NC v4.0.

Acknowledgements

We thank the anonymous reviewers for their valuable input on this article and our colleagues from the TALN team at LS2N for their proofreading and feedback. This work is part of the ANR DELICES project (ANR-19-CE38-0005) and was performed using HPC resources from GENCI-IDRIS (Grant 2022-[AD011013670]).

References

Wasi Ahmad, Xiao Bai, Soomin Lee, and Kai-Wei Chang. 2021. Select, extract and generate: Neural keyphrase generation with layer-wise coverage attention. In *Proceedings of the 59th Annual Meeting of the Association for Computational Linguistics and the 11th International Joint Conference on Natural Language Processing (Volume 1: Long Papers)*, pages 1389–1404, Online. Association for Computational Linguistics.

Isabelle Augenstein, Mrinal Das, Sebastian Riedel, Lakshmi Vikraman, and Andrew McCallum. 2017. SemEval 2017 task 10: ScienceIE - extracting keyphrases and relations from scientific publications. In *Proceedings of the 11th International Workshop on Semantic Evaluation (SemEval-2017)*, pages 546–555, Vancouver, Canada. Association for Computational Linguistics.

Florian Boudin. 2018. Unsupervised Keyphrase Extraction with Multipartite Graphs. In *Proceedings of the 2018 Conference of the North American Chapter of the Association for Computational Linguistics: Human Language Technologies, Volume 2 (Short Papers)*, pages 667–672, New Orleans, Louisiana. Association for Computational Linguistics.

Florian Boudin and Ygor Gallina. 2021. Redefining absent keyphrases and their effect on retrieval effectiveness. In *Proceedings of the 2021 Conference of the North American Chapter of the Association for Computational Linguistics: Human Language Technologies*, pages 4185–4193, Online. Association for Computational Linguistics.

Florian Boudin, Ygor Gallina, and Akiko Aizawa. 2020. Keyphrase generation for scientific document retrieval. In *Proceedings of the 58th Annual Meeting of the Association for Computational Linguistics*, pages 1118–1126, Online. Association for Computational Linguistics.

Hou Pong Chan, Wang Chen, Lu Wang, and Irwin King. 2019. Neural Keyphrase Generation via Reinforcement Learning with Adaptive Rewards. In *Proceedings of the 57th Annual Meeting of the Association for Computational Linguistics*, pages 2163–2174, Florence, Italy. Association for Computational Linguistics.

Jun Chen, Xiaoming Zhang, Yu Wu, Zhao Yan, and Zhoujun Li. 2018. Keyphrase Generation with Correlation Constraints. In *Proceedings of the 2018 Conference on Empirical Methods in Natural Language Processing*, pages 4057–4066, Brussels, Belgium. Association for Computational Linguistics.

Wang Chen, Yifan Gao, Jiani Zhang, Irwin King, and Michael R. Lyu. 2019. Title-Guided Encoding for Keyphrase Generation. *Proceedings of the AAAI Conference on Artificial Intelligence*, 33(01):6268–6275. Number: 01.

Md Faisal Mahbub Chowdhury, Gaetano Rossiello, Michael Glass, Nandana Mihindukulasooriya, and Alfio Gliozzo. 2022. Applying a Generic Sequence-to-Sequence Model for Simple and Effective Keyphrase Generation. *arXiv:2201.05302 [cs]*. ArXiv: 2201.05302.

J. Fagan. 1987. Automatic phrase indexing for document retrieval. In *Proceedings of the 10th Annual International ACM SIGIR Conference on Research and Development in Information Retrieval*, SIGIR '87, page 91–101, New York, NY, USA. Association for Computing Machinery.

Ygor Gallina, Florian Boudin, and Beatrice Daille. 2019. KPTimes: A large-scale dataset for keyphrase generation on news documents. In *Proceedings of the 12th International Conference on Natural Language Generation*, pages 130–135, Tokyo, Japan. Association for Computational Linguistics.

Zelalem Gero and Joyce C. Ho. 2019. Namedkeys: Unsupervised keyphrase extraction for biomedical documents. In *Proceedings of the 10th ACM International Conference on Bioinformatics, Computational Biology and Health Informatics*, BCB '19, page 328–337, New York, NY, USA. Association for Computing Machinery.

Anette Hulth. 2003. Improved automatic keyword extraction given more linguistic knowledge. In *Proceedings of the 2003 Conference on Empirical Methods in Natural Language Processing*, pages 216–223.

Steve Jones and Mark S. Staveley. 1999. Phrasier: A system for interactive document retrieval using keyphrases. In *Proceedings of the 22nd Annual International ACM SIGIR Conference on Research and Development in Information Retrieval*, SIGIR '99, page 160–167, New York, NY, USA. Association for Computing Machinery.

Su Nam Kim, Olena Medelyan, Min-Yen Kan, and Timothy Baldwin. 2010. SemEval-2010 task 5 : Automatic keyphrase extraction from scientific articles. In *Proceedings of the 5th International Workshop on Semantic Evaluation*, pages 21–26, Uppsala, Sweden. Association for Computational Linguistics.

Mike Lewis, Yinhan Liu, Naman Goyal, Marjan Ghazvininejad, Abdelrahman Mohamed, Omer Levy, Veselin Stoyanov, and Luke Zettlemoyer. 2020. BART: Denoising Sequence-to-Sequence Pre-training for Natural Language Generation, Translation, and Comprehension. In *Proceedings of the 58th Annual Meeting of the Association for Computational Linguistics*, pages 7871–7880, Online. Association for Computational Linguistics.

Kun Lu and Margaret E.I. Kipp. 2014. Understanding the retrieval effectiveness of collaborative tags and author keywords in different retrieval environments: An experimental study on medical collections. *Journal of the Association for Information Science and Technology*, 65(3):483–500.

Debanjan Mahata, Navneet Agarwal, Dibya Gautam, Amardeep Kumar, Swapnil Parekh, Yaman Kumar Singla, Anish Acharya, and Rajiv Ratn Shah. 2022. LDKP: A Dataset for Identifying Keyphrases from Long Scientific Documents. *arXiv:2203.15349 [cs]*. ArXiv: 2203.15349.

Rui Meng, Xingdi Yuan, Tong Wang, Sanqiang Zhao, Adam Trischler, and Daqing He. 2021. An empirical study on neural keyphrase generation. In *Proceedings of the 2021 Conference of the North American Chapter of the Association for Computational Linguistics: Human Language Technologies*, pages 4985–5007, Online. Association for Computational Linguistics.

Rui Meng, Sanqiang Zhao, Shuguang Han, Daqing He, Peter Brusilovsky, and Yu Chi. 2017. Deep keyphrase generation. In *Proceedings of the 55th Annual Meeting of the Association for Computational Linguistics (Volume 1: Long Papers)*, pages 582–592, Vancouver, Canada. Association for Computational Linguistics.

Aurélie Névéol, Rezarta Islamaj Doğan, and Zhiyong Lu. 2010. Author keywords in biomedical journal articles. In *AMIA annual symposium proceedings*, volume 2010, page 537. American Medical Informatics Association.

Rodrigo Nogueira, Wei Yang, Jimmy Lin, and Kyunghyun Cho. 2019. Document Expansion by Query Prediction. ArXiv:1904.08375 [cs].

Vahed Qazvinian, Dragomir R. Radev, and Arzucan Özgür. 2010. Citation summarization through keyphrase extraction. In *Proceedings of the 23rd International Conference on Computational Linguistics (Coling 2010)*, pages 895–903, Beijing, China. Coling 2010 Organizing Committee.

Alexander Thorsten Schutz. 2008. Keyphrase extraction from single documents in the open domain exploiting linguistic and statistical methods. Master's thesis, Digital Enterprise Research Institute, National University of Ireland, Galway.

Min Song, Il Yeol Song, Robert B. Allen, and Zoran Obradovic. 2006. Keyphrase extraction-based query expansion in digital libraries. In *Proceedings of the 6th ACM/IEEE-CS Joint Conference on Digital Libraries*, JCDL '06, page 202–209, New York, NY, USA. Association for Computing Machinery.

Ilya Sutskever, Oriol Vinyals, and Quoc V Le. 2014. Sequence to sequence learning with neural networks. In *Advances in Neural Information Processing Systems*, volume 27. Curran Associates, Inc.

Xiaojun Wan, Jianwu Yang, and Jianguo Xiao. 2007. Towards an iterative reinforcement approach for simultaneous document summarization and keyword extraction. In *Proceedings of the 45th Annual Meeting of the Association of Computational Linguistics*, pages 552–559, Prague, Czech Republic. Association for Computational Linguistics.

Hai Ye and Lu Wang. 2018. Semi-supervised learning for neural keyphrase generation. In *Proceedings of the 2018 Conference on Empirical Methods in Natural Language Processing*, pages 4142–4153, Brussels, Belgium. Association for Computational Linguistics.

Hongyi Yuan, Zheng Yuan, Ruyi Gan, Jiaxing Zhang, Yutao Xie, and Sheng Yu. 2022. BioBART: Pretraining and evaluation of a biomedical generative language model. In *Proceedings of the 21st Workshop on Biomedical Language Processing*, pages 97–109, Dublin, Ireland. Association for Computational Linguistics.

Hongyuan Zha. 2002. Generic summarization and keyphrase extraction using mutual reinforcement principle and sentence clustering. In *Proceedings of the 25th Annual International ACM SIGIR Conference on Research and Development in Information Retrieval*, SIGIR '02, page 113–120, New York, NY, USA. Association for Computing Machinery.

Chengxiang Zhai. 1997. Fast statistical parsing of noun phrases for document indexing. In *Fifth Conference on Applied Natural Language Processing*, pages 312–319, Washington, DC, USA. Association for Computational Linguistics.

A Training settings

- GPU type: V100 32Go
- Number of GPU: 4
- Trainer: Seq2SeqTrainer
- Text max size: 512
- Reference max size: 128
- Optimizer : AdamW
- Learning rate: 5×10^{-5}
- Other hyperparameters: Seq2SeqTrainer default values

Section Classification in Clinical Notes with Multi-task Transformers

Fan Zhang and **Itay Laish** and **Ayelet Benjamini** and **Amir Feder**
{zhanfan,itaylaish,ayelet,afeder}@google.com
Google Research

Abstract

Clinical notes are the backbone of electronic health records, often containing vital information not observed in other structured data. Unfortunately, the unstructured nature of clinical notes can lead to critical patient-related information being lost. Algorithms that organize clinical notes into distinct sections are often proposed in order to allow medical professionals to better access information in a given note. These algorithms, however, often assume a given partition over the note, and classify section types given this information. In this paper, we propose a multi-task solution for note sectioning, where a single model identifies context changes and labels each section with its medically-relevant title. Results on in-distribution (MIMIC-III) and out-of-distribution (private held-out) datasets reveal that our approach successfully identifies note sections across different hospital systems.

1 Introduction

The increasing role of free-text narrative in Electronic Health Records (EHR) is both a blessing and a curse. It allows much more nuanced information about patients' conditions being saved and documented (Uzuner et al., 2010; Jensen et al., 2012; Wang et al., 2018; Feder et al., 2020). However, the unstructured nature of this data can also make it unavailable to medical care givers interested in searching for specific patient-related information (Walsh, 2004; Ford et al., 2016).

To better organize free-form clinical notes and allow researchers and practitioners to quickly search over them, many solutions were proposed, mainly focusing on sectioning notes to correspond to headers described within the note (Pomares-Quimbaya et al., 2019). These solutions were often rule-based (Savova et al., 2010), identifying common section headers in the data. Unfortunately, this approach often failed to correctly classify sections across different hospital departments, care providers and EHR systems. For brevity throughout this paper, we refer these as different *data sources* or *distributions* interleaving. Alternatively, machine learning methods were proposed to classify individual text-spans and map them into a pre-existing list of possible sections. This approach successfully outperformed rule-based approaches, but was often not deployed because of its inability to identify section-boundaries.

With the recent success of transformer-based models in natural language understanding, we identify an opportunity to tackle the section boundary detection problem alongside section classification, and propose a unified solution. Our approach is based on pre-trained encoder-only transformer models, which were shown to produce superior results on natural language understanding (NLU) tasks broadly (Vaswani et al., 2017; Devlin et al., 2018), and specifically on clinically-relevant data (Alsentzer et al., 2019; Lee et al., 2020).

We start by exploring current section classification methods (§2). Then, we introduce our baseline, a marker-based section header extraction system, and describe how to use it to generate training labels for ML-based methods (§3). We then pose hypotheses for when should ML systems outperform rule-based approaches, and propose solutions based on the hypotheses (§4). We continue by proposing a dataset for training multi-task transformers from rule-based labels (§5) and demonstrate how such models can outperform rule-based approach on in-distribution and out-of-distribution data (§6). Finally, we conclude our work in light of our posed hypotheses (§7).

2 Related Work

Identifying section headers in free-form clinical notes is long identified as a crucial task for organizing patient-level data in biomedical informatics (Li et al., 2010). Both ML-based and rule-based solutions were proposed in the last decade to solve

Proceedings of the 13th International Workshop on Health Text Mining and Information Analysis (LOUHI), pages 54 - 59
December 7, 2022 ©2022 Association for Computational Linguistics

the problem (Pomares-Quimbaya et al., 2019). Unfortunately, existing solutions focus on solving the relatively narrowly-defined task of classifying predefined sections into section types, assuming that section borders are already given (Li et al., 2010; Tepper et al., 2012; Dai et al., 2015; Pomares-Quimbaya et al., 2019). In practice, however, we often observe complete notes, and are tasked with identifying distinct paragraphs and only then classifying them into individual sections.

Recently, there has been an influx of research demonstrating the power of pre-trained language models in solving multi-task problems (Peng et al., 2020; Radford et al., 2019; Wolf et al., 2020), including on long texts (Beltagy et al., 2020). Following this newly-formed conventional wisdom, we embrace this approach here, and propose an ML architecture that attempts to jointly detect section boundaries and classify individual sections.

3 Marker-based Section Header Extraction

We start by developing a *marker*-based section header extractor. This extractor will then be used for labeling our training data in §5 and as a baseline in §6. In this approach, a *marker* corresponds to a word that is usually used as the header of section. E.g. **PMH** is a typical marker word that represents the section Past Medical History. After examining patterns in the data, we discover hundreds of such markers in the MIMIC-III dataset (Johnson et al., 2016). Lines that start with these markers are extracted and are labeled as section headers. These headers mark the boundary between two sections and the text between two headers is then treated as one single section.

During our exploration, we recognized that there exists correlations between the type of the notes and the structure of the sections in the note. With that in regard, we customized our markers to the type of notes and certain markers will only be applied when the type of the note matches our definition. We identified 5 core note types that are most important for our usage: *History and Physical*, *Progress*, *Discharge summary*, *Consult* and *Operative*.

Building on the MIMIC-III dataset, we use an iterative approach to collect markers. A bootstrapping marker set is first developed on a sampled set of notes from the MIMIC-III dataset. The marker set is then used to extract sections on the sampled set and the extracted sections are then sent to experienced clinicians for rating. New markers are then added according to the errors collected from the raters and then used on a new set of randomly collected notes. This process is repeated until no more errors are reported from the raters. In practice, we found that this method shows both high precision and high coverage in recognizing the sections. However, this approach does not work well when we try to transfer it to a new dataset where the medical notes come from a different healthcare provider, where we see the recall numbers dropping significantly (see §6 for complete results).

By analyzing the errors, we are seeing the following patterns:

- Plurals. E.g. "complaint" and "complaints"
- Abbreviations. E.g. "ALL" for "allergy", "Hx" for "history".
- Mutation of marker orders. E.g. "PMH/PSH" and "PSH/PMH".
- Additional punctuations. E.g "** Marker **"
- Character splits, e.g. "P H Y S I C A L E X A M I N A T I O N".

By comparing with MIMIC-III, we observe that while the headers are semantically similar across different healthcare providers, many cases are actually non-identical and can therefore cause recall losses. Additionally, this approach does not take the context information into consideration, and is not able to recognize many cases above even if the section contents look similar to each other.

4 Section Classification Methods

To build solutions that are robust across different distributions or require minimum learning efforts to adapt, we need to understand what is the transferable knowledge that applies. Based on our experiences in building the marker-based approach, we have the following hypotheses:

- **Section titles are shared across different sources.** This means that we expect the same terminology is shared across different sources. For example, we would expect "assessment and plan" is a common terminology shared across different sources. There might be some slight variations, for example, "chief complaint" vs. "chief complaints".

- **Section contents are similar across different sources.** We are expecting that the same section type would have similar content even if they are from different healthcare systems.

- **Structure of the sections is different for different types of notes.** For example, we would expect the discharge summary notes to have a different set of sections in comparison to operative notes.

For the first hypothesis, we want to understand if we can build source-agnostic solutions by just expanding the markers used in the baseline. For the second hypothesis, we want to check if we can improve the accuracy of section type identification with additional information from the surrounding text of the section titles. For the last hypothesis, we propose to take advantage of the note type information within a multi-task framework.

4.1 Expanding section titles

We first explore the approach using the same mechanism as the baseline approach, where we identify section titles as section boundaries and categorize sections according to the marker types. Instead of the exact match used in the baseline approach, we modify the method to fuzzy-match with embedding-based similarity calculation. Here, we use embeddings from the Universal Sentence Encoder (Cer et al., 2018) to generate a sentence embedding for each section marker. Using the sentence embeddings, we calculate the cosine similarity and use it to filter out section markers. Using the dev set to select the best threshold in terms of both precision and recall, we find that 0.98 cosine similarity is the best for filtering potential markers.

4.2 Using context information

We conducted three types of experiments regarding the use of context information: (1) Section title only. For this, we only use the text of the target sentence itself as the input feature for our model and generate the input feature as <CLS><Target>. (2) Context information only. We exclude the section titles from the input feature of our model to see if we can achieve good enough performance with only context information. We generate the feature as <CLS><Text before><SEP><Text after>. (3) Title + Context. For this we use the entire segment of text including title + context for prediction and generate the feature as <CLS><Text before><SEP><Target><Text after>.

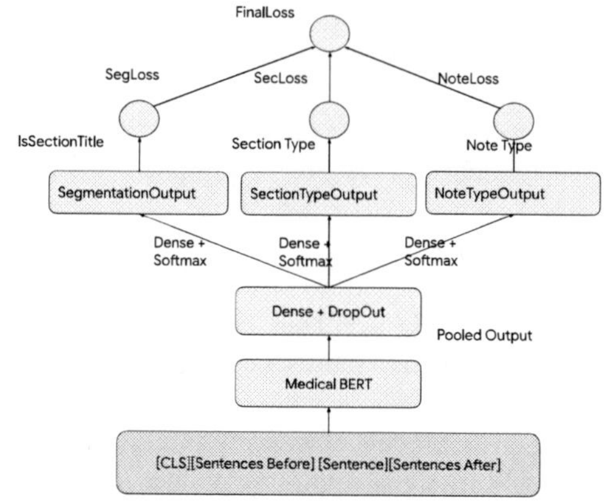

Figure 1: Structure of our multi-task BERT-based transformer model.

4.3 Multi-task BERT model

We propose the multi-task bert model[1] as shown in Fig 1[2]: We split the text into text spans according to line breaks and treat each text span as a training example. For each example, we create three separate losses for different tasks and use a combined loss as the final loss function.

- Segmentation Loss: This task does a binary classification regarding whether the target sentence is a section title or not.
- Section Type Loss: This task does a multi-class classification regarding the section type of the target sentence. We end up with a 19-way softmax by identifying 18 most important section type sand treat the rest as others. The details of thse 18 section types can be seen in Appendix A.
- Note Type Loss: This task predicts which type of the note the target sentence comes from. We end up with a 7-way softmax, including 5 core types as mentioned in Section 3 + 1 unspecified type for notes with no obvious structures + 1 others.

The combined loss is calculated as a weighted sum of all losses. We tested on our dev set and set an equal weight for each loss in our experiment. To verify whether the use of note type information is actually helpful, we added the experiment where we set the weight for note type loss to 0.

[1]For BERT, we are using medical-bert fine-tuned on pubmed data.

[2]Dense layers set as (128 - 32 - Final prediction towers) with 0.1 dropout

Method	Description	P	R
Embedding-based	Title only	0.82	1
BERT (target only)	Title only	**0.94**	**0.99**
BERT (context only)	Context Only	0.88	0.94
BERT (target + context)	Title + Context	**0.94**	**0.99**
BERT (no note loss)	Title + Context	**0.92**	**0.99**

Table 1: MIMIC-III (in-distribution) segmentation results. We only report segmentation results here as we found that the section type accuracy is usually high when we can recognize the correct section title.

Method	Description	P	R
MIMIC3 Markers	Baseline	0.98	0.65
Embedding-based	Title Only	0.66	0.84
BERT (target only)	Title Only	**0.72**	0.88
BERT (context only)	Context Only	0.66	0.80
BERT (target + context)	Title + Context	0.70	**0.95**

Table 2: out-of-distribution validation results

5 Data

To have enough data for training/evaluation, the output of the baseline system (Section 3) is used as the golden data. Due to the nature of the baseline algorithm, we can expect the generated data to have high precision/recall for training models on MIMIC-III and also high precision but low recall for validation on the held-out private dataset.

Test data For MIMIC-III data, we use the data described above. For the held-out private data, we use the same approach as described above and use all the extracted data as the test data. We randomly selected 500 notes for validation.

Data pre-processing The baseline approach uses the following rules for identifying the potential section titles which satisfy the following two constraints: (1) Sentence at the start of the text. (2) Sentence that ends with title endings (":", "-", "("). We followed similar ideas and relaxed this constraint in our data processing, where we split the text into spans of text when there is line break or a title ending. The section type information are then assigned the text spans according to the output of the baseline approach.

Training data We use MIMIC-III dataset as the training data. We randomly selected 4,000 notes for each of the five core note types, 4,000 notes where the note type is not specified and 8,000 notes randomly sampled from the entire dataset. As we don't have enough data for some categories, we end up having 20,000 notes with 3M text spans (among which there are 200k section titles). We split these examples to training/validation/testing in the ratio of 8:1:1.

6 Experiments and Results

We first conducted experiments on MIMIC-III dataset and Table 1 demonstrates the results. As our approaches are based on MIMIC-III markers and we are evaluating on the results extracted from the markers, we expected to see good recall performance for all our approaches. We are seeing that the embedding-based approach and BERT models that use title information were able to get a recall of more than 0.99. To our surprise, we also see that we are able to get a recall of 0.94 with just context information, proving that context information is useful even if used alone. However, we did not see better results with both title and context information, probably because that there exists limited headroom for improvement. In the meanwhile, we do see a small boost in precision with the inclusion of note type classification loss.

We applied the models trained on MIMIC-III and then to a new held-out dataset and results are shown in Table 2. The MIMIC3 markers-based approach was used as a baseline for comparison. We can see that while the markers-based approach still has a high precision due to its exact-match nature while its recall dropped to 0.65. With fuzzy title matching, the embedding-based approach improved the recall to 0.84 at the cost of dropping the precision to 0.66. Again, we see a reasonable performance with BERT + only context information. The BERT model with only title information reached a precision of 0.72 and recall of 0.88. With the addition of context information, the model's recall improves to 0.95 without much loss in precision.

7 Conclusion

In this work, we explored approaches for recognizing sections in free-form clinical notes. Our approach is based on the hypothesis that section content is similar across distributions and can be used to generate a robust section classifier. Our results demonstrate that our BERT-based model trained on MIMIC-III has very good performance on MIMIC-III and on our held-out private data, outperforming strong baselines.

References

Emily Alsentzer, John R Murphy, Willie Boag, Wei-Hung Weng, Di Jin, Tristan Naumann, and Matthew McDermott. 2019. Publicly available clinical bert embeddings. *arXiv preprint arXiv:1904.03323*.

Iz Beltagy, Matthew E Peters, and Arman Cohan. 2020. Longformer: The long-document transformer. *arXiv preprint arXiv:2004.05150*.

Daniel Cer, Yinfei Yang, Sheng-yi Kong, Nan Hua, Nicole Limtiaco, Rhomni St John, Noah Constant, Mario Guajardo-Cespedes, Steve Yuan, Chris Tar, et al. 2018. Universal sentence encoder for english. In *Proceedings of the 2018 conference on empirical methods in natural language processing: system demonstrations*, pages 169–174.

Hong-Jie Dai, Shabbir Syed-Abdul, Chih-Wei Chen, and Chieh-Chen Wu. 2015. Recognition and evaluation of clinical section headings in clinical documents using token-based formulation with conditional random fields. *BioMed research international*, 2015.

Jacob Devlin, Ming-Wei Chang, Kenton Lee, and Kristina Toutanova. 2018. Bert: Pre-training of deep bidirectional transformers for language understanding. *arXiv preprint arXiv:1810.04805*.

Amir Feder, Danny Vainstein, Roni Rosenfeld, Tzvika Hartman, Avinatan Hassidim, and Yossi Matias. 2020. Active deep learning to detect demographic traits in free-form clinical notes. *Journal of Biomedical Informatics*, 107:103436.

Elizabeth Ford, John A Carroll, Helen E Smith, Donia Scott, and Jackie A Cassell. 2016. Extracting information from the text of electronic medical records to improve case detection: a systematic review. *Journal of the American Medical Informatics Association*, 23(5):1007–1015.

Peter B Jensen, Lars J Jensen, and Søren Brunak. 2012. Mining electronic health records: towards better research applications and clinical care. *Nature Reviews Genetics*, 13(6):395–405.

Alistair EW Johnson, Tom J Pollard, Lu Shen, Li-wei H Lehman, Mengling Feng, Mohammad Ghassemi, Benjamin Moody, Peter Szolovits, Leo Anthony Celi, and Roger G Mark. 2016. Mimic-iii, a freely accessible critical care database. *Scientific data*, 3(1):1–9.

Jinhyuk Lee, Wonjin Yoon, Sungdong Kim, Donghyeon Kim, Sunkyu Kim, Chan Ho So, and Jaewoo Kang. 2020. Biobert: a pre-trained biomedical language representation model for biomedical text mining. *Bioinformatics*, 36(4):1234–1240.

Ying Li, Sharon Lipsky Gorman, and Noémie Elhadad. 2010. Section classification in clinical notes using supervised hidden markov model. In *Proceedings of the 1st ACM International Health Informatics Symposium*, pages 744–750.

Yifan Peng, Qingyu Chen, and Zhiyong Lu. 2020. An empirical study of multi-task learning on bert for biomedical text mining. In *Proceedings of the 19th SIGBioMed Workshop on Biomedical Language Processing*, pages 205–214.

Alexandra Pomares-Quimbaya, Markus Kreuzthaler, and Stefan Schulz. 2019. Current approaches to identify sections within clinical narratives from electronic health records: a systematic review. *BMC medical research methodology*, 19(1):1–20.

Alec Radford, Jeffrey Wu, Rewon Child, David Luan, Dario Amodei, Ilya Sutskever, et al. 2019. Language models are unsupervised multitask learners. *OpenAI blog*, 1(8):9.

Guergana K Savova, James J Masanz, Philip V Ogren, Jiaping Zheng, Sunghwan Sohn, Karin C Kipper-Schuler, and Christopher G Chute. 2010. Mayo clinical text analysis and knowledge extraction system (ctakes): architecture, component evaluation and applications. *Journal of the American Medical Informatics Association*, 17(5):507–513.

Michael Tepper, Daniel Capurro, Fei Xia, Lucy Vanderwende, and Meliha Yetisgen-Yildiz. 2012. Statistical section segmentation in free-text clinical records. In *Proceedings of the Eighth International Conference on Language Resources and Evaluation (LREC'12)*, pages 2001–2008.

Özlem Uzuner, Imre Solti, and Eithon Cadag. 2010. Extracting medication information from clinical text. *Journal of the American Medical Informatics Association*, 17(5):514–518.

Ashish Vaswani, Noam Shazeer, Niki Parmar, Jakob Uszkoreit, Llion Jones, Aidan N Gomez, Łukasz Kaiser, and Illia Polosukhin. 2017. Attention is all you need. *Advances in neural information processing systems*, 30.

Stephen H Walsh. 2004. The clinician's perspective on electronic health records and how they can affect patient care. *Bmj*, 328(7449):1184–1187.

Yanshan Wang, Liwei Wang, Majid Rastegar-Mojarad, Sungrim Moon, Feichen Shen, Naveed Afzal, Sijia Liu, Yuqun Zeng, Saeed Mehrabi, Sunghwan Sohn, et al. 2018. Clinical information extraction applications: a literature review. *Journal of biomedical informatics*, 77:34–49.

Thomas Wolf, Lysandre Debut, Victor Sanh, Julien Chaumond, Clement Delangue, Anthony Moi, Pierric Cistac, Tim Rault, Rémi Louf, Morgan Funtowicz, et al. 2020. Transformers: State-of-the-art natural language processing. In *Proceedings of the 2020 conference on empirical methods in natural language processing: system demonstrations*, pages 38–45.

A Appendix: Section Types

Table 3 shows a list of section types covered in this paper.

Section Type	Example Markers
CHIEF COMPLAINT	Chief Complaint, CC, Presenting Problem
PAST MEDICAL HISTORY	Pmh, Past Medical Problem
REVIEW OF SYSTEMS	ROS, Review of Systems
SOCIAL HISTORY	Family/Social History, Social Hx, SH
OTHER SUBJECTIVE	Subjective, health maintenance, Influenza vaccine screening
IMAGING	Image Result, IMAGING STUDIES
MEDICATION	Allergies/Medication List, med list, Infusions
PHYSICAL EXAMINATION	Physical Exam, Phys exam, PEx, Height And Weight
LAB RESULTS	Review of Laboratory Data, Labs and Reports, Blood Chemistry Studies
OTHER OBJECTIVE	Stress test, pathology
ASSESSMENT AND PLAN	A&P, Impression and Plan, Plan
PROBLEM LIST	Problem list, Problems (Active), Diagnoses
HOSPITAL COURSE	Brief history of hospital course, Hospital Summary
DISCHARGE TRANSFER DIAGNOSIS	Discharge/Transfer Diagnoses, Primary Diagnosis
DISCHARGE TRANSFER MEDICATION	Medications on discharge, Transfer Meds
FOLLOW UP	Discharge instructions and followup, Follow-up Plan, Followup Instructions
OTHER DISCHARGE INFORMATION	Discharge activity, Discharge Diet
INTERVAL EVENTS	Interval events, 24 hour events, o/n

Table 3: 18 core section types used in the study.

Building a Clinically-Focused Problem List From Medical Notes

Amir Feder* Itay Laish Shashank Agarwal Uri Lerner

Aviel Atias Cathy Cheung Peter Clardy Alon Cohen

Rachana Fellinger Hengrui Liu Lan Huong Nguyen Birju Patel

Natan Potikha Amir Taubenfeld Liwen Xu Seung Doo Yang

Ayelet Benjamini Avinatan Hassidim

Abstract

Clinical notes often contain useful information not documented in structured data, but their unstructured nature can lead to critical patient-related information being missed. To increase the likelihood that this valuable information is utilized for patient care, algorithms that summarize notes into a problem list have been proposed. Focused on identifying medically-relevant entities in the free-form text, these solutions are often detached from a canonical ontology and do not allow downstream use of the detected text-spans. Mitigating these issues, we present here a system for generating a canonical problem list from medical notes, consisting of two major stages. At the first stage, *annotation*, we use a transformer model to detect all clinical conditions which are mentioned in a single note. These clinical conditions are then grounded to a predefined ontology, and are linked to spans in the text. At the second stage, *summarization*, we develop a novel algorithm that aggregates over the set of clinical conditions detected on all of the patient's notes, and produce a concise patient summary that organizes their most important conditions.

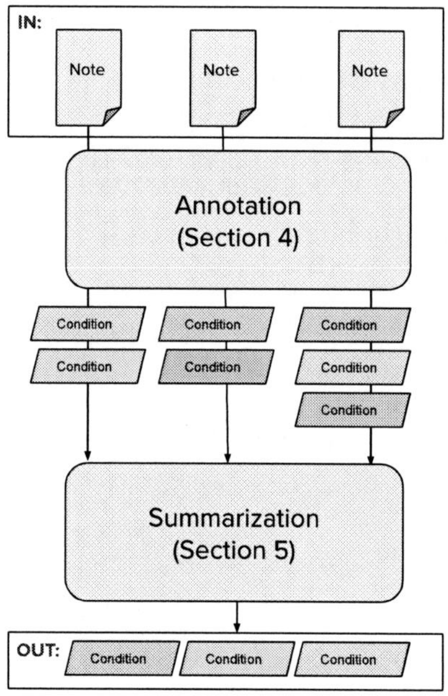

Figure 1: **System overview**: Conditions are extracted form each individual note at the *annotation* stage, and a single patient level list is generated from them at the *sumarization* stage.

1 Introduction

The pervasiveness of free-text narrative in Electronic Health Records (EHR) is both a blessing and a curse. It allows much more nuanced information about patients' conditions being saved and documented (Uzuner et al., 2010; Savova et al., 2010; Jensen et al., 2012; Wang et al., 2018; Feder et al., 2020). However, the unstructured nature of the data can also impede care givers' understanding of patient conditions (Walsh, 2004; Ford et al., 2016).

To allow care providers to better understand their patients' condition from medical notes, many machine learning (ML) models have been proposed (Uzuner et al., 2011; Jensen et al., 2012; Lee et al., 2020). These algorithms often solve a named-entity recognition (NER) task over the clinical notes, identifying text spans that correspond to clinical problems (Uzuner et al., 2011). While performance on such task has improved in the last decade (Wang et al., 2018), these models often do not link the identified entities to an ontology and are therefore sensitive to abbreviations, spelling errors and language ambiguity (Reátegui and Ratté, 2018; Gopinath et al., 2020; Gao et al., 2021). Moreover, these solutions operate at the note level, and are not able to aggregate a patient's overall medical problem list (Baumel et al., 2018). Both of these limitations decrease the utility of deploying these models in the real-world.

*Corresponding author (afeder@google.com).

Proceedings of the 13th International Workshop on Health Text Mining and Information Analysis (LOUHI), pages 60 - 68
December 7, 2022 ©2022 Association for Computational Linguistics

Another important limitation of many existing solutions is that they are built on top of recurrent neural networks designed for solving NER tasks which often do not fully utilize the nuanced linguistic signal (Wang et al., 2018). These approaches were shown to produce very good results on de-identification tasks on clinical notes (Hartman et al., 2020), but can fail when presented with tasks that demand better understanding of context (Devlin et al., 2019). The recent transition of the entire NLP community to pre-trained transformer-based models (Wu et al., 2020) thus offers an opportunity to further improve on existing condition extraction methods (Zhu et al., 2018).

In this paper, we take on the task of addressing these problems and limitations, and describe how to build an end-to-end system that is robust and trustworthy. Concretely, given a set of notes describing a single patient, our goal is to output a clinically-focused problem list. Our system consists of two major stages: (1) *Annotation* (§4): operating at the level of a single medical note, we detect all clinical conditions which are mentioned in the text. These clinical conditions are then grounded to a predefined set of entities, and are linked to text spans. More formally, the output of the *annotation* stage is a set of tuples, where each tuple is a clinical condition identifier, a character span and context metadata (e.g., the acuity and presence of the condition). (2) *Summarization* (§5): operating at patient-level, we consume the set of clinical conditions detected during the *annotation* stage, and produce a concise patient summary that organizes the conditions. Our system is backed by a tailored *Ontology* (§3), defined on-top of SNOMED-CT (Donnelly et al., 2006) used by both stages to model the clinical knowledge required for this task. See Figure 1 for an illustration of our system.

2 Related Work

Identifying patient-related information in medical notes is long recognized as a core task in clinical-NLP. As such, there exist standardized datasets and competitions (Uzuner, 2009; Savova et al., 2010; Jensen et al., 2012; Ford et al., 2016; Zhu et al., 2018). The task of identifying medical concepts in clinical notes was organized as a competition in i2b2 2010 (Uzuner et al., 2011). In i2b2 and in subsequent work, this task was defined as a named entity recognition (NER) task (Hartman et al., 2020), where individual words are classified as to whether they contain medical problems. Subsequently, a Named Entity Normalization (NEN) task, where entities are standardized into known medical concepts, was later added to the i2b2 (now n2c2) competitions (Luo et al., 2019). Solutions to the problem consequently followed the conventional NLP approaches to solving NER tasks. Recent approaches harness the transformer architecture, solving a token-level binary classification task (Peng et al., 2019; Yadav and Bethard, 2019; Si et al., 2019; Lee et al., 2020).

To connect identified text spans to an ontology, a common solution is to look for the most similar entity in a given knowledge graph. Knowledge graphs use a graph-structured data model to integrate data (Ehrlinger and Wöß, 2016). They are often used to store interlinked descriptions of entities—objects, events, situations or abstract concepts—while also encoding the semantics underlying the used terminology. They were shown to be very useful in the medical domain and are often used to encode medical knowledge (Lindberg et al., 1993; Donnelly et al., 2006; Lipscomb, 2000). Specifically, in the context of free-form text, as that in the clinical notes, graph structured data models can be used to map many alternative descriptions of the same condition into one canonical definition (Organization, 2015).

Finally, the task of aggregating patient-related information across multiple documents into one problem list in a single system was not, to the best of our knowledge, published in prior research. The focus of our work is building an end-to-end system that connects the text *annotation* with the *summarization* stage.

3 Ontology

Our system is based on a universe of entities (*ontology*). The *ontology* captures the clinical knowledge required for our system to provide a concise and clinically-focused problem list. This knowledge improves both the detection of clinical conditions in medical notes (§4), and the subsequent bucketing of related conditions (§5).

On the detection side, it is necessary for our algorithm to be aware of the ways in which clinical conditions may appear in medical notes. For example, "iddm" ("insulin dependent diabetes mellitus") is an alternative phrasing of "Diabetes mellitus type 1", and "Miller" may refer to "Miller Fisher syndrome". On the bucketing side, it is necessary to have knowledge about related conditions (e.g., "Biventricular congestive heart failure" is related to "Right heart failure") and about possible complications of certain conditions (e.g., "Diabetic nephropathy" is a complication of "Diabetes mellitus").

Failure to capture this knowledge may increase the redundancy at the problem list level, and might cause dilution of signals and features, which in turn results in poor quality. The *ontology* is therefore a fundamental building block that is being used across all the system stages, and the way it is created has critical quality implications. In this section, we describe the creation of our *ontology*. Instead of creating a full *ontology* end-to-end, we have opted to base our *ontology* on pre-existing datasets. We collected a set of *Ground Truth Problem List*, which were curated by clinicians, and examined the properties of each dataset against this ground truth. A useful *ontology* should demonstrate the following properties:

(i) **High coverage** of the entries in the ground truth Problem List's, and in the right granularity level.
(ii) **Easy to match** an entity from the ontology to the actual text in the medical note.
(iii) Entities should have **meaningful relationships** with other entities that are useful for reducing redundancy in the aggregated Problem List.

We considered multiple data sources, including: SNOMED-CT (Donnelly et al., 2006), MeSH (Lipscomb, 2000), ICD-10-CM (Organization, 2015), and UMLS (Lindberg et al., 1993).

3.1 ICD-10

.

ICD-10 is lacking some conditions (e.g., "Odynophagia") violating property ((i)); a single main entity is missing for some conditions (e.g. "Sepsis" and "Pneumonia" are associated with multiple unrelated entities), these conditions are cluttered across the dataset, making it more difficult to group mentions together (violating property ((iii))); and due to the verbose description of some entities (e.g. "K44.9 Diaphragmatic hernia without obstruction or gangrene"), it is hard to match an ontology entity to the text ("Hiatal hernia" in the previous example), in violation of property ((ii)).

3.2 MeSH

In MeSH we observed some significant recall losses. For example, "Hypertensive urgency" and "Generalized anxiety disorder" were missing, violating property ((i)).

3.3 UMLS

Since UMLS is a combination of multiple systems, the relationships and granularity it provides vary across entities. This makes all properties only partially satisfied.

3.4 SNOMED-CT

While SNOMED-CT was missing some entities (e.g., "Right eye glaucoma"), these could usually be compensated by other SNOMED concepts without any significant clinical impact, and overall, it outperformed on all three properties the other options considered.

We note that due to the uniqueness in structure, relation types and granularity of each ontology, any attempt of reconciliation is exposed to similar issues as observed in *UMLS*. Therefore we chose to base our solution on a single ontology source (SNOMED), where each entity in our ontology corresponds to exactly one SNOMED concept. This allows us to maintain the consistency and granularity of SNOMED concepts and relationships, and also allows us to incorporate new versions of SNOMED as they are released, which keeps the ontology up to date.

Additionally, in order to enhance the ability to match a SNOMED concept to text from medical notes, we enrich SNOMED concepts with the followings (using our NameMapper algorithm described in section 4.2):

1. The ICD-10 codes of all ICD-10 diagnoses for which the SNOMED concept is their closest concept in SNOMED.
2. Phrases which are alternative ways to mention the entity in medical notes.

For (1), we use two sources for mapping ICD-10 diagnoses to their closest SNOMED concept: (a) OHDSI-OMOP (Stang et al., 2010; Hripcsak et al., 2015); and (b) the *NameMapper* (details in section 4.2) algorithm, applied on a diagnosis' name in order to match it against the set of SNOMED terms. For (2), we consider clusters of phrases that are originated from various sources: MeSH, UMLS and manually-curated abbreviations. All phrases in a cluster refer to the same entity. We use NameMapper again, in order to match each phrase in a cluster against the full set of SNOMED terms. We add the entire cluster to the entity that corresponds to the closest SNOMED concept.

4 Annotation

The annotation stage is performed at the level of a single clinical note. At the end of this stage each mention of a condition in the text is exported in the form of $(ConditionID, (start, end), ContextInfo)$ tuple (where $start$, end refer to a char offset from the beginning of the note), $ConditionID$ is a unique entry in the ontology described in §3, and $ContextInfo$ includes extracted information about the condition, such as acuity, presence, etc.

We start this section by describing our detection (§4.1), and mapping (§4.2) algorithms.

4.1 ML model for surfacing candidates and context information

We extract condition spans from free text using an ML NER model. Later, we try to map these candidate spans into our ontology (§4.2).

Our model is a multi-task encoder-only transformer model (BERT; Devlin et al., 2019). Its main task is a 4-class classification task (using the labels **PROBLEM, BODY PART, QUALIFIER, PROCEDURE**), with additional two supplementary-tasks:

- Existences: For each of the four labels, whether it is PRESENT or ABSENT; e.g., in “ruled out cancer”, “cancer” is labeled as *ABSENT*.
- Relation: For both BODY_PARTS / QUALIFIERS, are they associated with the PROBLEM / PROCEDURE on their left, or on their right. E.g., in "diabetic foot ulcer", "ulcer" will have a LEFT_HAND_SIDE label. This information is later used to map the annotated term to the most accurate ontology entity.

This is the 2nd generation of NER models used in our system, our previous model was based on GloVe, Bi-LSTM and CRF (Hartman et al., 2020). On top of the CRF layer we placed three softmax layers to solve each of the three aforementioned tasks (this model is referred as *Bi-LSTM* below).

The BERT model described here showed superior performance (see table below). For BERT, we use a similar approach where we place three softmax layers on top of the pre-trained contextual embedding. The added layers are then fine-tuned on the MIMIC-III dataset (Johnson et al., 2016; using the same labels of the Bi-LSTM). We experimented with 3 pre-trained BERTs:

BERT-base from the original paper (Devlin et al., 2019).

BERT-small based on (Turc et al., 2019) – x2 more efficient than *BERT-base*.

PubMED BERT same architecture as *BERT-base*, pre-trained from scratch on MEDLINE/PubMed, using the original uncased word-piece tokenizer (Lee et al., 2020).

The labels are split $80\%/20\%$ for train/eval sets. The following table shows the results on the eval set. As can be seen, *PubMED BERT* surpasses the other models.

4.2 NameMapper – A graph traversal-based approach for ontology matching

In many cases, the hand-written text by clinicians in notes does not match the names of conditions in the ontology. To bridge this gap, and to increase the coverage of problems detected and matched by our algorithm in §4.1, we introduce a graph-based fuzzy text matcher called *NameMapper*. The NameMapper is used during the following stages of our system:

(i) **Ontology creation (§3)**: For mapping between entities in different ontologies (ICD-10→ SNOMED).

(ii) **Increase detection coverage**: Build a vocabulary used at the annotation stage for matching text spans to entities from the ontology.

(iii) **Mapping**: Map conditions spans generated in §4.1 to entities in our ontology.

NameMapper is essentially a string matching algorithm. It operates on text that is suspected to match a name of an entity in the ontology. It expands the string using different variations of each word within, and allows string manipulations using pre-defined operations. Each operation is associated with a cost. We use a graph (with these costs set as edge weights) to find the closest entity in the ontology to the input string. See an illustration in Figure 3). The process consists of three main stages:

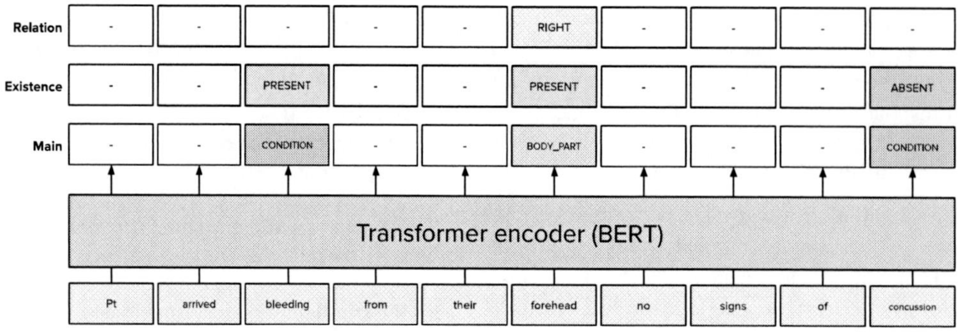

Figure 2: **Illustration of our multi-task encoder-only transformer**: Each token is labeled for *type classification(Main)*, *Existence* and *Relation*.

Parsing. We first break the input string into a (non-intersecting) sequence of name components of different types: e.g., connectors such as "due to" or "of" are modeled as a special type.

Generation. For each name component, we generate a set of alternatives, each alternative is associated with a cost. These alternatives represent different ways to refer to the same concept, e.g., "malignant tumor" → "cancer", "lung" → "pulmonary", "kidney" → "renal", "diabetes mellitus" → "diabetes"). These costs were manually curated. One could think of them as the conceptual distance between the two synonyms (e.g., replacing "ii" with "2" has a lower cost compared to replacing "infectious disease" with "infection"). The alternating names include the original string as it appears in the input name (up to lower-casing and some other default operations) and alternative wordings that are based on synonyms, dropping optional phrases, stemming and more. We manually curated those rules. For example, *"diabetes"* is a synonym of *diabetes mellitus*, *diabet* is the canonical form of *"diabetes"*.

Using the alternatives of each name component, we generate a list of alternative writings for the entire phrase. The alternatives include different combinations of the options created for the name components generated during the previous stage. This stage also allows phrase level transformations (with additional cost). For example, the connector "due to" allows a transformation of dropping itself and the possibility of swapping both of its sides: "coma due to diabetes mellitus" may generate alternatives such as "coma diabetes mellitus" and "diabetes mellitus coma" (each with a cost). The final (phrase level) cost is set to be the costs sum of all replacements and operations applied to the input string.

Selection. We output the ontology entity that matches the best candidate (lowest cost). For example, the terms "diabetic coma", "coma due to diabetes mellitus" and "diabetes mellitus coma" will all be mapped to the same ontology entity *"Coma due to diabetes mellitus"*, each with a cost.

5 Summarization

The *Annotation* Phase (§4) outputs the mentions of clinical conditions in the medical notes. The goal of the *Summarization* Phase is to take all the mentions across all the notes and generate a comprehensive and coherent problem list, optimized for the needs of clinicians who care for the patient. In addition to the mentions themselves, the Summarization Phase can use additional information in the patient's chart

	PROBLEM [3.8K]			BODY PART [1.4K]			QUALIFIER [0.7K]			PROCEDURE [0.4K]		
	P	R	F1	P	R	F1	P	R	F1	P	R	F1
Bi-LSTM	84.74	86.80	85.76	**75.52**	80.02	77.70	60.65	55.17	57.78	**71.83**	64.15	67.77
PubMED BERT	**87.64**	**89.69**	**88.65**	72.61	**86.79**	**79.07**	**65.27**	**62.18**	**63.69**	65.94	**76.58**	**70.86**
BERT base	86.47	87.19	86.83	74.22	82.51	78.15	61.24	57.77	59.45	64.99	67.80	66.36
BERT small	84.10	86.65	85.35	69.82	81.22	75.09	59.51	55.83	57.61	62.13	57.36	59.65

Table 1: ML model classification results

as well as general medical knowledge. We now describe the sequence of steps that make up the Summarization Phase.

5.1 Grouping

The first step is to collect all the clinical condition mentions related to the same condition. In this step, we drop conditions that the patient never had (e.g., mentions of known side effects of treatments, speculations written in the note etc.) using the existence signal generated by our annotator (§4.1).

5.2 Bucketing

In the next step, we group clinical conditions that are related to each other. For example, if we found mentions of *Systolic Heart Failure*, *Diastolic Heart Failure*, *Acute Heart Failure*, and *Acute Diastolic Heart Failure* in a patient's medical notes, we would bucket those mentions under a *Heart Failure* bucket, even if "Heart Failure" itself was not explicitly mentioned in the patient's record.

In the example above, *Heart Failure* is an anchor entity, used to bucket together more specific conditions as defined by the is-a relation of the SNOMED ontology (see Section 3). A bucket is defined as a collection of patient conditions composed of one or more anchor entities and their corresponding descendants (in the ontology).

Ideally, conditions inside a bucket should involve similar pathophysiologies, medications and therapies. Anchor entities should thus follow the Goldilocks Principle and be neither too broad nor too narrow. Overly broad anchor entities (e.g., *Heart disease*) represent conditions with very different pathophysiologies and therapies and therefore do not provide a good clinical view. Overly narrow anchor entities (e.g., *Systolic Heart Failure*, *Diastolic Heart Failure*, *Acute Heart Failure*, and *Acute Diastolic Heart Failure*) would make the Problem List overly long and redundant, reduc-

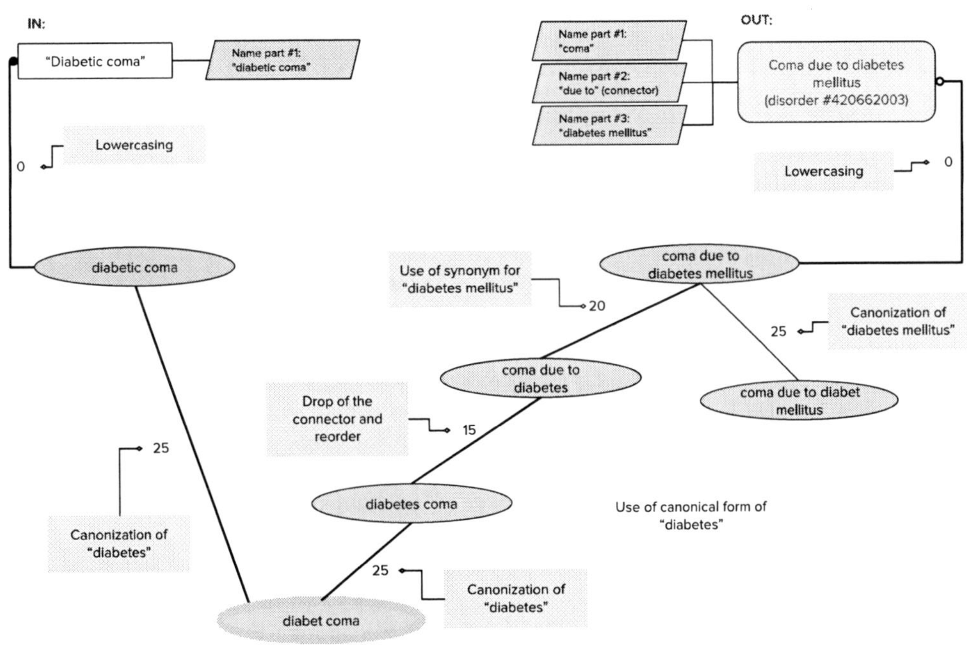

Figure 3: **Illustration of the NameMapper**: At stage **Parsing** the input *"Diabetic coma"* is parsed to its single name component (blue parallelogram). Additionally, the ontology entity *"Coma due to diabetes mellitus"* is parsed to *Coma, diabetes mellitus* and the connector *due to* (green parallelograms). Then the generation stage **Generation** will use the variations of each name component to create all possible permutations both for the input and the ontology entity(ies), given in blue and green ovals accordingly. Finally stage **Selection** will find the shortest path between the input and the ontology, that is *Coma due to diabetes mellitus* with the cost of 85 (in bold).

ing its usefulness. Identifying good anchor entities requires clinical expertise.

As part of the Bucketing step, we also determine the name of the bucket. This is typically the name of the anchor entity or entities around which the bucket is defined. However, if the bucket only contains more specific entities than the anchor entities, we give the bucket a more specific name.

For example, if the only conditions in the bucket anchored on *Heart Failure* were *Diastolic Heart Failure*, and *Acute Diastolic Heart Failure*, instead of naming the bucket *"Heart Failure"*, we would name it *"Diastolic Heart Failure"*, which is a more accurate description of all the entities that ended up in the bucket for this particular patient.

5.2.1 Secondary buckets

Some conditions are associated with other conditions typically but not always. For example, *Hyperglycemia* is often associated with *Diabetes Mellitus* but it is possible to have *nondiabetic Hyperglycemia*. In this case we consider *Diabetes Mellitus* as a secondary bucket of *Hyperglycemia*, meaning that if the *Diabetes Mellitus* bucket includes other conditions then it should also include *Hyperglycemia*, but if it is empty (does not exist) then *Hyperglycemia* should be its own bucket.

5.3 Bucket Presence

At the end of the bucketing step, we generate collections of clinical conditions that were mentioned in the patient's medical notes. However, the patient does not necessarily have all conditions that were mentioned. The typical reasons are mistakes in the Annotation Phase or actual uncertainty, e.g., a patient may have a mention indicating that Covid-19 is likely only to be ruled out in a later mention. Obviously, it is desirable to omit these conditions from the list, as the patient does not have them.

In the Condition Bucket Presence step, we determine if the patient is having, or has ever had each of the condition buckets. First, we make use of the Existence signal extracted during the Annotation phase (see Section 4.1), and drop the mentions classified as "ABSENT" by the algorithm based on the surrounding context. Since (as expected) the Annotator's existence classification is not always perfect, we apply an additional second level to improve the presence detection. To handle mistakes from the Annotation Phase we take into account the frequency with which the condition was mentioned in medical notes, the section where the condition was mentioned, and the credentials of the medical note's author. We are looking into using additional signals such as mentions in the notes of related conditions, documentations of conditions in the EHR structured Problem Lists, information about labs, vitals and medications, and many others.

5.4 Classification

Over time, a patient is expected to have many clinical conditions. However, not all conditions are active when the patient is reviewed, and if they all were to be displayed, the list would quickly become overwhelmingly long and not particularly useful. Imagine reading a chart of a patient that caught a common cold a few years ago who is also diabetic; more details about diabetes and related conditions should be surfaced, and the information about the common cold's occurrence in the far past would be no longer relevant today, and therefore should be skipped to avoid unnecessary clutter. Since our Annotation phase also detects symptoms and procedures seen in the patient's medical notes, the length of the generated Problem List can become extremely long. It is thus important for us to strive for conciseness, and avoid information overload that could distract physicians from the important active conditions.

To make the Problem List easier to comprehend, we classify the clinical conditions into four categories: *Active Conditions*, *Historical Conditions*, *Procedures*, and *Symptoms*.

SNOMED classifies every entity into a type which includes disorders, findings, and procedures. We consider the SNOMED types of all the entities in a bucket to determine the type of the bucket: a bucket with disorders is a Conditions bucket, a bucket with findings is a Symptoms bucket and so on. We have additional logic for mixed buckets, e.g., a mixture of disorders and findings is considered a Conditions bucket.

In the next step we classify the Conditions buckets into Active and Historical category. We do this by first classifying individual conditions included in a bucket separately and then again classifying the entire bucket based on the whole collection: a bucket with at least one Active condition is Active, otherwise it is Historical. A chronic lifelong condition such as *Type 1 Diabetes Mellitus* is always considered *Active*. The remaining conditions are considered *Active* and then moved to *Historical* if a mention confirming their presence wasn't seen for a long time, or if a mention was found explicitly indicating that the condition was resolved. The duration after which a non-lifetime condition is automatically classified as *Historical* (because it was not mentioned again as present) varies, and is part of our curated knowledge gathered with assistance of expert clinicians.

5.5 Ranking

Finally, we rank the conditions in each category so that the most clinically important conditions are displayed first. Our ranking function accounts for the severity and recency clinical conditions to determine the order. More severe and more recent conditions are ranked higher to highlight the conditions that might require more attention from physicians.

5.6 Summarization evaluation

Each step in the *Summarization* Phase is evaluated separately so that we are able to test those steps in isolation. At the same time, we also test the overall pipeline by evaluating the resulting Problem List holistically. In addition to evaluating metrics such as precision and recall, we also measure the usefulness of the Problem List, which captures the effects of steps such as Bucketing, Classification, and Ranking.

6 Conclusion

In this work, we present an end-to-end system for summarizing a patient's problem list directly from their entire collection of medical notes. This system aggregates over identified conditions in each note, producing a concise list mapped to a canonical ontology and without duplicated conditions. Building on recent improvements in natural language understanding models, especially encoder-only transformers, we show how NLP models can be used as part of an holistic system. We hope that our work will spur more research on how to utilize NLP for better, more robust and trustworthy, health informatics systems.

References

Tal Baumel, Jumana Nassour-Kassis, Raphael Cohen, Michael Elhadad, and Noémie Elhadad. 2018. Multi-label classification of patient notes: case study on icd code assignment. In *Workshops at the thirty-second AAAI conference on artificial intelligence*.

Jacob Devlin, Ming-Wei Chang, Kenton Lee, and Kristina Toutanova. 2019. Bert: Pre-training of deep bidirectional transformers for language understanding. In *Proceedings of the 2019 Conference of the North American Chapter of the Association for Computational Linguistics: Human Language Technologies, Volume 1 (Long and Short Papers)*, pages 4171–4186.

Kevin Donnelly et al. 2006. Snomed-ct: The advanced terminology and coding system for ehealth. *Studies in health technology and informatics*, 121:279.

Lisa Ehrlinger and Wolfram Wöß. 2016. Towards a definition of knowledge graphs. *SEMANTiCS (Posters, Demos, SuCCESS)*, 48(1-4):2.

Amir Feder, Danny Vainstein, Roni Rosenfeld, Tzvika Hartman, Avinatan Hassidim, and Yossi Matias. 2020. Active deep learning to detect demographic traits in free-form clinical notes. *Journal of Biomedical Informatics*, 107:103436.

Elizabeth Ford, John A Carroll, Helen E Smith, Donia Scott, and Jackie A Cassell. 2016. Extracting information from the text of electronic medical records to improve case detection: a systematic review. *Journal of the American Medical Informatics Association*, 23(5):1007–1015.

Yanjun Gao, Dmitriy Dligach, Leslie Christensen, Samuel Tesch, Ryan Laffin, Dongfang Xu, Timothy Miller, Ozlem Uzuner, Matthew M Churpek, and Majid Afshar. 2021. A scoping review of publicly available language tasks in clinical natural language processing. *arXiv preprint arXiv:2112.05780*.

Divya Gopinath, Monica Agrawal, Luke Murray, Steven Horng, David Karger, and David Sontag. 2020. Fast, structured clinical documentation via contextual autocomplete. In *Machine Learning for Healthcare Conference*, pages 842–870. PMLR.

Tzvika Hartman, Michael D Howell, Jeff Dean, Shlomo Hoory, Ronit Slyper, Itay Laish, Oren Gilon, Danny Vainstein, Greg Corrado, Katherine Chou, et al. 2020. Customization scenarios for de-identification of clinical notes. *BMC medical informatics and decision making*, 20(1):1–9.

George Hripcsak, Jon D Duke, Nigam H Shah, Christian G Reich, Vojtech Huser, Martijn J Schuemie, Marc A Suchard, Rae Woong Park, Ian Chi Kei Wong, Peter R Rijnbeek, et al. 2015. Observational health data sciences and informatics (ohdsi): opportunities for observational researchers. *Studies in health technology and informatics*, 216:574.

Peter B Jensen, Lars J Jensen, and Søren Brunak. 2012. Mining electronic health records: towards better research applications and clinical care. *Nature Reviews Genetics*, 13(6):395–405.

Alistair EW Johnson, Tom J Pollard, Lu Shen, Li-wei H Lehman, Mengling Feng, Mohammad Ghassemi, Benjamin Moody, Peter Szolovits, Leo Anthony Celi, and Roger G Mark. 2016. Mimic-iii, a freely accessible critical care database. *Scientific data*, 3(1):1–9.

Jinhyuk Lee, Wonjin Yoon, Sungdong Kim, Donghyeon Kim, Sunkyu Kim, Chan Ho So, and Jaewoo Kang. 2020. Biobert: a pre-trained biomedical language representation model for biomedical text mining. *Bioinformatics*, 36(4):1234–1240.

Donald AB Lindberg, Betsy L Humphreys, and Alexa T McCray. 1993. The unified medical language system. *Yearbook of medical informatics*, 2(01):41–51.

Carolyn E Lipscomb. 2000. Medical subject headings (mesh). *Bulletin of the Medical Library Association*, 88(3):265.

Yen-Fu Luo, Weiyi Sun, and Anna Rumshisky. 2019. Mcn: a comprehensive corpus for medical concept normalization. *Journal of biomedical informatics*, 92:103132.

World Health Organization. 2015. International classification of diseases, tenth revision, (icd-10).

Yifan Peng, Shankai Yan, and Zhiyong Lu. 2019. Transfer learning in biomedical natural language processing: an evaluation of bert and elmo on ten benchmarking datasets. *arXiv preprint arXiv:1906.05474*.

Ruth Reátegui and Sylvie Ratté. 2018. Comparison of metamap and ctakes for entity extraction in clinical notes. *BMC medical informatics and decision making*, 18(3):13–19.

Guergana K Savova, James J Masanz, Philip V Ogren, Jiaping Zheng, Sunghwan Sohn, Karin C Kipper-Schuler, and Christopher G Chute. 2010. Mayo clinical text analysis and knowledge extraction system (ctakes): architecture, component evaluation and applications. *Journal of the American Medical Informatics Association*, 17(5):507–513.

Yuqi Si, Jingqi Wang, Hua Xu, and Kirk Roberts. 2019. Enhancing clinical concept extraction with contextual embeddings. *Journal of the American Medical Informatics Association*, 26(11):1297–1304.

Paul E Stang, Patrick B Ryan, Judith A Racoosin, J Marc Overhage, Abraham G Hartzema, Christian Reich, Emily Welebob, Thomas Scarnecchia, and Janet Woodcock. 2010. Advancing the science for active surveillance: rationale and design for the observational medical outcomes partnership. *Annals of internal medicine*, 153(9):600–606.

Iulia Turc, Ming-Wei Chang, Kenton Lee, and Kristina Toutanova. 2019. Well-read students learn better: On the importance of pre-training compact models. *arXiv preprint arXiv:1908.08962*.

Özlem Uzuner. 2009. Recognizing obesity and comorbidities in sparse data. *Journal of the American Medical Informatics Association*, 16(4):561–570.

Özlem Uzuner, Imre Solti, and Eithon Cadag. 2010. Extracting medication information from clinical text. *Journal of the American Medical Informatics Association*, 17(5):514–518.

Özlem Uzuner, Brett R South, Shuying Shen, and Scott L DuVall. 2011. 2010 i2b2/va challenge on concepts, assertions, and relations in clinical text. *Journal of the American Medical Informatics Association*, 18(5):552–556.

Stephen H Walsh. 2004. The clinician's perspective on electronic health records and how they can affect patient care. *Bmj*, 328(7449):1184–1187.

Yanshan Wang, Liwei Wang, Majid Rastegar-Mojarad, Sungrim Moon, Feichen Shen, Naveed Afzal, Sijia Liu, Yuqun Zeng, Saeed Mehrabi, Sunghwan Sohn, et al. 2018. Clinical information extraction applications: a literature review. *Journal of biomedical informatics*, 77:34–49.

Stephen Wu, Kirk Roberts, Surabhi Datta, Jingcheng Du, Zongcheng Ji, Yuqi Si, Sarvesh Soni, Qiong Wang, Qiang Wei, Yang Xiang, et al. 2020. Deep learning in clinical natural language processing: a methodical review. *Journal of the American Medical Informatics Association*, 27(3):457–470.

Vikas Yadav and Steven Bethard. 2019. A survey on recent advances in named entity recognition from deep learning models. *arXiv preprint arXiv:1910.11470*.

Henghui Zhu, Ioannis Ch Paschalidis, and Amir Tahmasebi. 2018. Clinical concept extraction with contextual word embedding. *arXiv preprint arXiv:1810.10566*.

Specializing Static and Contextual Embeddings in the Medical Domain Using Knowledge Graphs: Let's Keep It Simple

Hicham El Boukkouri[1], **Olivier Ferret**[2], **Thomas Lavergne**[1], **Pierre Zweigenbaum**[1]
[1]Université Paris-Saclay, CNRS, LISN, Orsay, France,
[2]Université Paris-Saclay, CEA, List, F-91120, Palaiseau, France,
{elboukkouri,lavergne,pz}@lisn.fr, olivier.ferret@cea.fr

Abstract

Domain adaptation of word embeddings has mainly been explored in the context of retraining general models on large specialized corpora. While this usually yields good results, we argue that knowledge graphs, which are used less frequently, could also be utilized to enhance existing representations with specialized knowledge. In this work, we aim to shed some light on whether such knowledge injection could be achieved using a basic set of tools: graph-level embeddings and concatenation. To that end, we adopt an incremental approach where we first demonstrate that static embeddings can indeed be improved through concatenation with in-domain *node2vec* representations. Then, we validate this approach on contextual models and generalize it further by proposing a variant of BERT that incorporates knowledge embeddings within its hidden states through the same process of concatenation. We show that this variant outperforms plain retraining on several specialized tasks, then discuss how this simple approach could be improved further. Both our code and pre-trained models are open-sourced for future research. In this work, we conduct experiments that target the medical domain and the English language.

1 Introduction

The popularization of transfer learning, particularly in the context of pre-training language models to serve as encoders in downstream tasks, has led to an ever-expanding set of methods for representing textual data: e.g. ELMo (Peters et al., 2018), BERT (Devlin et al., 2019), XLNet (Yang et al., 2019). While these models continuously push forward the expected level of performance on so-called "general domain" tasks (e.g. GLUE[1]), they usually lag behind when it comes to more specialized areas like the medical domain (see BLUE[2] and BLURB[3] benchmarks). As a result, there is a growing interest in finding ways in which these out-of-the-box representations can be specialized, with most efforts focusing on retraining general models on specialized corpora: e.g. ClinicalBERT (Alsentzer et al., 2019), BioBERT (Lee et al., 2020), and BioMed-RoBERTa (Gururangan et al., 2020). However, pre-trained language models have also been shown to benefit from external knowledge injection, with approaches like LIBERT (Lauscher et al., 2020), KnowBERT (Peters et al., 2019), and KEPLER (Wang et al., 2021b) in the general domain, or (Hao et al., 2020) and (Lu et al., 2021) in the medical domain. Yet, these efforts usually involve complex modifications to the architecture of underlying models and/or their pre-training procedure, which may convey the impression that knowledge injection cannot be achieved in simpler ways.

In this work, we propose a simple approach to embedding specialization that relies on knowledge graph embeddings and concatenation. We argue that by building a simple but strong baseline first, we lay the foundation for future improvements that may be easily achieved by upgrading to more sophisticated knowledge embeddings or combination methods. In practice, we show that medical concept embeddings obtained from an in-domain knowledge graph can be combined through concatenation with word representations to effectively construct specialized "meta-embeddings" (Yin and Schütze, 2016). Moreover, in the specific case of contextual embeddings, we show that these concept embeddings can be combined either externally, with a general-domain model, or internally, during the pre-training of a specialized model, to achieve varying levels of model specialization. All our models are trained and evaluated in pairs, and in exactly the same conditions, to highlight to the greatest extent the impact of our strategies.

[1]https://gluebenchmark.com/leaderboard

[2]https://github.com/ncbi-nlp/BLUE_Benchmark#baselines

[3]https://microsoft.github.io/BLURB/leaderboard.html

Proceedings of the 13th International Workshop on Health Text Mining and Information Analysis (LOUHI), pages 69 - 80
December 7, 2022 ©2022 Association for Computational Linguistics

Our contributions are the following:

- We build two sets of knowledge representations by applying *node2vec* (Grover and Leskovec, 2016) to concepts from MeSH (biomedical) and SNOMED CT (clinical).
- We construct specialized meta-embeddings by concatenating fastText embeddings (Bojanowski et al., 2017) with the *node2vec* vectors. We show that this improves the performance of both general and medical domain representations on several medical tasks.
- We conduct the same experiments with contextual BERT and CharacterBERT (El Boukkouri et al., 2020) representations, and show similar improvements on most evaluation tasks with a slight edge for the character-based model.
- We generalize the meta-embedding approach to the pre-training of contextual models by introducing a 'Knowledge Injection Module' that combines incoming hidden states from a Transformer layer (Vaswani et al., 2017) with external knowledge representations through the same process of concatenation.
- We retrain both original and modified versions of BERT and CharacterBERT on a medical corpus and show that the modified models perform better on several medical tasks.
- We propose improvements to our methods and share our code and pre-trained models to facilitate future attempts at enhancing word embeddings using knowledge graphs.

Our experiments are conducted on general and medical corpora in the English language. Generalization to other cases is left for future work.

2 Related Work

Our approach is related to the similar but usually distinct topics of knowledge injection and domain adaptation. In fact, most attempts at domain adaptation do not aim to inject external knowledge directly into models but rather indirectly, through retraining on specialized corpora, as this is known to bring significant improvements when such in-domain corpora are available (Si et al., 2019). On the other hand, research concerned with knowledge injection usually tackles the problem within the same domain. For instance, SemBERT (Zhang et al., 2020), COMET (Bosselut et al., 2019), ERNIE (Zhang et al., 2019), K-BERT (Liu et al., 2020), and KEPLER all inject general knowledge into general-domain models. Similar efforts in the medical domain (Hao et al., 2020; He et al., 2020a; Michalopoulos et al., 2021; Lu et al., 2021) directly inject medical knowledge during medical pre-training. In this work, we first set out to determine whether the performance of general-domain models, both static and contextual, can be improved solely using specialized knowledge embeddings, then only do we incorporate this approach into the usual model adaptation via pre-training.

Methods that utilize knowledge graphs, for instance (Roy and Pan, 2021; Sharma et al., 2019; Chang et al., 2021), can be broadly grouped into two categories: those that use the structured data directly and those that encode this data into numerical representations. Instances of direct utilization include KG-BERT (Yao et al., 2019) where triples (`concept_1`, `relation`, `concept_2`) are used to inject BERT with medical information through auxiliary tasks like knowledge graph completion and triple classification. Entity linking in (Yuan et al., 2021) or more specialized tasks in (He et al., 2020c) are also used as auxiliary tasks for performing such injection. While these methods can be effective, we argue that an indirect approach is desirable as it presents the specialized knowledge in the same format as the word embeddings, thus reformulating knowledge injection as a meta-embedding problem.

Meta-embeddings combine two or more underlying sets of embeddings into a single final representation. There are many approaches to meta-embeddings like Dynamic Meta Embeddings (DME, Kiela et al. (2018)) where each embedding is projected down to the same dimension before being used in a linear combination, or Word Prisms (He et al., 2020b), which further improve upon DMEs by enforcing desirable orthogonality properties during training. In this work, we use a simple but strong baseline for meta-embeddings—concatenation—which ensures that both word and knowledge information is accessible at all times. More sophisticated approaches, although likely to improve our overall performance, are left for future work.

3 Knowledge Graph Representations

In order to use concatenation to specialize word embeddings with a knowledge base, we first need to be able to convert this knowledge base into dense numerical representations. There are several meth-

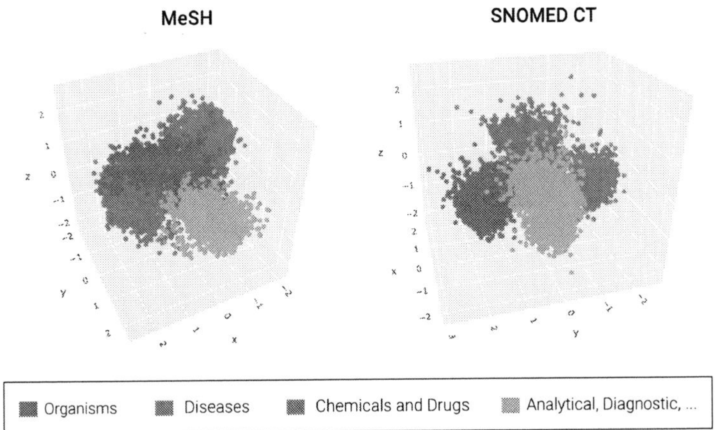

Figure 1: PCA of MeSH and SNOMED embeddings for four categories of medical concepts.

ods for embedding knowledge graphs (e.g. RotatE (Sun et al., 2019), TuckER (Balazevic et al., 2019)), and these usually produce multifaceted relation-dependent concept representations. However, for simplicity, we only consider a single relation which enables us to use a graph-level method instead, namely *node2vec* (Grover and Leskovec, 2016).

3.1 UMLS, MeSH, and SNOMED CT

The Unified Medical Language System (UMLS) (Lindberg et al., 1993) includes a meta-thesaurus that contains multiple subsets (called vocabularies) that organize specific groups of medical concepts according to a large number of varied relationships (e.g. `active_ingredient_of`, `associated_with`, `branch_of`). Among the many vocabularies in the UMLS, we use the Medical Subject Headings (MeSH)[4], which mainly organizes concepts from the biomedical domain, as well as the Systematized Nomenclature Of Medicine - Clinical Terms (SNOMED_CT),[5] which also has a coverage of the clinical domain. Given both vocabularies, we query[6] the UMLS and recover all pairs of Concept Unique Identifiers (CUI) for concepts related through the `is_a` relation (e.g. `Chronic Bronchitis` is a `Chronic disease`). Although many more types of relations are available, we focus on the single most frequent type `is_a`, which also allows us to extract a single graph and use a graph-level method like *node2vec*.

[4]`https://www.nlm.nih.gov/mesh/meshhome.html`

[5]`https://www.nlm.nih.gov/healthit/snomedct/index.html`

[6]SQL scripts are available in our code repository.

3.2 Dense Representations with *node2vec*

The *node2vec* (Grover and Leskovec, 2016) method effectively applies a *word2vec* (Mikolov et al., 2013) objective to learn node representations from a set of node sequences that are generated randomly using a flexible type of random walks on the knowledge graph. Running the official `Python` implementation[7] with default parameters allows us to learn 256-dimensional dense representations for each node of the provided graphs. This step yields 29,738 CUI embeddings for MeSH concepts and 389,872 CUI embeddings for SNOMED with 15,418 overlapping CUIs having both a MeSH and SNOMED representation. The visualization of these embeddings using a PCA (see Figure 1) shows that this method is able to separate different categories of medical concepts in different subspaces, which suggests some level of encoded medical knowledge.

Using *node2vec* Embeddings in Practice For each possible CUI, we concatenate both sets of knowledge embeddings and use zero-padding when a CUI does not appear in either MeSH or SNOMED. This produces a final 512-dimensional knowledge representation for each concept. However, using these representations in practice requires locating concept mentions in texts, which refers to the task of concept normalization. This normalization aims to identify the various linguistic forms that a given concept can take, which we perform in our case by running an exact string matching between the reference linguistic forms from the

[7]`https://github.com/aditya-grover/node2vec`

UMLS[8] and the target texts. Ultimately, the tokens from each mention are assigned the *node2vec* representation of their concept, with out-of-mention tokens getting an empty zero-valued vector instead.

4 Embedding-Specialization Methods

4.1 Static Representations

To determine whether word embeddings can be successfully specialized using in-domain knowledge representations, we first conduct experiments on static embeddings. In particular, we learn word representations using fastText[9] (Bojanowski et al., 2017) and then attempt to specialize these representations by concatenating fastText and node2vec vectors at the token level. We consider the following corpora for learning word embeddings:

Gigaword (Graff et al., 2003): a newswire corpus constructed from many sources including the New York Times. This is a general domain corpus with ≈ 1 billion tokens.

PubMed (MEDLINE): scientific abstracts from the biomedical literature.[10] This is a medical domain corpus with ≈ 2 billion tokens.

MIMIC (Johnson et al., 2016): clinical notes from several hospitals.[11] This is a medical domain corpus with ≈ 0.5 billion tokens.

4.2 Contextual Representations

We also experiment with contextual embeddings, namely: BERT (Devlin et al., 2019) and CharacterBERT (El Boukkouri et al., 2020).[12] The former is included as a strong baseline for transformer-based embeddings and the latter is included as it produces word-level representations and seems to perform well in the medical domain. Furthermore, considering these two models allows us to have a larger sample size for measuring the impact of our strategies on transformer-based models.

We specialize contextual embeddings in two ways: either externally, via token-level concatenation similar to static embeddings; or internally, by introducing the following specialization layers.

[8]These synonyms are available in the MRCONSO table.

[9]Training scripts are available in our code repository.

[10]Available at: https://www.nlm.nih.gov/databases/download/pubmed_medline.html

[11]Available at: https://physionet.org/content/mimiciii-demo/1.4/

[12]We use the "base-uncased" versions of these models.

Knowledge Injection Modules (KIM) These are small layers that generalize the idea of concatenating word and knowledge embeddings to the internal states of a transformer-based model. When placed after a given layer, this module concatenates the hidden representations from that layer h_i with their corresponding knowledge representations KG_i. Then, it projects this concatenation to recover a set of "enhanced states" $\mathbb{h}_i$ with the same dimensionality as the original hidden representations. Since this operation may lose some of the information from the original hidden states, we compute a mixture of the enhanced and original states with trainable parameters $\alpha \in [0, 1]$ and $\beta = 1 - \alpha$. The final output $\boldsymbol{h}_i$ is fed to the next layer. In summary:

$$\boldsymbol{h}_i = \alpha\, \mathbb{h}_i + \beta\, \mathrm{h}_i$$

where $\mathbb{h}_i = [\mathrm{h}_i; \mathrm{KB}_i]\, W + b$ and W, b are respectively the weight matrix and bias of the linear projection operation (see Figure 2).

Our KIMs are loosely related to Adapter Modules (Houlsby et al., 2019; Wang et al., 2021a) but are conceptually simpler and only focus on incorporating external representations into the hidden states of transformer-based models.

5 Experiments

5.1 Embedding Models

Our final embeddings come in five configurations:

Random: randomly initialized 256-dimensional static embeddings used as a baseline for static word representations.

***Model*:** either 256-dimensional static embeddings of the form "fastText(*corpus*)" where *corpus* is one of the corpora presented in section 4.1, or a 768-dimensional BERT or CharacterBERT model.

[*Model*, node2vec]: token-level concatenation of *Model* with the pre-trained 512-dimensional node2vec representations from Section 3.2.

***Model*(medical):** a transformer model adapted via pre-training on a large medical corpus that consists of ≈ 0.5 billion tokens from MIMIC-III clinical notes and ≈ 0.5 billion tokens from abstracts extracted from PMC-OA[13] biomedical articles.

[13]https://www.ncbi.nlm.nih.gov/pmc/tools/openftlist/

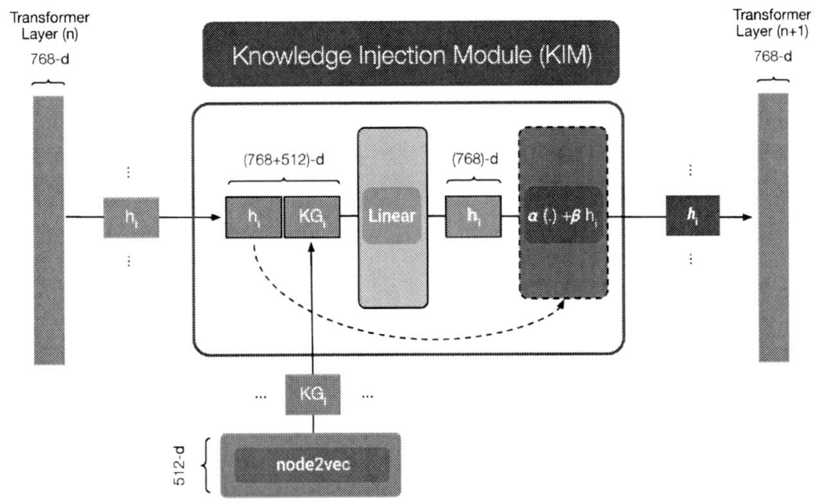

Figure 2: Detailed view of a Knowledge Injection Module (KIM) between two Transformer layers. Given an incoming hidden (h_i) and knowledge representation (KG_i), the module concatenates both vectors ($[\mathrm{h}_i; \mathrm{KG}_i]$), applies a linear projection down to the original size ($\mathbb{h}_i$), then computes a mixture of the enhanced and original states using parameters $\alpha \in [0, 1]$ and $\beta = 1 - \alpha$. The output ($\boldsymbol{h}_i$) is ultimately fed to the next Transformer layer.

Enhanced*Model*(medical): same as the configuration above but this time, the architecture is changed to use a KIM after each transformer layer, as well as either the WordPiece embeddings (Wu et al., 2016) for BERT, or Character-CNN (Peters et al., 2018) for CharacterBERT.

For the last two configurations, we follow a standard pre-training procedure comprising Masked Language Modeling (MLM) and Next Sentence Prediction (NSP), and adapt the implementation from El Boukkouri et al. (2020) while keeping the same hyper-parameters.[14]

5.2 Evaluation Tasks

Insights from model evaluation can be misleading, especially when only a few tasks are considered. To conduct a thorough evaluation of our models, we consider multiple tasks from both the biomedical and clinical domains (see Table 1):

i2b2 This is the i2b2/VA 2010 clinical concept extraction task (Uzuner et al., 2011), which is a sequence labeling task that aims to detect three categories of clinical entities: PROBLEM (e.g. "headache"), TREATMENT (e.g. "oxycodone") and TEST (e.g. "MRI"). The exact match F1-score is used as an evaluation metric.

BC5-Disease/Chemical These are two sequence labeling tasks from BioCreative V CDR (Li et al., 2016), which respectively aim to detect DISEASE (e.g. "hepatitis") and CHEMICAL (e.g. "corticosteroid") entities. The exact F1 is used as a metric.

DDI This is a relation extraction task from SemEval 2013 - Task 9.2. (Herrero-Zazo et al., 2013), which focuses on classifying drug-drug interactions into five categories: ADVISE (DDI-advise), EFFECT (DDI-effect), MECHANISM (DDI-mechanism), INTERACTION (DDI-int), and DDI-false for no interaction. The micro-averaged F1 over all four non-negative classes is used as a metric.

ChemProt This is a relation extraction task from BioCreative VI (Krallinger et al., 2017), which focuses on classifying chemical-protein relations into six categories: ACTIVATOR (CPR:3), INHIBITOR (CPR:4), AGONIST (CPR:5), ANTAGONIST (CPR:6), SUBSTRATE (CPR:9) and FALSE for no relation. The micro-averaged F1-score over non-negative classes is used as a metric.

BIOSSES This is a small sentence similarity dataset in the biomedical domain (Soğancıoğlu et al., 2017). The Pearson correlation of predicted and gold similarities is used as a metric.

ClinicalSTS This is a clinical sentence similarity task from the OHNLP Challenge 2018 (Wang et al., 2018). It uses Pearson correlation as well.

[14]Specifically, we use the parameters at this URL.

	i2b2	BC5-Disease	BC5-Chemical	ChemProt	DDI	BIOSSES	ClinicalSTS	MEDNLI
Train	22,263	4,182	5,203	4,154	2,937	64	600	11,232
Val.	5,565	4,244	5,347	2,416	1,004	16	150	1,395
Test	45,009	4,424	5,385	3,458	979	20	318	1,422

Table 1: Number of examples (entities, positive relations, or samples) for each evaluation task.

MedNLI This is a clinical natural language inference task (Romanov and Shivade, 2018), which aims to classify pairs of sentences into three categories: ENTAILMENT, CONTRADICTION, and NEUTRAL. The classification accuracy is used as a metric.

5.3 Evaluation Architectures

We use different architectures depending on the model and fine-tuning tasks at hand.

Sequence Labeling The architecture for tagging uses an encoder followed by a classification layer and a CRF (Lafferty et al., 2001). The encoder changes according to the type of input embeddings: **fastText** and **[fastText, node2vec]** are fed to a Bi-LSTM,[15] variants of **BERT** are their own encoders, and variants of **[BERT, node2vec]** concatenate knowledge (node2vec) embeddings with token (BERT) representations and feed it forward.

Classification The architecture for relation extraction is similar but requires a summarized representation at the example level to be fed to a classification layer. Here again, **fastText** and **[fastText, node2vec]** are fed to a Bi-LSTM, but this time, the output is average-pooled to produce a single feature vector. With variants of **BERT**, the pooler output is used. Finally, when using variants of **[BERT, node2vec]**, the knowledge representations are average-pooled before being concatenated with the pooler representation.

Sentence Similarity For STS tasks, we use a different approach for static and contextual embeddings. For **static embeddings**, we compute a bag-of-word representation for each sentence, then measure the cosine similarity between the two representations. When **contextual embeddings** are involved, we treat the task as a regression problem and use the same encoder as for classification.

Natural Language Inference For NLI tasks, we require a summarized representation at the sentence-pair level that we can ultimately feed to a classification layer. For **static embeddings**, we compute an average-pooled Bi-LSTM representation for the first sentence u as well as for the second one v, then compute a global feature vector $[u, v, |u - v|, u * v]$ following the approach of InferSent (Conneau et al., 2017). When using variants of **BERT**, we simply use the pooler representation as these models can accept sentence pairs. Finally, with variants of **[BERT, node2vec]**, we concatenate the pooler output with InferSent-style features computed from the node2vec vectors.

[15] All future mentions of a Bi-LSTM refer to a 3-layer network with 50% recurrent dropout and an output size of 512.

5.4 Evaluation Method

Optimization All parameters (including static and knowledge embeddings) are fine-tuned using the following hyper-parameters:

- **Validation Ratio:** when no validation set is available, we use 20% of the training data.
- **Epochs:** we run 15 epochs for all tasks, except for BIOSSES and ClinicalSTS for which we run 100 and 50 epochs respectively.
- **Batch Size:** we use batches of 32 examples.
- **Optimizer & Learning Rate:** we use the Adam optimizer (Kingma and Ba, 2015) with a learning rate of 1e-3 for non-transformer weights and a learning rate of 3e-5 for transformer weights. We also use a weight decay of 10% and a linear schedule with a 10% warmup for transformer weights.

Model Ensembles To account for some of the randomnesses during fine-tuning, we evaluate each model on each task using 10 different random seeds. Given these single models, we compute ensembles using a majority vote, except for STS tasks where we use the average similarity instead. Then, to account for the variance of the ensembles as well, we compute 10 different ensembles by excluding a single seed from the ensemble set and repeating this process. The average ensemble score is then used as the final performance for the model.

Statistical Significance We use Almost Stochastic Order (ASO) tests from Dror et al. (2019) in an attempt to rigorously compare our models. In this framework, the test takes a set of scores from

	i2b2	bc5-disease	bc5-chemical	chemprot	ddi	biosses	clinical_sts	mednli
Random	83.42	74.12	75.91	51.72	64.92	**64.42**	62.09	**70.68**
[Random, node2vec]	**84.57**	**80.93**	**84.91**	**53.66**	**65.14**	59.83	**63.72**	70.36
fastText(Gigaword)	84.70	76.79	82.76	52.06	62.83	**82.43**	**71.93**	69.66
[fastText(Gigaword), node2vec]	**84.99**	**80.86**	**86.50**	**52.64**	**64.44**	53.55	65.95	**70.08**
fastText(PubMed)	85.16	79.71	88.67	**54.62**	66.17	**91.49**	**72.10**	70.51
[fastText(PubMed), node2vec]	**85.49**	**82.62**	**89.45**	54.36	**67.11**	62.32	67.46	**70.82**
fastText(MIMIC)	85.49	78.92	84.93	51.54	67.12	**76.94**	**72.42**	**71.74**
[fastText(MIMIC), node2vec]	**86.01**	**80.70**	**86.38**	**52.90**	**67.59**	51.72	69.17	70.96
BERT	**88.16**	79.56	88.63	**71.75**	**79.95**	81.94	**84.71**	**79.75**
[BERT, node2vec]	87.76	**80.66**	**88.88**	71.36	79.60	**85.93**	84.34	79.30
BERT(medical)	**89.45**	81.88	**90.67**	**71.96**	**79.79**	89.14	**84.21**	**83.66**
EnhancedBERT(medical)	89.40	**83.48**	90.36	71.22	78.74	**92.12**	83.59	83.03
CharacterBERT	**88.08**	80.90	88.73	70.61	79.42	90.58	84.49	78.85
[CharacterBERT, node2vec]	87.81	**81.63**	**89.39**	**71.01**	**81.23**	**91.03**	**84.89**	**79.19**
CharacterBERT(medical)	**89.82**	83.60	92.07	**73.63**	**80.67**	87.52	83.63	**84.66**
EnhancedCharacterBERT(medical)	89.76	**85.05**	**92.08**	73.01	79.39	**92.65**	**84.42**	84.46

Table 2: Performance of model ensembles on evaluation tasks from the medical domain. Results are displayed in pairs: baseline model on the top line and specialized version (either through concatenation or KIM) on the bottom line. The colors show statistical significance, with bluer colors meaning the specialized models improve more significantly over the baselines and redder colors showing a more significant degradation in performance.

a model A and a model B, then returns a value $\epsilon \in [0, 1]$ that quantifies the stochastic order between A and B, with $\epsilon = 0$ meaning $A \succeq B$, $\epsilon = 1$ meaning $B \succeq A$, and $\epsilon = 0.5$ meaning that no stochastic order can be found for A and B.

6 Results and Discussion

For better visibility and given the large number of experiments, we present our results in pairs composed of a baseline and a specialized version of that baseline. We report the performances of each model pair as a set of two consecutive rows with the baseline on top (see Table 2). We also emphasize in bold the best performance on each task (column) and color the specialized version according to its ASO distance (ϵ) to the baseline model.[16]

Random vs. [Random, node2vec] It is interesting to note that randomly initialized static embeddings manage to achieve reasonable results, sometimes even outperforming pre-trained fastText representations (see Random vs. Gigaword or PubMed on MedNLI). However, given the random nature of these vectors, we can easily expect in-domain knowledge representations to be able to improve the performance on downstream specialized tasks. While this is verified in most situations, we note a degradation on BIOSSES and MedNLI. This could point to situations where external knowledge is not relevant to the task at hand.

fastText(X) vs. [fastText(X), node2vec] Overall, using concatenation to combine knowledge representations with fastText embeddings seems to result in consistent gains, notably on tagging and classification tasks (see the top-left section of the table). Moreover, these results seem to hold regardless of the domain of origin, as word embeddings trained on Gigaword (general domain), PubMed (biomedical domain), and MIMIC (clinical domain) all seem to benefit from this combination. However, we can see that the results on STS are significantly worse with drops of up to 30 points of correlation on BIOSSES with fastText(Gigaword). This may be due to the "bag-of-word + cosine similarity"

[16] Colors range from red ($\epsilon = 0$) for a significant degradation, to blue ($\epsilon = 1$) for a significant improvement.

approach not being suited for meta-embeddings made of both word and knowledge representations, especially since the node2vec vectors are rather sparse (most concepts do not have both a MeSH and SNOMED representation) and twice as large as the word representations.

BERT vs. [BERT, node2vec] When looking at the results for contextual embeddings, we can see several instances where the concatenation with *node2vec* proves to be beneficial. However, there seems to be a discrepancy where sometimes this concatenation does improve the CharacterBERT baseline on one hand but impairs the BERT baseline on the other (see ChemProt and DDI). A closer look at these cases shows that plain CharacterBERT performs slightly lower than plain BERT in these situations, which may mean that the knowledge representations compensate for any information that may be missing in the baseline CharacterBERT model, relative to the task.

BERT(medical) vs. EnhancedBERT(medical) The addition of KIMs seems to give variable results depending on the evaluation task. In fact, we can see that EnhancedBERT and Enhanced-CharacterBERT respectively lose 1.05 and 1.28 F1 relative to their baselines on the DDI task, however, we also see that these same models gain 1.6 and 1.45 F1 on the BC5-Disease task. Incidentally, the BC5 tasks are interesting as they use the exact same corpus but focus on two different types of entities: DISEASE and CHEMICAL. Therefore, given that EnhancedBERT(medical) performs better than BERT(medical) on BC5-Disease and worse on BC5-Chemical, we can safely assume that this is not due to the KIMs being particularly harmful but rather to the information available in the knowledge representations being, relative to what is already available in the base model, more relevant for the first task than for the second one. Consequently, we may assume that the KIMs can successfully incorporate external information into a model but that the downstream performance may depend on the relevance of this information for any given task.

Observed Trends All in all, we notice that the best models remain either BERT or CharacterBERT-based models and that the addition of external knowledge to static representations is not sufficient to make them outperform their contextual counterparts. This is globally true with a few exceptions. In fact, we may observe in the case of sequence labeling tasks (i2b2 and BC5-Disease/Chemical) that the addition of knowledge is often beneficial for static models. The matter is more complex for contextual models however, where the benefits are less clear but for which it may still be desirable to use external knowledge as any potential degradation seems to be relatively minor. In the case of relation classification tasks (ChemProt and DDI), leveraging external knowledge is once again positive for static models but seems to be harmful to some Transformer-based models (especially BERT). Finally, for semantic similarity and inference tasks (BIOSSES, ClinicalSTS, and MedNLI), we may not recommend using our methods as any existing gains are relatively small when compared to the potential losses, although there may be some benefit for contextual models. Overall, we can see that our knowledge enhancement methods, either by external concatenation or through KIMs, always benefit CharacterBERT with appreciable gains in performance: choosing CharacterBERT with KIMs ensures obtaining the highest performance or being very close to it.

	all	no	some	full	homog
fastText(PubMed)	+3.3	-4.6	+4.5	+5.1	+6.2
CharacterBERT	+0.3	-1.7	+0.6	+0.9	+1.1
EnhancedCBERT	+1.4	-1.7	+1.8	+2.1	+2.5

Table 3: Variations (percentages) of True Positives for the BC5-Disease task according to the coverage of the gold entities by concepts of our knowledge graph.

Contribution of the Knowledge Graph To measure the contribution of external knowledge, specifically in the case of sequence labeling tasks, we compute, for each gold entity of the test set, the average change in *true positives* brought by the use of the knowledge embeddings. To dig a bit deeper, we compute this change in buckets with varying the degrees of coverage of gold entities by a concept of the knowledge graph: **no** coverage; **some** coverage; all the tokens are **fully** covered; and finally, a full and **homog**eneous coverage (i.e. same CUI everywhere). We display the results for BC5-Disease and three different models in Table 3: fastText(PubMed) and CharacterBERT, which both rely on token-level concatenations, and Enhanced-CharacterBERT (EnhancedCBERT), which leverages KIMs. While the over**all** contribution is positive, we can see that this effect increases with the coverage of gold entities by the knowledge base.

Moreover, when the coverage is null, the impact becomes negative, emphasizing the importance of choosing a complete and adequate knowledge base when using such knowledge injection methods.

7 Conclusion and Future Work

In this paper, we focused on exploring the extent to which specialized information from a knowledge graph could be injected into existing word embeddings using a very simple set of tools: graph embeddings and concatenation. While focusing on the medical domain in the English language, we conducted multiple evaluations on tasks ranging from entity recognition to sentence similarity. These evaluations demonstrated that concatenation with in-domain graph representations can be a simple yet effective approach to model specialization, with significant gains on several tasks. Moreover, applying the same process of concatenation within transformer-based contextual models proved beneficial as well, with notable improvements using Knowledge Injection Modules (KIMs) on several downstream tasks.

As mentioned in Section 3.1, many more types of relations beyond `is_a` could be used to improve the quality of the generated knowledge representations. An interesting path for future work may be to use recent meta-embedding methods like Word Prisms to learn multifaceted knowledge representations from multiple underlying representations corresponding to two or more types of relations.

8 Acknowledgements

This work was funded by the French National Research Agency (ANR) under the ADDICTE project (ANR-17-CE23-0001) and was performed using HPC resources from GENCI-IDRIS (Grant AD011011698R1).

References

Emily Alsentzer, John Murphy, William Boag, Wei-Hung Weng, Di Jindi, Tristan Naumann, and Matthew McDermott. 2019. Publicly available clinical BERT embeddings. In *Proceedings of the 2nd Clinical Natural Language Processing Workshop*, pages 72–78, Minneapolis, Minnesota, USA. Association for Computational Linguistics.

Ivana Balazevic, Carl Allen, and Timothy Hospedales. 2019. TuckER: Tensor factorization for knowledge graph completion. In *Proceedings of the 2019 Conference on Empirical Methods in Natural Language Processing and the 9th International Joint Conference on Natural Language Processing (EMNLP-IJCNLP)*, pages 5185–5194, Hong Kong, China. Association for Computational Linguistics.

Piotr Bojanowski, Edouard Grave, Armand Joulin, and Tomas Mikolov. 2017. Enriching word vectors with subword information. *Transactions of the Association for Computational Linguistics*, 5:135–146.

Antoine Bosselut, Hannah Rashkin, Maarten Sap, Chaitanya Malaviya, Asli Celikyilmaz, and Yejin Choi. 2019. COMET: Commonsense transformers for automatic knowledge graph construction. In *Proceedings of the 57th Annual Meeting of the Association for Computational Linguistics*, pages 4762–4779, Florence, Italy. Association for Computational Linguistics.

David Chang, Eric Lin, Cynthia Brandt, and Richard Andrew Taylor. 2021. Incorporating domain knowledge into language models by using graph convolutional networks for assessing semantic textual similarity: Model development and performance comparison. *JMIR medical informatics*, 9(11):e23101.

Alexis Conneau, Douwe Kiela, Holger Schwenk, Loïc Barrault, and Antoine Bordes. 2017. Supervised learning of universal sentence representations from natural language inference data. In *Proceedings of the 2017 Conference on Empirical Methods in Natural Language Processing*, pages 670–680, Copenhagen, Denmark. Association for Computational Linguistics.

Jacob Devlin, Ming-Wei Chang, Kenton Lee, and Kristina Toutanova. 2019. BERT: Pre-training of deep bidirectional transformers for language understanding. In *Proceedings of the 2019 Conference of the North American Chapter of the Association for Computational Linguistics: Human Language Technologies, Volume 1 (Long and Short Papers)*, pages 4171–4186, Minneapolis, Minnesota. Association for Computational Linguistics.

Rotem Dror, Segev Shlomov, and Roi Reichart. 2019. Deep dominance - how to properly compare deep neural models. In *Proceedings of the 57th Annual Meeting of the Association for Computational Linguistics*, pages 2773–2785, Florence, Italy. Association for Computational Linguistics.

Hicham El Boukkouri, Olivier Ferret, Thomas Lavergne, Hiroshi Noji, Pierre Zweigenbaum, and Jun'ichi Tsujii. 2020. CharacterBERT: Reconciling ELMo and BERT for word-level open-vocabulary representations from characters. In *Proceedings of the 28th International Conference on Computational Linguistics*, pages 6903–6915, Barcelona, Spain (Online). International Committee on Computational Linguistics.

David Graff, Junbo Kong, Ke Chen, and Kazuaki Maeda. 2003. English gigaword. *Linguistic Data Consortium, Philadelphia*, 4(1):34.

Aditya Grover and Jure Leskovec. 2016. node2vec: Scalable feature learning for networks. In *Proceedings of the 22nd ACM SIGKDD international conference on Knowledge discovery and data mining*, pages 855–864.

Suchin Gururangan, Ana Marasović, Swabha Swayamdipta, Kyle Lo, Iz Beltagy, Doug Downey, and Noah A. Smith. 2020. Don't stop pretraining: Adapt language models to domains and tasks. In *Proceedings of the 58th Annual Meeting of the Association for Computational Linguistics*, pages 8342–8360, Online. Association for Computational Linguistics.

Boran Hao, Henghui Zhu, and Ioannis Paschalidis. 2020. Enhancing clinical BERT embedding using a biomedical knowledge base. In *Proceedings of the 28th International Conference on Computational Linguistics*, pages 657–661, Barcelona, Spain (Online). International Committee on Computational Linguistics.

Bin He, Di Zhou, Jinghui Xiao, Xin Jiang, Qun Liu, Nicholas Jing Yuan, and Tong Xu. 2020a. BERT-MK: Integrating graph contextualized knowledge into pre-trained language models. In *Findings of the Association for Computational Linguistics: EMNLP 2020*, pages 2281–2290, Online. Association for Computational Linguistics.

Jingyi He, Kc Tsiolis, Kian Kenyon-Dean, and Jackie Chi Kit Cheung. 2020b. Learning efficient task-specific meta-embeddings with word prisms. In *Proceedings of the 28th International Conference on Computational Linguistics*, pages 1229–1241, Barcelona, Spain (Online). International Committee on Computational Linguistics.

Yun He, Ziwei Zhu, Yin Zhang, Qin Chen, and James Caverlee. 2020c. Infusing Disease Knowledge into BERT for Health Question Answering, Medical Inference and Disease Name Recognition. In *Proceedings of the 2020 Conference on Empirical Methods in Natural Language Processing (EMNLP)*, pages 4604–4614, Online. Association for Computational Linguistics.

María Herrero-Zazo, Isabel Segura-Bedmar, Paloma Martínez, and Thierry Declerck. 2013. The DDI corpus: An annotated corpus with pharmacological substances and drug–drug interactions. *Journal of biomedical informatics*, 46(5):914–920.

Neil Houlsby, Andrei Giurgiu, Stanislaw Jastrzebski, Bruna Morrone, Quentin De Laroussilhe, Andrea Gesmundo, Mona Attariyan, and Sylvain Gelly. 2019. Parameter-efficient transfer learning for NLP. In *International Conference on Machine Learning*, pages 2790–2799. PMLR.

Alistair EW Johnson, Tom J Pollard, Lu Shen, H Lehman Li-Wei, Mengling Feng, Mohammad Ghassemi, Benjamin Moody, Peter Szolovits, Leo Anthony Celi, and Roger G Mark. 2016. MIMIC-III, a freely accessible critical care database. *Scientific data*, 3(1):1–9.

Douwe Kiela, Changhan Wang, and Kyunghyun Cho. 2018. Dynamic meta-embeddings for improved sentence representations. In *Proceedings of the 2018 Conference on Empirical Methods in Natural Language Processing*, pages 1466–1477, Brussels, Belgium. Association for Computational Linguistics.

Diederik P Kingma and Jimmy Ba. 2015. Adam: A Method for Stochastic Optimization. In *ICLR*.

Martin Krallinger, Obdulia Rabal, Saber A Akhondi, Martın Pérez Pérez, Jesús Santamaría, Gael Pérez Rodríguez, et al. 2017. Overview of the bioCreative VI chemical-protein interaction track. In *Proceedings of the sixth BioCreative challenge evaluation workshop*, volume 1, pages 141–146.

John D. Lafferty, Andrew McCallum, and Fernando C. N. Pereira. 2001. Conditional random fields: Probabilistic models for segmenting and labeling sequence data. In *Proceedings of the Eighteenth International Conference on Machine Learning (ICML 2001)*, pages 282–289, Williams College, Williamstown, MA, USA. Morgan Kaufmann.

Anne Lauscher, Ivan Vulić, Edoardo Maria Ponti, Anna Korhonen, and Goran Glavaš. 2020. Specializing unsupervised pretraining models for word-level semantic similarity. In *Proceedings of the 28th International Conference on Computational Linguistics*, pages 1371–1383, Barcelona, Spain (Online). International Committee on Computational Linguistics.

Jinhyuk Lee, Wonjin Yoon, Sungdong Kim, Donghyeon Kim, Sunkyu Kim, Chan Ho So, and Jaewoo Kang. 2020. BioBERT: a pre-trained biomedical language representation model for biomedical text mining. *Bioinformatics*, 36(4):1234–1240.

Jiao Li, Yueping Sun, Robin J Johnson, Daniela Sciaky, Chih-Hsuan Wei, Robert Leaman, Allan Peter Davis, Carolyn J Mattingly, Thomas C Wiegers, and Zhiyong Lu. 2016. BioCreative v CDR task corpus: a resource for chemical disease relation extraction. *Database*, 2016.

D. A. Lindberg, B. L. Humphreys, and A. T. McCray. 1993. The unified medical language system. *Methods of information in medicine*, 32(4):281–291.

Weijie Liu, Peng Zhou, Zhe Zhao, Zhiruo Wang, Qi Ju, Haotang Deng, and Ping Wang. 2020. K-BERT: Enabling language representation with knowledge graph. *Proceedings of the AAAI Conference on Artificial Intelligence*, 34(03):2901–2908.

Qiuhao Lu, Dejing Dou, and Thien Huu Nguyen. 2021. Parameter-efficient domain knowledge integration from multiple sources for biomedical pre-trained language models. In *Findings of the Association for Computational Linguistics: EMNLP 2021*, pages 3855–3865, Punta Cana, Dominican Republic. Association for Computational Linguistics.

George Michalopoulos, Yuanxin Wang, Hussam Kaka, Helen Chen, and Alexander Wong. 2021. Umls-BERT: Clinical Domain Knowledge Augmentation of Contextual Embeddings Using the Unified Medical Language System Metathesaurus. In *Proceedings of the 2021 Conference of the North American Chapter of the Association for Computational Linguistics: Human Language Technologies*, pages 1744–1753, Online. Association for Computational Linguistics.

Tomas Mikolov, Kai Chen, Greg Corrado, and Jeffrey Dean. 2013. Efficient estimation of word representations in vector space. *arXiv preprint arXiv:1301.3781*.

Matthew E. Peters, Mark Neumann, Mohit Iyyer, Matt Gardner, Christopher Clark, Kenton Lee, and Luke Zettlemoyer. 2018. Deep contextualized word representations. In *Proceedings of the 2018 Conference of the North American Chapter of the Association for Computational Linguistics: Human Language Technologies, Volume 1 (Long Papers)*, pages 2227–2237, New Orleans, Louisiana. Association for Computational Linguistics.

Matthew E. Peters, Mark Neumann, Robert Logan, Roy Schwartz, Vidur Joshi, Sameer Singh, and Noah A. Smith. 2019. Knowledge enhanced contextual word representations. In *Proceedings of the 2019 Conference on Empirical Methods in Natural Language Processing and the 9th International Joint Conference on Natural Language Processing (EMNLP-IJCNLP)*, pages 43–54, Hong Kong, China. Association for Computational Linguistics.

Alexey Romanov and Chaitanya Shivade. 2018. Lessons from natural language inference in the clinical domain. In *Proceedings of the 2018 Conference on Empirical Methods in Natural Language Processing*, pages 1586–1596, Brussels, Belgium. Association for Computational Linguistics.

Arpita Roy and Shimei Pan. 2021. Incorporating medical knowledge in BERT for clinical relation extraction. In *Proceedings of the 2021 Conference on Empirical Methods in Natural Language Processing*, pages 5357–5366, Online and Punta Cana, Dominican Republic. Association for Computational Linguistics.

Soumya Sharma, Bishal Santra, Abhik Jana, Santosh Tokala, Niloy Ganguly, and Pawan Goyal. 2019. Incorporating Domain Knowledge into Medical NLI using Knowledge Graphs. In *Proceedings of the 2019 Conference on Empirical Methods in Natural Language Processing and the 9th International Joint Conference on Natural Language Processing (EMNLP-IJCNLP)*, pages 6092–6097, Hong Kong, China. Association for Computational Linguistics.

Yuqi Si, Jingqi Wang, Hua Xu, and Kirk Roberts. 2019. Enhancing clinical concept extraction with contextual embeddings. *Journal of the American Medical Informatics Association*, 26(11):1297–1304.

Gizem Soğancıoğlu, Hakime Öztürk, and Arzucan Özgür. 2017. BIOSSES: a semantic sentence similarity estimation system for the biomedical domain. *Bioinformatics*, 33(14):i49–i58.

Zhiqing Sun, Zhi-Hong Deng, Jian-Yun Nie, and Jian Tang. 2019. RotatE: Knowledge Graph Embedding by Relational Rotation in Complex Space. In *International Conference on Learning Representations*.

Özlem Uzuner, Brett R South, Shuying Shen, and Scott L DuVall. 2011. 2010 i2b2/VA challenge on concepts, assertions, and relations in clinical text. *Journal of the American Medical Informatics Association*, 18(5):552–556.

Ashish Vaswani, Noam Shazeer, Niki Parmar, Jakob Uszkoreit, Llion Jones, Aidan N Gomez, Ł ukasz Kaiser, and Illia Polosukhin. 2017. Attention is all you need. In *Advances in Neural Information Processing Systems*, volume 30. Curran Associates, Inc.

Ruize Wang, Duyu Tang, Nan Duan, Zhongyu Wei, Xuanjing Huang, Jianshu Ji, Guihong Cao, Daxin Jiang, and Ming Zhou. 2021a. K-Adapter: Infusing Knowledge into Pre-Trained Models with Adapters. In *Findings of the Association for Computational Linguistics: ACL-IJCNLP 2021*, pages 1405–1418, Online. Association for Computational Linguistics.

Xiaozhi Wang, Tianyu Gao, Zhaocheng Zhu, Zhengyan Zhang, Zhiyuan Liu, Juanzi Li, and Jian Tang. 2021b. KEPLER: A Unified Model for Knowledge Embedding and Pre-trained Language Representation. *Transactions of the Association for Computational Linguistics*, 9:176–194.

Yanshan Wang, Naveed Afzal, Sijia Liu, Majid Rastegar-Mojarad, Liwei Wang, Feichen Shen, Sunyang Fu, and Hongfang Liu. 2018. Overview of the BioCreative/OHNLP challenge 2018 task 2: clinical semantic textual similarity. *Proceedings of the BioCreative/OHNLP Challenge*, 2018.

Yonghui Wu, Mike Schuster, Zhifeng Chen, Quoc V Le, Mohammad Norouzi, Wolfgang Macherey, Maxim Krikun, Yuan Cao, Qin Gao, Klaus Macherey, et al. 2016. Google's neural machine translation system: Bridging the gap between human and machine translation. *arXiv preprint arXiv:1609.08144*.

Zhilin Yang, Zihang Dai, Yiming Yang, Jaime Carbonell, Russ R Salakhutdinov, and Quoc V Le. 2019. XLNet: Generalized autoregressive pretraining for language understanding. In *Advances in Neural Information Processing Systems*, volume 32. Curran Associates, Inc.

Liang Yao, Chengsheng Mao, and Yuan Luo. 2019. KG-BERT: BERT for knowledge graph completion. *arXiv preprint arXiv:1909.03193*.

Wenpeng Yin and Hinrich Schütze. 2016. Learning word meta-embeddings. In *Proceedings of the 54th Annual Meeting of the Association for Computational*

Linguistics (Volume 1: Long Papers), pages 1351–1360, Berlin, Germany. Association for Computational Linguistics.

Zheng Yuan, Yijia Liu, Chuanqi Tan, Songfang Huang, and Fei Huang. 2021. Improving biomedical pretrained language models with knowledge. In *Proceedings of the 20th Workshop on Biomedical Language Processing*, pages 180–190, Online. Association for Computational Linguistics.

Zhengyan Zhang, Xu Han, Zhiyuan Liu, Xin Jiang, Maosong Sun, and Qun Liu. 2019. ERNIE: Enhanced language representation with informative entities. In *Proceedings of the 57th Annual Meeting of the Association for Computational Linguistics*, pages 1441–1451, Florence, Italy. Association for Computational Linguistics.

Zhuosheng Zhang, Yuwei Wu, Hai Zhao, Zuchao Li, Shuailiang Zhang, Xi Zhou, and Xiang Zhou. 2020. Semantics-aware bert for language understanding. *Proceedings of the AAAI Conference on Artificial Intelligence*, 34(05):9628–9635.

BioSimCSE: BioMedical Sentence Embeddings using Contrastive learning

Kamal Raj Kanakarajan, Bhuvana Kundumani, Abhijith Neil Abraham
Malaikannan Sankarasubbu
SAAMA AI Research Lab, Chennai, India
{kamal.raj, bhuvana.kundumani, abhigith.abraham, malaikannan.sankarasubbu}@saama.com

Abstract

Sentence embeddings in the form of fixed-size vectors that capture the information in the sentence as well as the context are critical components of Natural Language Processing systems. With transformer model based sentence encoders outperforming the other sentence embedding methods in the general domain, we explore the transformer based architectures to generate dense sentence embeddings in the biomedical domain. In this work, we present BioSimCSE, where we train sentence embeddings with domain specific transformer based models with biomedical texts. We assess our model's performance with zero-shot and fine-tuned settings on Semantic Textual Similarity (STS) and Recognizing Question Entailment (RQE) tasks. Our BioSimCSE model using BioLinkBERT achieves state of the art (SOTA) performance on both tasks.

1 Introduction

Word embeddings or vector representations of words generated by neural network architectures, capture the semantic relationships between words much better than traditional methods such as one hot encoding, bag of words, and so on. When dealing with large texts in real-world situations, it is essential to capture the semantic relationship between words as well as between sentences. Thus, rich sentence embeddings that capture the overall sentence semantics are critical for NLP systems. Sentence embeddings play a significant role in various NLP tasks such as information retrieval, semantic search, intent detection, natural language inference tasks.

In recent years, pre-trained models with transformer architecture for the general domain have grown in popularity. The advent of BERT Devlin et al. (2018) based models has made generating high-quality vector representations for natural language text much more manageable. These embeddings act as feature inputs for several downstream tasks. However, these models only generate word-level embeddings, from which we can derive sentence-level embeddings by averaging over the word-level embeddings. Another method is to use a cross encoder network from BERT to directly fine-tune for the task.Although this approach outperforms the averaging approach, it is computationally expensive and unsuitable for semantic similarity search and clustering tasks.

The biomedical domain with its corpora, significantly different from the general domain corpora, needs sophisticated and domain-specific models for effective knowledge representation. In this paper, we adapt SimCSE Gao et al. (2021), a state-of-the-art contrastive learning-based sentence embedding method, and release BioSimCSE - a biomedical domain-specific sentence embedding model.

In summary, our contributions are

1. We train and release[1] biomedical sentence embeddings with supervised and unsupervised training objectives from SimCSE.
2. We evaluate our models on biomedical STS and RQE tasks and demonstrate that our BioSimCSE achieves outstanding outcomes in both zero-shot and fine-tuned settings.

2 Background

Transformer-based language representations produced by Universal Sentence Encoder Cer et al. (2018) and BERT has aided NLP practitioners and researchers in various NLP tasks. Using BERT, sentence embeddings can either be generated by averaging the context embeddings of the last few layers or from the output of the last layer. SBERT Reimers and Gurevych (2019) shows that sentence embeddings produced by averaging word-level embeddings from BERT-like transformer models are unsuitable for standard similarity measurements

[1]https://github.com/kamalkraj/BioSimCSE

Proceedings of the 13th International Workshop on Health Text Mining and Information Analysis (LOUHI), pages 81 - 86
December 7, 2022

such as cosine similarity. SBERT uses the siamese network Schroff et al. (2015), a modified BERT network for the generation of fixed-size sentence embeddings. Though SBERT significantly reduces the time during inference and produces quality sentence embeddings, it follows a supervised approach. It heavily relies on labelled data to train sentence embedding models that might not be suitable for domains without labelled corpora.

Natural Language Inference (NLI) datasets are commonly used for supervised training of sentence embeddings models. The Multi-Genre Natural Language Inference (MultiNLI) Williams et al. (2018) corpus is mainly used to train general domain sentence embeddings. MultiNLI has 433k sentence pairs that have textual entailment information annotated. In the biomedical domain, large corpora of text are publicly available as research papers and articles. However, the availability of annotated datasets is lower than that of the general domain, and the number of samples is also low. Medical Natural Language Inference (MedNLI) Romanov and Shivade (2018) and Radiology Natural Language Inference (RadNLI) Miura et al. (2021a) Miura et al. (2021b) are biomedical NLI datasets; merging both yields only 15K sentence pairs for supervised training.

Recent research has explored different training objectives to derive sentence embeddings in an unsupervised manner. Before widespread adoption of transfomer-based models, Skipthought vectors Kiros et al. (2015) and Quick thoughts Logeswaran and Lee (2018) use unsupervised learning to derive sentence representations from unlabeled data with encoder-decoder and encoder architectures respectively. Semantic Re-tuning with Contrastive Tension (CT) Carlsson et al. (2021), BERT-flow Li et al. (2020), Transformer-based Sequential Denoising Auto-Encoder (TSDAE) Wang et al. (2021) and Simple Contrastive Learning of Sentence Embeddings (SimCSE) Gao et al. (2021) propose methods to generate sentence embeddings using a unsupervised approach with different training objectives. A domain like biomedical, where supervised datasets are unavailable, has to rely on unsupervised techniques. In this work, we selected SimCSE because its unsupervised training is comparable to that of its supervised competitors for training sentence embeddings; in addition, SimCSE performed better in our initial experiment with the other unsupervised training objectives described above.

3 Methods

The training objective for SimCSE Gao et al. (2021) utilises the contrastive learning approach, which has a cross-entropy loss function with in-batch negatives. In Unsupervised learning, positive pairs are made by giving the same sentence to the pre-trained encoder twice with regular dropout as noise, all other sentences in a mini-batch act as negative pairs. The NLI dataset has a contradiction hypothesis for each premise and its entailment hypothesis. For supervised sentence embeddings training, SimCSE uses entailment pairs as positive cases and adds matching contradiction pairs and other sentences in the mini-batch as negatives. As in the original SimCSE implementation, we use the `[CLS]` (first token of the sequence) as sentence embedding. Unsupervised SimCSE uses `[CLS]` with an MLP layer (only used in training), and supervised SimCSE uses `[CLS]` with MLP.

We initialize our sentence embeddings models from state-of-the-art transformer model, PubMedBERT Gu et al. (2020) and BioLinkBERT Yasunaga et al. (2022) from Biomedical Language Understanding and Reasoning Benchmark (BLURB) Gu et al. (2020) for our experiments. We excluded ELECTRA Clark et al. (2020) variants BioELECTRA Kanakarajan et al. (2021) and BioM Alrowili and Shanker (2021) from the BLURB because the quality of embeddings created by ELECTRA due to its Replaced Token Detection pre-training task is poor as shown in COCO-LM Meng et al. (2021).

Using biomedical corpora detailed in 3.1, we train biomedical domain-specific unsupervised and supervised sentence transformer models. The sentence embeddings training is done only with the model base architecture - 12 layers of transformers block with a hidden dimension of 768 and multi-head attention over 12 layers. Hyper-parameters used for training are provided in Appendix A. The trained sentence embeddings are then evaluated in zero-shot and fine-tuned settings on the three datasets outlined in 4.2.

3.1 Training data

We obtained 1 million sentences randomly sampled from PubMed Central (PMC) [2] published as of April 2022. Pubmed Parser Achakulvisut et al. (2020) is used to extract the abstracts and SciSpacy Neumann et al. (2019) for sentence tokenization. This data is used for unsupervised model training.

[2]https://www.ncbi.nlm.nih.gov/pmc/

The MedNLI dataset comprises sentence pairs annotated for contradiction, neutrality, and entailment by physicians from the Past Medical History section of MIMIC-III Johnson et al. (2016) clinical notes. The dataset contains 11,232 training, 1,395 validation, and 1,422 test cases. The RadNLI dataset contains annotated sentence pairings from the MIMIC-CXR database Johnson et al. (2019). The dataset includes a validation set of 480 and a test set of 480 pairings. For supervised model training, we merge the training, validation, and test sets from these two datasets.

4 Evaluation

We use STS and RQE tasks in the biomedical domain to evaluate the performance of our BioSimCSE sentence embeddings model. The datasets used for evaluation are detailed in 4.2. The similarity between two sentence embeddings is determined using cosine similarity. We determine a threshold in cosine similarity using the development set to classify entailment or not for RQE (binary classification) dataset. Using Spearman's correlation, we evaluate STS in line with the original SimCSE research. For RQE, accuracy is used. We evaluate BioSimCSE sentence embeddings under zero-shot and fine-tuned settings.

4.1 Evaluation Settings

In a zero-shot setting, the trained supervised and unsupervised BioSimCSE model is used to derive the sentence embeddings directly and evaluate the tasks. In the fine-tuned setting, With task-specific datasets, we further fine-tune the supervised and unsupervised trained BioSimCSE models to adapt better to the task's requirements for the sentence embeddings. For fine-tuning, the sentence embeddings (u, v) for each pair of sentences are derived. Using mean squared loss as the objective function for STS datasets and contrastive loss for question entailment datasets, we optimize cosine similarity between (u, v). Hyper-parameters used for fine-tuning are provided in Appendix A. The fine-tuned sentence embeddings are evaluated only with the corresponding task used for fine-tuning.

The results for zero-shot and fine-tuned are shown in table 1. We also train cross encoder, in which the transformer model takes two sentences and predicts a similarity score or a classification label, as described in the BERT Devlin et al. (2018) paper. This is the standard approach for STS and RQE (Binary classification) tasks. Results for cross encoder is available in table 2. We only compare our models to biomedical-specific models because recent research Gu et al. (2020) has shown that models pretrained with biomedical domain-specific corpora perform significantly better than general-domain language models for Biomedical NLP tasks.

4.2 Evaluation Data

BIOSSES dataset provides a collection of 100 similar sentence pairs manually annotated in the biomedical domain. We use the train-test split from BLURB Gu et al. (2020), 64 pairs for training, 16 pairs for validation and the remaining 20 pairs for testing. ClinicalSTS is the STS task in the clinical domain, the latest version provided by n2c2 2019 challenge Wang et al. (2020) has 1641 samples for training and a test set of 412 samples. We use the test set for evaluation, and we have split 1641 samples into 80% train and 20% validation set. Finding entailment between two questions in the context of QA is the objective of RQE. 8,588 training pairs and 302 testing pairs in the initial release Abacha and Demner-Fushman (2016). We use the MEDIQA 2019 Challenge Ben Abacha et al. (2019) test set as the testing pair and the original as the development set.

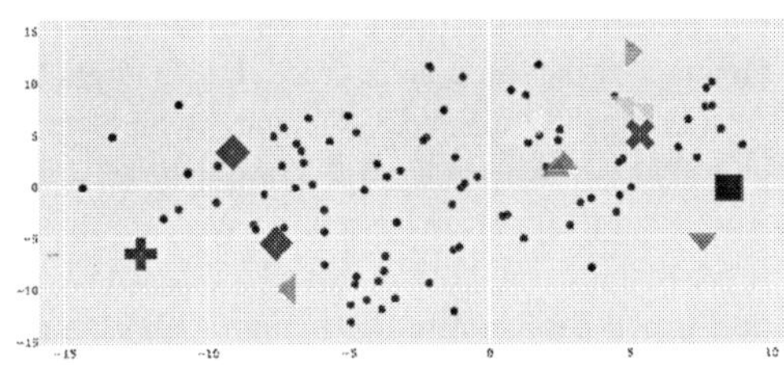

Figure 1: The t-SNE of sentence representation of BioLinkBERT before training with SimCSE

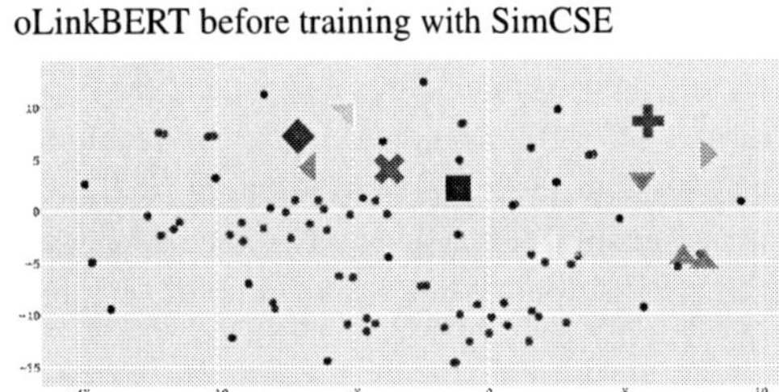

Figure 2: : The t-SNE of sentence representations after with training Unsupervised SimCSE. Similar pairs are denoted by identical shapes. The points are drawn from ClinicalSTS's most semantically comparable sentence pairings (with 5-score labels).

		Zero shot			fine-tuned		
		BIOSSES	**ClinicalSTS**	**RQE**	**BIOSSES**	**ClinicalSTS**	**RQE**
Sent2vec	BioSentVec	77.98	48.72	51.56	-	-	-
$\text{BioSimCSE}_{\text{Supervised}}$	PubMedBERT	83.13	72.17	53.04	85.91	77.87	56.52
	BioLinkBERT	90.32	76.42	54.35	92.73	81.35	57.39
$\text{BioSimCSE}_{\text{Unsupervised}}$	PubMedBERT	90.61	80.67	51.94	93.57	81.16	56.52
	BioLinkBERT	94.55	81.02	56.61	**96.37**	**83.76**	**60.04**

Table 1: Results on BIOSSES, ClinicalSTS and RQE test sets as described in 4. Metric, Spearman's correlation for BIOSSES and ClinicalSTS and accuracy for RQE.

	BIOSSES	**ClinicalSTS**	**RQE**
PubMedBERT	89.94	79.28	51.73
BioLinkBERT	91.75*	80.42	53.47

Table 2: Results on cross encoder architecture. * Current state of the art (SOTA). Metric, Spearman's correlation for BIOSSES and ClinicalSTS and accuracy for RQE.

5 Results

The $\text{BioSimCSE}_{\text{unsupervised}}$ BioLinkBERT model achieves remarkable results on all three datasets during the zero-shot evaluation. The zero-shot performs even better than BioLinkBERT fine-tuned with task-specific data using cross-encoder architecture. From the t-sne of sentence representation Figure 2, we can see that the similar sentence pairs (denoted by the same shapes) are closely aligned after training the BioLinkBERT model with SimCSE and also the average cosine similarity increased from 86.5 to 90.1 for the same. We can also observe that the transformer-based models have outperformed BioSentVec Chen et al. (2019), a non-transformer-based model with a large margin for both BIOSSES and ClinicalSTS. BioSentVec utilizes word vectors and n-grams approach to generate sentence embeddings using sent2vec Pagliardini et al. (2018) model. From the results, we can see that the supervised training of SimCSE is not practical as the unsupervised training, as the biomedical domain has a limited no.of samples in the NLI dataset.

When fine-tuned with task-specific data $\text{BioSimCSE}_{\text{unsupervised}}$ BioLinkBERT model sets new state-of-the-art results for all three datasets. For BIOSSES, Spearman's correlation is improved by +4.62 points, compared to the previous SOTA of 91.75. For ClinicalSTS the current SOTA is by BioELECTRA Kanakarajan et al. (2021) 82.11, BioSimCSE improve the SOTA by +1.65 points. BioSimCSE improve the RQE baseline 54.1 accuracy score Abacha et al. (2019) by +5.94 points and sets a new SOTA. We have omitted the RQE SOTA result from PANLP Zhu et al. (2019) (accuracy of 74.9), as this score is achieved using multitask and ensemble methods.

Performance on the evaluation datasets has steadily improved for both BioLinkBERT and PubMedBERT following training with SimCSE compared to the cross-encoder approach.

6 Conclusion

In our work, we have explored SimCSE for training sentence embeddings in the biomedical domain. We utilize the publicly available biomedical literature and NLI dataset for training the network in an unsupervised and supervised fashion and further fine-tune them with the task-specific datasets to adapt better to the task's requirements. Our BioSimCSE model has achieved SOTA on all three evaluation datasets. Our results demonstrate that SimCSE unsupervised training objectives can be able to train high-quality biomedical domain-specific sentence embeddings. We make the code and weights available for all of our models for reproducibility.

Limitations

In our experiments, we have only considered transformer base size models, whereas the Original SimCSE work evaluated both base and large size models. The sample sizes of the datasets used to evaluate sentence embeddings are limited. However, these are the biomedical domain's only sentence pair datasets. After training with SimCSE, the models have only been evaluated on sentence pair similarity/classification tasks and not on any classification of single sentence tasks.

References

Asma Ben Abacha and Dina Demner-Fushman. 2016. Recognizing question entailment for medical question answering. *AMIA ... Annual Symposium proceedings. AMIA Symposium*, 2016:310–318.

Asma Ben Abacha, Chaitanya P. Shivade, and Dina Demner-Fushman. 2019. Overview of the mediqa 2019 shared task on textual inference, question entailment and question answering. In *BioNLP@ACL*.

Titipat Achakulvisut, Daniel Acuna, and Konrad Kording. 2020. Pubmed parser: A python parser for pubmed open-access xml subset and medline xml dataset xml dataset. *Journal of Open Source Software*, 5(46):1979.

Sultan Alrowili and Vijay Shanker. 2021. BioM-transformers: Building large biomedical language models with BERT, ALBERT and ELECTRA. In *Proceedings of the 20th Workshop on Biomedical Language Processing*, pages 221–227, Online. Association for Computational Linguistics.

Asma Ben Abacha, Chaitanya Shivade, and Dina Demner-Fushman. 2019. Overview of the MEDIQA 2019 shared task on textual inference, question entailment and question answering. In *Proceedings of the 18th BioNLP Workshop and Shared Task*, pages 370–379, Florence, Italy. Association for Computational Linguistics.

Fredrik Carlsson, Amaru Cuba Gyllensten, Evangelia Gogoulou, Erik Ylipää Hellqvist, and Magnus Sahlgren. 2021. Semantic re-tuning with contrastive tension. In *International Conference on Learning Representations*.

Daniel Cer, Yinfei Yang, Sheng yi Kong, Nan Hua, Nicole Limtiaco, Rhomni St. John, Noah Constant, Mario Guajardo-Cespedes, Steve Yuan, Chris Tar, Yun-Hsuan Sung, Brian Strope, and Ray Kurzweil. 2018. Universal sentence encoder.

Qingyu Chen, Yifan Peng, and Zhiyong Lu. 2019. Biosentvec: creating sentence embeddings for biomedical texts. *2019 IEEE International Conference on Healthcare Informatics (ICHI)*.

Kevin Clark, Minh-Thang Luong, Quoc V. Le, and Christopher D. Manning. 2020. ELECTRA: Pre-training text encoders as discriminators rather than generators. In *ICLR*.

Jacob Devlin, Ming-Wei Chang, Kenton Lee, and Kristina Toutanova. 2018. Bert: Pre-training of deep bidirectional transformers for language understanding. *arXiv preprint arXiv:1810.04805*.

Tianyu Gao, Xingcheng Yao, and Danqi Chen. 2021. Simcse: Simple contrastive learning of sentence embeddings.

Yu Gu, Robert Tinn, Hao Cheng, Michael Lucas, Naoto Usuyama, Xiaodong Liu, Tristan Naumann, Jianfeng Gao, and Hoifung Poon. 2020. Domain-specific language model pretraining for biomedical natural language processing.

Alistair E. W. Johnson, Tom J. Pollard, Seth J. Berkowitz, Nathaniel R. Greenbaum, Matthew P. Lungren, Chih ying Deng, Roger G. Mark, and Steven Horng. 2019. Mimic-cxr, a de-identified publicly available database of chest radiographs with free-text reports. *Scientific Data*, 6.

Alistair EW Johnson, Tom J Pollard, Lu Shen, Li-wei H Lehman, Mengling Feng, Mohammad Ghassemi, Benjamin Moody, Peter Szolovits, Leo Anthony Celi, and Roger G Mark. 2016. Mimic-iii, a freely accessible critical care database. *Scientific data*, 3(1):1–9.

Kamal raj Kanakarajan, Bhuvana Kundumani, and Malaikannan Sankarasubbu. 2021. BioELECTRA:pretrained biomedical text encoder using discriminators. In *Proceedings of the 20th Workshop on Biomedical Language Processing*, pages 143–154, Online. Association for Computational Linguistics.

Ryan Kiros, Yukun Zhu, Ruslan Salakhutdinov, Richard S. Zemel, Antonio Torralba, Raquel Urtasun, and Sanja Fidler. 2015. Skip-thought vectors.

Bohan Li, Hao Zhou, Junxian He, Mingxuan Wang, Yiming Yang, and Lei Li. 2020. On the sentence embeddings from pre-trained language models.

Lajanugen Logeswaran and Honglak Lee. 2018. An efficient framework for learning sentence representations. In *International Conference on Learning Representations*.

Yu Meng, Chenyan Xiong, Payal Bajaj, Saurabh Tiwary, Paul Bennett, Jiawei Han, and Xia Song. 2021. Coco-lm: Correcting and contrasting text sequences for language model pretraining.

Yasuhide Miura, Yuhao Zhang, Emily Tsai, Curtis Langlotz, and Dan Jurafsky. 2021a. Improving factual completeness and consistency of image-to-text radiology report generation. In *Proceedings of the 2021 Conference of the North American Chapter of the Association for Computational Linguistics: Human Language Technologies*, pages 5288–5304, Online. Association for Computational Linguistics.

Yasuhide Miura, Yuhao Zhang, Emily Tsai, Curtis Langlotz, and Dan Jurafsky. 2021b. Radnli: A natural language inference dataset for the radiology domain.

Mark Neumann, Daniel King, Iz Beltagy, and Waleed Ammar. 2019. ScispaCy: Fast and Robust Models for Biomedical Natural Language Processing. In *Proceedings of the 18th BioNLP Workshop and Shared Task*, pages 319–327, Florence, Italy. Association for Computational Linguistics.

Matteo Pagliardini, Prakhar Gupta, and Martin Jaggi. 2018. Unsupervised learning of sentence embeddings using compositional n-gram features. In *Proceedings of the 2018 Conference of the North American Chapter of the Association for Computational*

Linguistics: Human Language Technologies, Volume 1 (Long Papers), pages 528–540, New Orleans, Louisiana. Association for Computational Linguistics.

Nils Reimers and Iryna Gurevych. 2019. Sentence-bert: Sentence embeddings using siamese bert-networks.

Alexey Romanov and Chaitanya Shivade. 2018. Lessons from natural language inference in the clinical domain.

Florian Schroff, Dmitry Kalenichenko, and James Philbin. 2015. Facenet: A unified embedding for face recognition and clustering. *2015 IEEE Conference on Computer Vision and Pattern Recognition (CVPR)*.

Kexin Wang, Nils Reimers, and Iryna Gurevych. 2021. Tsdae: Using transformer-based sequential denoising auto-encoder for unsupervised sentence embedding learning.

Y. Wang, S. Fu, F. Shen, S. Henry, O. Uzuner, and H. Liu. 2020. The 2019 n2c2/OHNLP Track on Clinical Semantic Textual Similarity: Overview. *JMIR Med Inform*, 8(11):e23375.

Adina Williams, Nikita Nangia, and Samuel Bowman. 2018. A broad-coverage challenge corpus for sentence understanding through inference. In *Proceedings of the 2018 Conference of the North American Chapter of the Association for Computational Linguistics: Human Language Technologies, Volume 1 (Long Papers)*, pages 1112–1122. Association for Computational Linguistics.

Michihiro Yasunaga, Jure Leskovec, and Percy Liang. 2022. Linkbert: Pretraining language models with document links. In *Association for Computational Linguistics (ACL)*.

Wei Zhu, Xiaofeng Zhou, Keqiang Wang, Xun Luo, Xiepeng Li, Yuan Ni, and Guo Tong Xie. 2019. Panlp at mediqa 2019: Pre-trained language models, transfer learning and knowledge distillation. In *BioNLP@ACL*.

A Example Appendix

The learning rate and batch size for training SimCSE supervised, and unsupervised models are the same as the original work. Our search for hyperparameters also shows that these give the best results. Both supervised, and unsupervised training was done using 512 batch sizes and learning rates 1e-5 and 5e-5, respectively. For the unsupervised model, we train the model with 1 million and 2 million examples, and we use zero-shot sentence similarity to measure how well it does. For one epoch, the model was trained. Adding more data after 1 million does not make a big difference in performance compared to the cost of training the model. The sequence length is restricted to 128 tokens in all our experiments. We use the SimCSE[3] implementation that the authors made available as open source to train our sentence embeddings. All the experiments are done using a single NVIDIA Titan RTX (24GB VRAM) GPU.

Table 3 lists all of the hyperparameters used for task specific fine-tuning of sentence embedding and cross encoder fine-tuning.

	Hyperparameters
Epochs	3-20
Learning rate	1e-5, 2e-5, 5e-5
Batch size	8, 16

Table 3: Sentence embeddings and cross encoder fine-tuning hyperparameters

Figure 3 shows how the similarity of sentence pairs is computed using the cosine similarity metric. The standard cross encoder architecture used with transformer models for sentence pair tasks is shown in Figure 4.

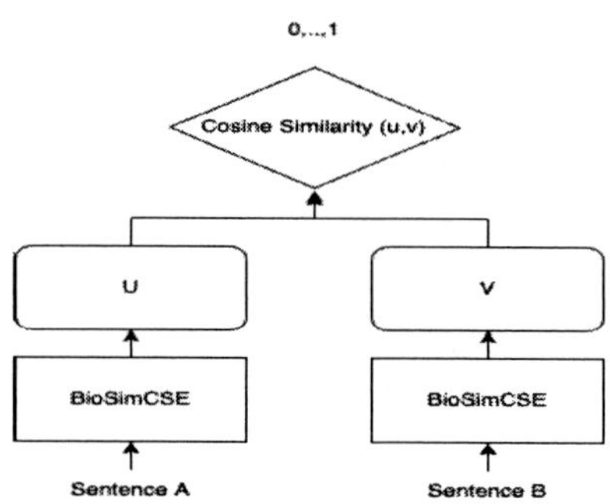

Figure 3: Finding similarity of sentence pair using BioSimCSE model

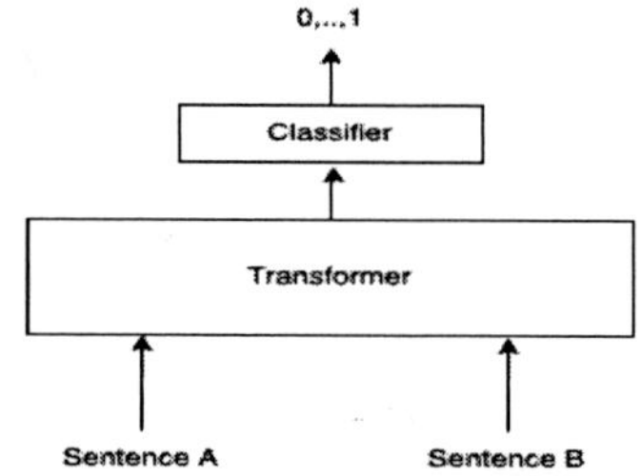

Figure 4: Cross encoder fine-tuning for sentence pair regression/classification

[3]https://github.com/princeton-nlp/SimCSE

Proxy-based Zero-Shot Entity Linking by Effective Candidate Retrieval

Maciej Wiatrak[1,*], **Eirini Arvaniti**[1], **Angus Brayne**[1], **Jonas Vetterle**[1,2], **Aaron Sim**[1]
[1]BenevolentAI [2]Moonfire Ventures
London, United Kingdom
{maciej.wiatrak, eirini.arvaniti, angus.brayne, aaron.sim}@benevolent.ai
jonas@moonfire.com

Abstract

A recent advancement in the domain of biomedical Entity Linking is the development of powerful two-stage algorithms – an initial *candidate retrieval* stage that generates a shortlist of entities for each mention, followed by a *candidate ranking* stage. However, the effectiveness of both stages are inextricably dependent on computationally expensive components. Specifically, in candidate retrieval via dense representation retrieval it is important to have *hard* negative samples, which require repeated forward passes and nearest neighbour searches across the entire entity label set throughout training. In this work, we show that pairing a proxy-based metric learning loss with an adversarial regularizer provides an efficient alternative to hard negative sampling in the candidate retrieval stage. In particular, we show competitive performance on the recall@1 metric, thereby providing the option to leave out the expensive candidate ranking step. Finally, we demonstrate how the model can be used in a zero-shot setting to discover out of knowledge base biomedical entities.

1 Introduction

The defining challenge in biomedical Entity Linking (EL) is performing classification over a large number of entity labels with limited availability of labelled mention data, in a constantly evolving knowledge base. For instance, while the Unified Medical Language System (UMLS) knowledge base (Bodenreider, 2004) contains millions of unique entity labels, the EL training data in the biomedical domain as a whole is notoriously scarce, particularly when compared to the general domain – Wikipedia, for instance, is powerful as *both* a Knowledge base and a source of matching entities and mentions. Furthermore, biomedical knowledge bases are evolving rapidly with new entities being added constantly. Given this knowledge base evolution and scarcity of training data it is crucial that biomedical entity linking systems can scale efficiently to large entity sets, and can discover or discern entities outside of the knowledge base and training data.

Recent methods in the general entity linking domain (Logeswaran et al., 2019; Wu et al., 2020) address the data issue with zero-shot entity linking systems that use entity descriptions to form entity representations and generalise to entities without mentions. A particularly powerful architecture was initially proposed by Humeau et al. (2019) and further improved by Wu et al. (2020). It consists of a two-stage approach: 1) candidate retrieval in a dense space performed by a *bi-encoder* (Wu et al., 2020) which independently embeds the entity mention and its description, and 2) candidate ranking performed by a *cross-encoder* which attends across both the mention and entity description (Logeswaran et al., 2019). In this work we focus on the former, which is traditionally optimised with the cross-entropy (CE) loss and aims to maximise the similarity between the entity mention and its description relative to the similarities of incorrect mention-description pairs. In practice, the large number of knowledge base entities necessitates the use of negative sampling to avoid the computational burden of comparing each mention to all of the entity descriptions. However, if the sampled distribution of negatives is not reflective of the model distribution, the performance may be poor. Recently, Zhang and Stratos (2021) showed that using hard negatives - the highest scoring incorrect examples - results in bias reduction through better approximation of the model distribution. Collecting hard negatives is computationally expensive, as it requires periodically performing inference and retrieving approximate nearest neighbours for each mention.

At the ranking stage, negative sampling is not required, as the number of candidates usually does

* Corresponding author.

Proceedings of the 13th International Workshop on Health Text Mining and Information Analysis (LOUHI), pages 87 - 99
December 7, 2022 ©2022 Association for Computational Linguistics

not exceed 64. However, the state-of-the-art cross-encoder model used for ranking is very expensive to run, scaling quadratically with the input sequence length. This highlights the need for efficient and performant candidate retrieval models capable of disambiguating mentions without the need for the expensive ranking step.

In this paper, we propose and evaluate a novel loss for the candidate retrieval model, which breaks the dependency between the positive and negative pairs. Our contributions are: (1) a novel loss which significantly outperforms the benchmark cross-entropy loss on the candidate retrieval task when using random negatives, and performs competitively when using hard negatives. (2) We design and apply an adversarial regularization method, based on the Fast Gradient Sign Method (Goodfellow et al., 2015), which is designed to simulate hard negative samples without expensively mining them. (3) We construct a biomedical dataset for out of knowledge base detection evaluation using the MedMentions corpus and show that our model can robustly identify mentions that lack a corresponding entry in the knowledge base, while maintaining high performance on the retrieval task.

Our main testing ground is the biomedical entity linking dataset MedMentions (Mohan and Li, 2019), which utilizes UMLS as its knowledge base. Additionally, to confirm that our method works also in the general, non-biomedical domain, we evaluate it on the Zero-Shot Entity Linking (ZESHEL) dataset proposed in Logeswaran et al. (2019). We focus on the retrieval task with the recall@1 metric, because we are aiming to predict the entity directly without requiring the additional expensive ranking stage. Our results show that both the proposed loss and regularization improve performance, achieving state-of-the-art results on recall@1 and competitive performance on recall@64 on both datasets. Finally, we demonstrate that our model can robustly identify biomedical out of knowledge base entities, without requiring any changes to the training procedure.

2 Related Work

Zero-Shot Entity Linking There is a plethora of work on zero-shot entity linking methods leveraging the bi-encoder architecture (Wu et al., 2020) for candidate retrieval. These include novel scoring functions between the input and the label (Humeau et al., 2019; Luan et al., 2021; Khattab and Zaharia, 2020), cross-domain pretraining methods (Varma et al., 2021), training and inference optimisation techniques (Bhowmik et al., 2021) and effective entity representation methods (Ma et al., 2021). Our work instead focuses on optimising the candidate retriever's loss function.

The impact of hard negatives on the entity linking model performance has also been investigated (Gillick et al., 2019; Zhang and Stratos, 2021). Notably, Zhang and Stratos (2021) develop analytical tools to explain the role of hard negatives and evaluate their model on the zero-shot entity linking task. We draw on this work, but move away from the CE loss towards a novel contrastive proxy-based loss.

Finally, there is a body of work on zero-shot entity linking in the biomedical domain using clustering (Angell et al., 2021; Agarwal et al., 2021). Our method does not consider the affinities between mentions directly and links them independently. Therefore, we do not study entity discovery.

An important aspect of biomedical entity linking systems is the detection of "unlinkable" mentions that lack a corresponding entry in the Knowledge Base - referred to as NIL detection. Methods for this task can be grouped into four main strategies (Shen et al., 2014; Sevgili et al., 2020): (1) label a mention as NIL when the corresponding candidate retriever does not return any candidate entities (Tsai and Roth, 2016), (2) assign the NIL label to mentions whose corresponding top-ranked entity does not exceed some score threshold (Bunescu and Pasca, 2006; Gottipati and Jiang, 2011; Lazic et al., 2015), (3) train a classifier that predicts whether the top-ranked entity for a given mention is correct (Moreno et al., 2017), (4) explicitly introduce a NIL class to the candidate ranking model (Kolitsas et al., 2018). A downside of the final approach is that knowledge of the NIL mention distribution is required at training time. In this work we tune a NIL score threshold (2) on a validation set. Detecting unlinkable mentions is particularly important in the biomedical domain, where the knowledge bases are rapidly evolving.

Proxy-based Losses State-of-the-art entity linking models such as BLINK (Wu et al., 2020) leverage metric learning loss during training to make mentions similar to its assigned entity representations. Metric learning losses could be divided into two categories, pair-based and proxy-based losses (Kim et al., 2020). Pair-based losses can leverage semantic relations between data points, here mentions. However, training them can be highly computationally expensive. On the other hand, proxy-based losses are significantly less computationally complex. This is done by establishing a proxy for each class and trying to increase the similarity between data points and its assigned proxies. Therefore, avoiding comparing the mentions to each other in favour of comparing the mentions to their proxies. We draw heavily on proxy-based losses (Movshovitz-Attias et al., 2017; Kim et al., 2020) from metric learning by treating entity descriptions as the proxies. We establish a proxy for each entity, creating mention-proxy (i.e. entity) pairs, and optimise the model to embed the mention close to its assigned proxy. The loss proposed here is similar to the Proxy-NCA loss of Movshovitz-Attias et al. (2017). Our modification is the use of the Softplus function, similar to Kim et al. (2020), to avoid a vanishing gradient for the true mention-proxy pair.

Adversarial Regularization Entity linking systems often rely on careful mining of hard negative examples to boost their performance (Gillick et al., 2019; Zhang and Stratos, 2021) at the expense of increased computational complexity. The model needs update hard negatives for each mention periodically. A potential alternative to hard negative mining is training on adversarial examples (Szegedy et al., 2013; Goodfellow et al., 2015) - synthetic data points designed to induce the model to making incorrect predictions, such that they are more challenging. Adversarial training can be seen as data augmentation and can help reduce overfitting. Goodfellow et al. (2015) introduced a simple method for generating adversarial examples, called Fast Gradient Sign Method (FGSM), which we build upon in this work. FGSM creates adversarial examples by applying small perturbations to the original inputs - often the word embeddings for NLP problems. FGSM has been used successfully as a regulariser in supervised and semi-supervised NLP tasks (Miyato et al., 2016; Pan et al., 2021). Here, we follow a similar approach and use FGSM to augment our training pairs with adversarial positive and negative examples.

3 Task formulation

In the **Entity Linking task** we are provided with a list of documents $D \in \mathcal{D}$, where each document has a set of mentions $M_D = \{m_1, m_2, \ldots, m_{N_D}\}$. The task is to link each mention m_i to an entity e_i, where each entity belongs to the Knowledge Base (KB) $\mathcal{E}$. In this work we focus specifically on the problem of biomedical zero-shot entity linking. The setup for the zero-shot task is the same as for entity linking introduced above, except that the set of entities present in the test set is not present in the training set, i.e. $\mathcal{E}_{\text{train}} \cap \mathcal{E}_{\text{test}} = \emptyset$ with $\mathcal{E}_{\text{train}} \cup \mathcal{E}_{\text{test}} = \mathcal{E}$. We focus specifically on the **Candidate Retrieval** task, where the goal is given a mention m_i, reduce the pool of potential candidate entities from a KB to a smaller subset. Candidate retrieval is crucial for biomedical entity linking because of the large size of knowledge bases. In this work we use the bi-encoder architecture for candidate retrieval. Finally, in addition to the *in-KB* entity linking task, where you only consider entities inside the KB, we also consider an **out of KB** scenario, where the task is to map mentions to the augmented set of labels $\mathcal{E} \cup$ NIL, with NIL indicating the absence of a corresponding KB entity.

4 Methods

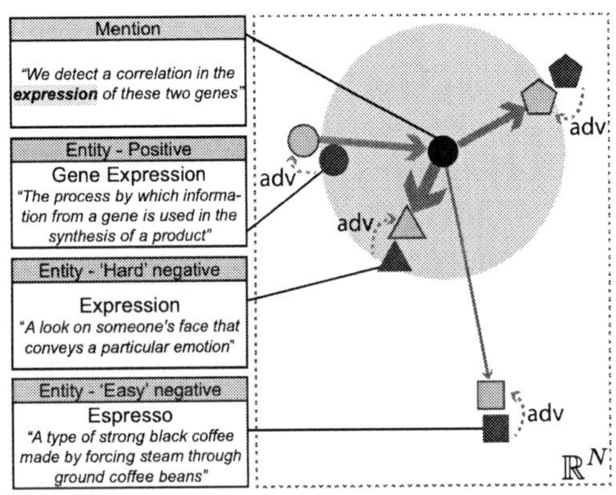

Figure 1: Overview of our proxy-based entity linking method. The mention and entity embeddings are encoded into a joint embedding space. During training, the magnitude of the gradients of the Proxy loss function with respect to the embedding coordinates is a function of the similarity between the mention and the entities (proxies). The gradients are represented by arrows whose widths indicate their magnitude. The adv-labelled dotted arrows are the Fast Gradient Sign Method adversarial perturbations. The blue circle symbolizes the margin δ.

In this section, we review the categorical CE loss, used by current state-of-the-art models, in the context of entity linking (Wu et al., 2020; Zhang and Stratos, 2021). We then compare it to our proposed Proxy-based loss. Finally, we describe and motivate our regularization approach.

4.1 Loss

Given a set of data points corresponding to mention representations $m \in M$ and to a set of proxies corresponding to entities $e \in \mathcal{E}$, the categorical CE loss is defined as:

$$L_{\text{CE}}(m, P) := -\log\left(\frac{\exp(s(m, e^+))}{\sum_{e \in P} \exp(s(m, e))}\right), \tag{1}$$

where $s(\cdot, \cdot)$ denotes a similarity function (e.g. cosine similarity or dot product), e^+ is the positive proxy for mention representation m, P^- is a set of negative proxies used as negative samples, and $P = \{e^+\} \cup P^-$.

The gradient of the CE loss with respect to $s(m, e)$ is given by:

$$\frac{\partial L_{\text{CE}}}{\partial s(m, e)} = \begin{cases} -1 + \dfrac{\exp(s(m, e^+))}{\sum_{e \in P} \exp(s(m, e))}, & e = e^+ \\ \dfrac{\exp(s(m, e^-))}{\sum_{e \in P} \exp(s(m, e))}, & e \in P^- \end{cases} \tag{2}$$

In practice training is performed with negative sampling. If the negatives are sampled randomly, often the exponential term for the positive entity is much larger than that of the negative samples and the gradients vanish. When $s(m, e^+) \gg s(m, e^-)\ \forall e^- \in P^-$ then $\partial L_{\text{CE}} / \partial s(m, e) \to 0$. This behaviour is desirable when training with the full distribution of negative pairs, but stifles learning in the noisier sampling setup. A common approach is the use of hard negatives (Gillick et al., 2019; Zhang and Stratos, 2021), which increases performance over training with random negatives at the cost of increased computational complexity.

On the other hand, contrastive metric learning losses (Bromley et al., 1993; Chopra et al., 2005; Hadsell et al., 2006) alleviate the vanishing gradients problem by decoupling the positive and negative loss terms. Proxy-based contrastive losses, such as Proxy-NCA (Movshovitz-Attias et al., 2017), aim to increase the similarity between a data point x and its assigned proxy e^+, while decreasing the similarity between x and its negative proxies $e^- \in P^-$. As demonstrated in (Kim et al., 2020), a downside of Proxy-NCA is that the scale of its gradient is constant for positive samples. This issue is alleviated by the Proxy Anchor loss (Kim et al., 2020), whose gradient reflects the relative hardness of both positive and negative pairs, resulting in improved model performance.

Drawing inspiration from the proxy-based metric learning losses described above, we formulate our Proxy-based (Pb) candidate retrieval loss as follows:

$$\begin{aligned} L_{\text{Pb}}(m, P) = {} & \log(1 + \exp(-\alpha(s(m, e^+) - \delta)) \\ & + \log(1 + \sum_{e^- \in P^-} \exp(\alpha(s(m, e^-) + \delta)), \end{aligned} \tag{3}$$

where we use the same notation as in Eq. 1. In addition, α is a hyperparameter controlling how strongly positive and negative samples pull and push each other, and δ is a margin. If α and δ are large, the model will be strongly penalized for the positive pair being too far from each other, and conversely the negative pair for being too close to each other. If α and δ are small, the model will receive weaker feedback. The Softplus function, a smooth approximation of the ReLU, introduces an additional margin beyond which the model stops penalising both positive and negative pairs, thus reducing overfitting. The gradient of our Proxy-based loss function is given by:

$$\frac{\partial L_{\text{Pb}}}{\partial s(m, e)} = \begin{cases} \dfrac{-\alpha \exp(-\alpha s^+)}{1 + \exp(-\alpha s^+)}, & e = e^+ \\ \dfrac{\alpha \exp(\alpha s^-)}{1 + \sum_{e^- \in P^-} \exp(\alpha s -)}, & e \in P^- \end{cases} \tag{4}$$

where $s^+ = s(m, e^+) - \delta$, $s^- = s(m, e^-) + \delta$. This gradient reflects the relative hardness of negative examples, decoupled from the positive pair, which makes it less sensitive to the choice of negative sampling scheme.

4.2 Regularization

Our regularization approach is based on a simple adversarial training technique, called *Fast Gradient Sign Method* (FGSM) (Goodfellow et al., 2015). The idea of FGSM is to generate adversarial examples according to the following equation:

$$x_{\text{adv}} = x + \epsilon * \text{sign}(\nabla_x L(x, y)) \tag{5}$$

where x is the original training example, y its corresponding label, L the loss function that is minimised during model training, and ϵ a small number defining the magnitude of the perturbation.

FGSM applies a small perturbation to the input example that should not change the label of the resulting example x_{adv}. However, Goodfellow et al. (2015) demonstrated that even infinitesimal perturbations can cause drastic changes to the model output when carefully designed. This effect is due to the locally linear nature of neural networks in combination with the high dimensionality of their inputs. Moreover, it is the direction, rather than the magnitude, of the perturbation that matters the most. In FGSM the direction is determined by the gradient of the loss function with respect to the model input - x is pushed in the direction of highest loss increase given its true label y.

In the context of entity linking task, we are interested in generating examples adversarial to the learned metric, in other words hard negative and hard positive examples for a given mention m. To this end, we applied the following perturbations to the entity encoder input embeddings $z = \text{input\_embed}(e)$:

$$z^-_{\text{adv}} = z^- + \epsilon * \text{sign}(\nabla_{z^-} s(m, e^-)) \quad (6)$$
$$z^+_{\text{adv}} = z^+ - \epsilon * \text{sign}(\nabla_{z^+} s(m, e^+)) \quad (7)$$

where m is the anchor mention and z^-, z^+ are the encoder input embeddings of negative and positive entities e^-, e^+ correspondingly.

Given N negative entities for a mention m, the generated adversarial entity embeddings $P_{\text{adv}} = \{z^-_{\text{adv\_1}}, \ldots, z^-_{\text{adv\_N}}, z^+_{\text{adv}}\}$ are used as additional training examples, giving rise to an auxiliary loss term that encourages the model to be invariant to local adversarial perturbations. Thus, the final objective we are trying to minimise becomes:

$$L_{\text{Pb}}(m, P) + \lambda L_{\text{Pb}}(m, P_{\text{adv}}) \quad (8)$$

where λ is a hyperparameter controlling the relative contributions of the two losses.

5 Experiments

5.1 Datasets

MedMentions This is is a biomedical entity-linking dataset consisting of over 4,000 PubMed abstracts (Mohan and Li, 2019). As recommended by the authors, we use the ST21PV subset, which has around 200,000 mentions in total. A large number of mentions in both the validation and test splits are zero-shot, meaning their ground truth label is not present in the training data. We do not carry out any additional preprocessing on the dataset. Finally, for the knowledge base (KB), we follow the framework in Varma et al. (2021) and use the UMLS 2017AA version filtered by the types present in the ST21PV subset. The final KB includes approximately 2.36M entities.

	MedMentions			Zero-Shot EL		
	Train	Val	Test	Train	Val	Test
Mentions	120K	40K	40K	49K	10K	10K
Entities	19K	8K	8K	333K	90K	70K
% Entities seen	100	57.5	57.5	100	0	0

Table 1: Statistics of datasets used. "% Entities seen" signifies the percentage of ground truth entities seen during training.

To evaluate our models in the NIL detection setting, we have created a new dataset based on MedMentions. In this dataset, we have assigned mentions corresponding to 11 entity types a NIL label and removed them from the Knowledge Base. Details on the dataset statistics and removed entity types can be found in the Appendix.

Zero-Shot Entity Linking dataset ZESHEL, a general domain dataset was constructed by Logeswaran et al. (2019) from Wikias[1]. It consists of 16 independent Wikias. The task is to link mentions in each document to a Wikia-specific entity dictionary with provided entity descriptions. The dataset is zero-shot, meaning there is no overlap in entities between training, validation and test sets.

5.2 Input Representation and Model Architecture

Similarly to Wu et al. (2020); Zhang and Stratos (2021); Varma et al. (2021) our candidate retriever is a bi-encoder consisting of two independent BERT transformers. We use the bi-encoder to encode a textual mention and an entity description independently then obtain a similarity score between them.

Namely, Given a mention and its surrounding context τ_m and an entity τ_e, we obtain dense vector representations $\mathbf{y_m} = \text{red}(T_1(\tau_m))$ and $\mathbf{y_e} = \text{red}(T_2(\tau_e))$, where T_1 and T_2 are the two independent transformers of the bi-encoder and $\text{red}(\cdot)$ is a function that reduces the output of a transformer into a single vector. We use a mean pooling operation for the function $\text{red}(\cdot)$.

As in Wu et al. (2020); Zhang and Stratos (2021); Varma et al. (2021) we use the dot product to score the mention $\mathbf{y_m}$ against an entity vector $\mathbf{y_e}$ when

[1]https://wikia.com

using the CE loss. For our Proxy-based loss we use cosine similarity.

In this, work, we focus on entity linking by efficient candidate retrieval, but we also include the ranker results using the highest scoring candidate entities in the Appendix, where we also include more details on entity, mention and context modelling.

5.3 Training & Evaluation Details

In all our experiments we used the transformer architecture (Vaswani et al., 2017) for the encoders. Namely, we used BERT (Devlin et al., 2019), initialised with appropriate pre-trained weights: SapBERT (Liu et al., 2021) for MedMentions and the uncased BERT-base (Devlin et al., 2019) for ZESHEL. For FGSM regularization, we apply adversarial perturbations to the composite token embeddings (i.e. sum of word, position and segment embeddings) used as input to BERT. We apply our regularization to both Proxy-based and CE. For information on hyperparameter tuning please refer to the Appendix. We tune all of our experiments on the validation set and report results on the test set. Due to hardware limitations, the training was conducted on a single V100 GPU machine with 16 GB of GPU memory. The limited GPU capacity, in particular, memory, posed a challenge by constraining us to using a relatively low number of negatives when training a retriever.

5.3.1 Candidate Retriever

The retriever model is optimised with the Proxy-based loss (3) and benchmark CE loss (1) for fair comparison. We evaluate the retriever on the micro-averaged recall@1 and recall@64 metrics, where in our setup recall@1 is equivalent to accuracy. Here we focus on the recall@1 metric, which is highly relevant for efficient candidate retrieval models that do not necessitate running an expensive cross-encoder for candidate ranking. We use two negative sampling techniques: (1) Random, where the negatives are sampled uniformly at random from all entities in the knowledge base, and (2) Mixed-p: p percent of the negatives are hard, the rest are random. This is motivated by the results shown in Zhang and Stratos (2021). We set the p to 50%.

Hard negative mining Retrieving hard negatives requires running the model in the inference mode over the entire KB. Then, for each mention, the most similar (i.e. hard) negatives are sampled according to a scoring function. Here, we use FAISS (Johnson et al., 2019) for obtaining hard negatives given a mention and an index of entity embeddings from the KB.

Running a forward pass over the entire KB at regular intervals can be costly and time-consuming as the KB often amounts to millions of entities. Moreover, the computational complexity of retrieving hard negatives may grow exponentially depending on the scoring function. For example, the traditionally used scoring function also leveraged in this work, where the mention and entity are both represented with a single embedding requires $\mathcal{O}(me)$ approximate nearest neighbour searches, where m and e are the number of mentions and entities respectively. However, employing an alternative scoring function such as the *sum-of-max* used in Zhang and Stratos (2021) which requires comparing a set of mention embeddings with a set of entity embeddings results in $\mathcal{O}(mexy)$ where x and y is the number of mention vector and entity vector embeddings. In Zhang and Stratos (2021) x and y are set to 128, the number of maximum tokens in the mention and entity input sequence.

This highlights the computational cost of hard negative mining and underlines the need for both methods which can work effectively with random samples as well as more efficient hard negative mining strategies. In this work we propose a method for the former.

Biomedical Out of Knowledge Base Detection For the biomedical NIL detection scenario training proceeds exactly as in the in-KB setting. We train models with the Proxy-based loss with different margins, and also a model with the CE loss. In each case, we use a validation set that includes NIL mentions to select an appropriate threshold for the retrieval model. Mentions whose corresponding top-ranked entity does not achieve this score are assigned the NIL label. We choose the threshold that maximises the F1 score for NIL entities in the validation set. We then apply this threshold to detect NIL mentions in the test set.

6 Results

We present the results for candidate retrieval and benchmark our models against suitable methods. We name our method Proxy-based Entity Linking (PEL-Pb). We also report the results of a version of our model which uses the CE (PEL-CE) loss on

	# Neg.	recall@1	recall@64
Angell et al. (2021)	-	50.8	85.3
Agarwal et al. (2021)	-	72.3	95.6
Varma et al. (2021)	100	71.7	-
PEL-CE	32 (mixed)	72.1	95.5
	64 (mixed)	72.1	95.6
	64 (random)	55.7	94.0
PEL-Pb	32 (mixed)	71.6	93.3
	64 (mixed)	**72.6**	95.0
	64 (random)	63.3	**95.9**
PEL-CE + FGSM	32 (mixed)	72.3	95.5
PEL-Pb + FGSM	32 (mixed)	72.4	93.7

Table 2: Candidate retrieval results on the MedMentions dataset. CE and Pb refers to cross-entropy and proxy-based losses respectively. All experiments were run with mixed random and hard negatives "(mixed)", or only "(random)" negatives. The bold figures represent the best score for each recall metric. Note that FGSM PEL variants were only run with 32 negatives due to GPU memory constraints.

	Random		Mixed	
	recall@1	recall@64	recall@1	recall@64
Wu et al. (2019)*	-	81.80	46.5	84.8
Agarwal et al. (2021)	38.6	84.0	50.4	85.1
Ma et al. (2021)	45.4	**90.8**	-	-
Zhang and Stratos (2021)	-	87.62	-	**89.6**
PEL-CE	44.1	84.8	52.5	87.2
PEL-Pb	48.9	85.2	53.1	86.0
PEL-CE + FGSM	44.1	85.2	53.2	87.2
PEL-Pb + FGSM	**49.7**	85.6	**54.2**	86.6

Table 3: Candidate retrieval results on the ZESHEL dataset. CE and Pb refers to cross-entropy and proxy-based losses respectively. The negative to positive sample ratio for all PEL runs is 32. The bold figures represent the best score for each sampling strategy (random vs. mixed random and hard). The highlighted figure represents the best overall score across strategies. * we use the results reported in Zhang and Stratos (2021) for random negatives and Ma et al. (2021) for mixed negatives.

all experiments for comparison.

6.1 MedMentions

Table 2 shows that all approaches using bi-encoder transformer models strongly outperform the N-Gram TF-IDF proposed in Angell et al. (2021) for recall@1 and also recall@64. We also observe the strong positive effect of including hard negatives during model training. The effect is particularly strong for the CE loss, where recall@1 increases by 17% compared with training on random negatives. We believe that such difference is partly due to the large size of the KB MedMentions KB, amounting to 2.36M entities, which contributes to the importance of hard negative mining. For the Proxy-based loss, including hard negatives increases recall@1 by 9%, achieving state-of-the-art performance of 72.6%. Adding FGSM regularisation boosted performance, as can be seen from the experiments with 32 negatives (the largest number of negatives we could fit into GPU memory when applying FGSM). However, it did not exceed the performance of the unregularized model with 64 negative samples.

	NIL			All classes incl. NIL	
	auPR	Precision	Recall	Recall@1	Recall@64
Pb (m=0)	83.7	81.2	71.0	**72.6**	**90.4**
Pb (m=0.01)	84.4	81.6	71.5	72.5	90.2
Pb (m=0.05)	85.8	83.3	73.5	72.4	89.9
Pb (m=0.1)	**87.6**	**85.2**	**79.2**	69.4	85.7
CE	32.3	31.8	74.0	64.4	76.1

Table 4: NIL detection results on the MedMentions dataset. auPR, precision and recall are reported exclusively for the NIL class, whereas micro-averaged recall@1 and recall@64 are reported for all classes including NIL. Pb: Proxy-based with margin m, CE: Cross-Entropy.

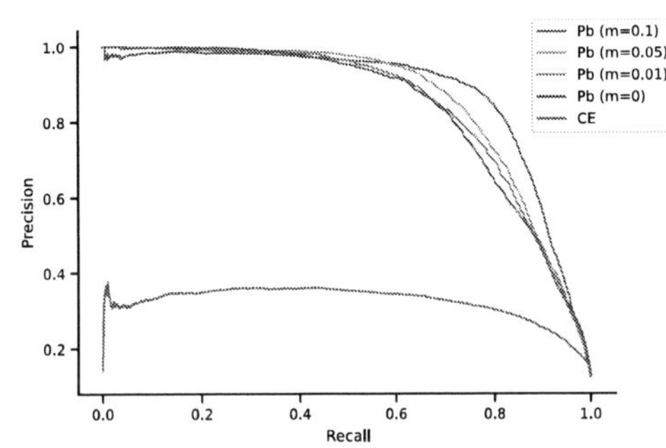

Figure 2: Precision Recall curves for NIL detection on the MedMentions dataset. Pb: Proxy-based with margin m, CE: Cross-Entropy.

Biomedical Out of Knowledge Base detection We also evaluated our proposed loss function on NIL detection. All models trained with the Proxy-based loss significantly outperform the CE-based model in terms of both precision and recall (Figure 2). The CE loss does not encourage low scores in absolute value for negatives examples, but rather encourages scores that are lower than the scores of positive examples. As we can see from the results, CE training fails to assign low scores to NIL mentions, as these are out-of-distribution negatives and thus have not been compared to positive examples during model training. Our Proxy-based loss does not suffer from this issue, even with a margin of 0.

We believe that this is accomplished by the decoupling of the positive and negative loss terms, such that low *absolute* score values are encouraged for negative examples.

Furthermore, the higher the Proxy-based margin the better the model's performance with respect to detecting NIL mentions. At the same time, Proxy-based models with lower margins perform better at the overall recall metrics (Figure 2). These metrics are computed with respect to all classes including the NIL class. Given that the performance differences among models with different margins are minimal, a practitioner could choose how to set the margin considering the trade-off between NIL detection and overall model performance. To our knowledge, we are the first to propose a method for NIL detection using the bi-encoder architecture.

6.2 Zero-Shot Entity Linking dataset

Based on the candidate retrieval results in Table 3, we can conclude six key points. (1) Proxy-based models (Pb) outperform their Cross-Entropy (CE) counterparts across all considered settings for recall@1. In particular, our Proxy-based model using hard negatives and FGSM regularization achieves state-of-the-art recall@1 on this dataset. This highlights the gain that we get by breaking the dependency between positive and negative pairs. (2) Including hard negatives always boosts model performance. This is particularly evident on the recall@1 metric. The model trained with CE loss strongly depends on hard negatives, with recall@1 increasing by 8% compared to training with random negatives. For the Proxy-based loss the increase is 4%, as the model already performs competitively when trained with random negatives. This showcases the importance of hard negative sampling for the CE loss. Hard negatives provide the model with much more meaningful feedback and avoid the threat of vanishing gradients (Eq. 2). (3) The difference between Pb and CE models becomes much smaller for recall@64. Trivially, as k increases, recall@k for all models will converge towards 1. Additionally, as k increases to above the number of hard negatives, the model's ability to distinguish the hard negatives from the positive will not be seen in the metric. (4) CE models marginally outperform Pb models with hard negatives at recall@64. Hard negatives consistently have a larger impact on CE compared to Pb also at recall@64 (2), while the benefits of Pb have been nullified as discussed in (3). (5) Alternative methods leveraging the CE loss and different model architectures such as MuVER (Ma et al., 2021) and SOM (Zhang and Stratos, 2021) outperform the bi-encoder based approach at recall@64. However, both MuVER and SOM are more complex models tuned for achieving high recall@64, whereas the main focus of our approach is high recall@1 in the pursuit of avoiding the additional ranking stage. Pb outperforms the only single stage entity linking model Agarwal et al. (2021) across the board. (6) FGSM regularization boosts the results of both Proxy-based and CE models, demonstrating its promise as a general method for regularizing the retrieval model.

7 Discussion & Future Work

We have proposed and evaluated a novel proxy-based loss for biomedical candidate retrieval. Additionally, we have adopted an adversarial regularization technique designed to simulate hard negatives, and shown that both our loss and regularization boost performance on the recall@1 metric. We have also constructed a biomedical dataset for NIL detection and demonstrated that our candidate retrieval model can robustly identify biomedical NIL entities, while maintaining high overall performance. These are important advances towards closing the gap between the two-stage approach that include an expensive cross-encoder and a candidate retriever-only setup.

Notably, our work highlights the importance of hard negative sampling when optimising the candidate generator with the CE loss. Random negative sampling together with CE loss can result in the problem becoming trivial, for example the randomly sampled negative entity having a different type. However, accessing hard negative examples during model training can be challenging, particularly when the knowledge base is large and entity representations are frequently updated.

Considering this, we recommend to employ our Proxy-based loss for the candidate retrieval task in three different scenarios: (1) training with random negatives, (2) optimising for recall@1, (3) detecting NIL entities. Moreover, we also recommend leveraging FGSM regularisation in any setup and both retrieval and ranking tasks.

An interesting approach would be to attempt to approximate hard negatives without frequent updates of the entity representations. This could potentially be done by keeping the entity encoder frozen, or exploring alternative relatedness measures which does not require frequently running the model over the whole knowledge base. Finally, there is a plethora of work on proxy-based (Movshovitz-Attias et al., 2017; Kim et al., 2020) and pair-based losses (Bromley et al., 1993; Chopra et al., 2005; Schroff et al., 2015; Dong and Shen, 2018), usually discussed in the computer vision and metric learning literature. Improving the candidate retrieval is a crucial step towards high-performing and efficient entity linking systems that can be easily applied in real-world settings.

Limitations

There are several limitations of our work. Firstly, we only demonstrate the advantages of our proposed method when computing hard negatives is computationally expensive, which is the case with large knowledge bases and expensive scoring methods. If computing hard negatives is not a bottleneck, one may use negative sampling with the baseline CE loss. However, biomedical knowledge bases typically contain a huge number of entities. Secondly, in our experiments we were limited to single GPU machines with at most 16GB of GPU memory. This prevented us from including more than 64 negatives samples in the standard setup and 32 negative samples when using FGSM regularization, which could potentially be benefit model performance. Thirdly, we acknowledge that some comparison to related work is missing, in particular, Zhang and Stratos (2021). We were not able to reproduce the results cited in the paper using the publicly available code. Finally, our work is limited to proxy-based metric learning losses. More space could be devoted to the topic of how one could utilise metric learning more broadly for biomedical entity linking. We leave this for future work.

Ethics Statement

The BERT-based models fine-tuned in this work and datasets are publicly available. We will also make our code as well as the biomedical out of knowledge base detection dataset publicly available.

The task of entity linking is often crucial for downstream applications, such as relation extraction, hence potential biases at the entity lining stage can have significant harmful downstream consequences. One source of such biases are the pre-trained language models fine-tuned in this work. There is a considerable body of work devoted to the topic of biases in language models. One way the entity linking systems can be particularly harmful is when they commit or propagate errors in the language models, knowledge bases, mention detection across certain populations such as races or genders. Because of the high ambiguity across biomedical mentions and entities in the knowledge base, it is important that the users investigate the output prediction of the entity linking system and often take is a suggestion, rather than gold standard. Finally, we highlight that linking the entity to its entry in the knowledge base and out of knowledge base detection can be analogous to surveillance and tracking in the computer vision domain, which comes with substantial ethical considerations.

Acknowledgements

We thank Dane Corneil, Georgiana Neculae and Juha Iso-Sipilä for helpful feedbacks and the anonymous reviewers for constructive comments on the manuscript.

References

Dhruv Agarwal, Rico Angell, Nicholas Monath, and Andrew McCallum. 2021. Entity linking and discovery via arborescence-based supervised clustering. *arXiv preprint arXiv:2109.01242*.

Rico Angell, Nicholas Monath, Sunil Mohan, Nishant Yadav, and Andrew McCallum. 2021. Clustering-based inference for biomedical entity linking. In *Proceedings of the 2021 Conference of the North American Chapter of the Association for Computational Linguistics: Human Language Technologies*, pages 2598–2608, Online. Association for Computational Linguistics.

Rajarshi Bhowmik, Karl Stratos, and Gerard de Melo. 2021. Fast and effective biomedical entity linking using a dual encoder. *ArXiv*, abs/2103.05028.

Olivier Bodenreider. 2004. The Unified Medical Language System (UMLS): integrating biomedical terminology. *Nucleic acids research*, 32(Database issue):D267–D270.

Jane Bromley, Isabelle Guyon, Yann LeCun, Eduard Säckinger, and Roopak Shah. 1993. Signature verification using a "siamese" time delay neural network. In *Proceedings of the 6th International Conference on Neural Information Processing Systems*, NIPS'93, page 737–744, San Francisco, CA, USA. Morgan Kaufmann Publishers Inc.

Razvan Bunescu and Marius Pasca. 2006. Using encyclopedic knowledge for named entity disambiguation.

Association for Computational Linguistics, European Chapter.

S. Chopra, R. Hadsell, and Y. LeCun. 2005. Learning a similarity metric discriminatively, with application to face verification. In *2005 IEEE Computer Society Conference on Computer Vision and Pattern Recognition (CVPR'05)*, volume 1, pages 539–546 vol. 1.

Jacob Devlin, Ming-Wei Chang, Kenton Lee, and Kristina Toutanova. 2019. BERT: Pre-training of deep bidirectional transformers for language understanding. In *Proceedings of the 2019 Conference of the North American Chapter of the Association for Computational Linguistics: Human Language Technologies, Volume 1 (Long and Short Papers)*, pages 4171–4186, Minneapolis, Minnesota. Association for Computational Linguistics.

Xingping Dong and Jianbing Shen. 2018. Triplet loss in siamese network for object tracking. In *Proceedings of the European Conference on Computer Vision (ECCV)*.

Daniel Gillick, Sayali Kulkarni, Larry Lansing, Alessandro Presta, Jason Baldridge, Eugene Ie, and Diego Garcia-Olano. 2019. Learning dense representations for entity retrieval. *CoNLL 2019 - 23rd Conference on Computational Natural Language Learning, Proceedings of the Conference*, pages 528–537.

Ian Goodfellow, Jonathon Shlens, and Christian Szegedy. 2015. Explaining and harnessing adversarial examples. In *International Conference on Learning Representations (ICLR)*.

Swapna Gottipati and Jing Jiang. 2011. Linking entities to a knowledge base with query expansion. Association for Computational Linguistics.

Raia Hadsell, Sumit Chopra, and Yann LeCun. 2006. Dimensionality reduction by learning an invariant mapping. In *2006 IEEE Computer Society Conference on Computer Vision and Pattern Recognition (CVPR'06)*, volume 2, pages 1735–1742. IEEE.

Samuel Humeau, Kurt Shuster, Marie-Anne Lachaux, and J. Weston. 2019. Poly-encoders: Transformer architectures and pre-training strategies for fast and accurate multi-sentence scoring. *arXiv: Computation and Language*.

Jeff Johnson, Matthijs Douze, and Hervé Jégou. 2019. Billion-scale similarity search with GPUs. *IEEE Transactions on Big Data*, 7(3):535–547.

Omar Khattab and Matei Zaharia. 2020. *ColBERT: Efficient and Effective Passage Search via Contextualized Late Interaction over BERT*, page 39–48. Association for Computing Machinery, New York, NY, USA.

Sungyeon Kim, Dongwon Kim, Minsu Cho, and Suha Kwak. 2020. Proxy anchor loss for deep metric learning. In *Proceedings of the IEEE/CVF Conference on Computer Vision and Pattern Recognition (CVPR)*.

Nikolaos Kolitsas, Octavian-Eugen Ganea, and Thomas Hofmann. 2018. End-to-end neural entity linking. *arXiv preprint arXiv:1808.07699*.

Nevena Lazic, Amarnag Subramanya, Michael Ringgaard, and Fernando Pereira. 2015. Plato: A selective context model for entity resolution. *Transactions of the Association for Computational Linguistics*, 3:503–515.

Fangyu Liu, Ehsan Shareghi, Zaiqiao Meng, Marco Basaldella, and Nigel Collier. 2021. Self-alignment pretraining for biomedical entity representations. In *Proceedings of the 2021 Conference of the North American Chapter of the Association for Computational Linguistics: Human Language Technologies*, pages 4228–4238, Online. Association for Computational Linguistics.

Lajanugen Logeswaran, Ming-Wei Chang, Kenton Lee, Kristina Toutanova, Jacob Devlin, and Honglak Lee. 2019. Zero-shot entity linking by reading entity descriptions. In *Proceedings of the 57th Annual Meeting of the Association for Computational Linguistics*, pages 3449–3460, Florence, Italy. Association for Computational Linguistics.

Yi Luan, Jacob Eisenstein, Kristina Toutanova, and Michael Collins. 2021. Sparse, dense, and attentional representations for text retrieval. *Transactions of the Association for Computational Linguistics*, 9:329–345.

Xinyin Ma, Yong Jiang, Nguyen Bach, Tao Wang, Zhongqiang Huang, Fei Huang, and Weiming Lu. 2021. MuVER: Improving first-stage entity retrieval with multi-view entity representations. In *Proceedings of the 2021 Conference on Empirical Methods in Natural Language Processing*, pages 2617–2624, Online and Punta Cana, Dominican Republic. Association for Computational Linguistics.

Takeru Miyato, Andrew M Dai, and Ian Goodfellow. 2016. Adversarial training methods for semi-supervised text classification. *arXiv preprint arXiv:1605.07725*.

Sunil Mohan and Donghui Li. 2019. Medmentions: A large biomedical corpus annotated with UMLS concepts. *CoRR*, abs/1902.09476.

Jose G Moreno, Romaric Besançon, Romain Beaumont, Eva D'hondt, Anne-Laure Ligozat, Sophie Rosset, Xavier Tannier, and Brigitte Grau. 2017. Combining word and entity embeddings for entity linking. In *European Semantic Web Conference*, pages 337–352. Springer.

Yair Movshovitz-Attias, Alexander Toshev, Thomas K. Leung, Sergey Ioffe, and Saurabh Singh. 2017. No fuss distance metric learning using proxies. In *Proceedings of the IEEE International Conference on Computer Vision (ICCV)*.

Lin Pan, Chung-Wei Hang, Avirup Sil, Saloni Potdar, and Mo Yu. 2021. Improved text classification via contrastive adversarial training. *arXiv preprint arXiv:2107.10137*.

Florian Schroff, Dmitry Kalenichenko, and James Philbin. 2015. Facenet: A unified embedding for face recognition and clustering. In *2015 IEEE Conference on Computer Vision and Pattern Recognition (CVPR)*, pages 815–823.

Ozge Sevgili, Artem Shelmanov, Mikhail Arkhipov, Alexander Panchenko, and Chris Biemann. 2020. Neural entity linking: A survey of models based on deep learning. *arXiv preprint arXiv:2006.00575*.

Wei Shen, Jianyong Wang, and Jiawei Han. 2014. Entity linking with a knowledge base: Issues, techniques, and solutions. *IEEE Transactions on Knowledge and Data Engineering*, 27(2):443–460.

Christian Szegedy, Wojciech Zaremba, Ilya Sutskever, Joan Bruna, Dumitru Erhan, Ian Goodfellow, and Rob Fergus. 2013. Intriguing properties of neural networks. *arXiv preprint arXiv:1312.6199*.

Chen-Tse Tsai and Dan Roth. 2016. Cross-lingual wikification using multilingual embeddings. In *Proceedings of the 2016 Conference of the North American Chapter of the Association for Computational Linguistics: Human Language Technologies*, pages 589–598.

Maya Varma, Laurel Orr, Sen Wu, Megan Leszczynski, Xiao Ling, and Christopher Ré. 2021. Cross-domain data integration for named entity disambiguation in biomedical text. In *Findings of the Association for Computational Linguistics: EMNLP 2021*, pages 4566–4575, Punta Cana, Dominican Republic. Association for Computational Linguistics.

Ashish Vaswani, Noam Shazeer, Niki Parmar, Jakob Uszkoreit, Llion Jones, Aidan N. Gomez, undefinedukasz Kaiser, and Illia Polosukhin. 2017. Attention is all you need. In *Proceedings of the 31st International Conference on Neural Information Processing Systems*, NIPS'17, page 6000–6010, Red Hook, NY, USA. Curran Associates Inc.

Ledell Wu, Fabio Petroni, Martin Josifoski, Sebastian Riedel, and Luke Zettlemoyer. 2019. Zero-shot Entity Linking with Dense Entity Retrieval.

Ledell Wu, Fabio Petroni, Martin Josifoski, Sebastian Riedel, and Luke Zettlemoyer. 2020. Scalable zero-shot entity linking with dense entity retrieval. In *Proceedings of the 2020 Conference on Empirical Methods in Natural Language Processing (EMNLP)*, pages 6397–6407, Online. Association for Computational Linguistics.

Wenzheng Zhang and Karl Stratos. 2021. Understanding hard negatives in noise contrastive estimation. In *Proceedings of the 2021 Conference of the North American Chapter of the Association for Computational Linguistics: Human Language Technologies*, pages 1090–1101, Online. Association for Computational Linguistics.

Appendices

A Context and Mention Modelling

We represent a mention and its surrounding context, τ_m, as a sequence of word piece tokens

$$[\text{CLS}]\ \text{ctxt}_\text{l}\ [\text{M}_s]\ \text{mention}\ [\text{M}_e]\ \text{ctxt}_\text{r}\ [\text{SEP}]$$

where mention, ctxt_l and ctxt_r are the word-piece tokens of the mention, left and right context, and $[\text{M}_s]$ and $[\text{M}_e]$ are special tokens marking the start and end of a mention respectively.

Due to the differences in available data, we represent entities differently for ZESHEL and MedMentions. On ZESHEL, we represent entities with a sequence of word piece tokens

$$[\text{CLS}]\ \text{title}\ [\text{ENT}]\ \text{description}\ [\text{SEP}]$$

where [ENT] is a special separator token. In contrast, when training on the MedMentions dataset we represent an entity by the sequence

$$[\text{CLS}]\ \text{title}\ [\text{SEP}]\ \text{types}\ [\text{SEP}]\ \text{description}\ [\text{SEP}]$$

Descriptions of entities were sourced from UMLS.

B Candidate ranker setup and results

To evaluate the impact of our candidate retriever model on the downstream task of candidate ranking, we also conducted ranking experiments on both datasets.

		# Candidates	Ranker	Accuracy
ZESHEL	Wu et al. (2020)	64	Base	61.3
	Wu et al. (2020)	64	Large	63.0
	Zhang and Stratos (2021)	64	Base	66.7
	Zhang and Stratos (2021)	64	Large	**67.1**
	PEL-Pb	16	Base	62.8
	PEL-Pb + FGSM	16	Base	64.6
MedMentions	Bhowmik et al. (2021)*	-	-	68.4
	Angell et al. (2021)	-	-	72.8
	Varma et al. (2021)	10	Base	**74.6**
	PEL-Pb	16	Base	74.0
	PEL-Pb + FGSM	16	Base	**74.6**
	Angell et al. (2021) + post-processing	-	-	74.1
	Varma et al. (2021) + post-processing	10	Base	**74.8**

Table 5: Ranker results on the ZESHEL and MedMentions datasets. * uses the full MedMentions dataset, rather than the ST21PV subset used by other models reported in the table and recommended by MedMentions authors'.

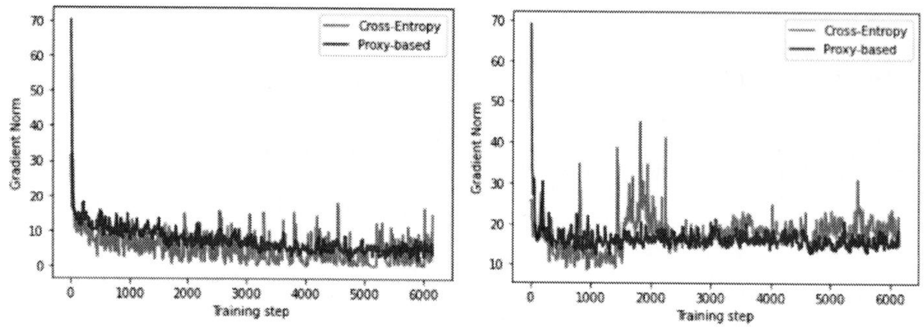

Figure 3: Comparison of smoothed gradient norms over training steps using two losses, CE and Proxy-based. The left plot visualizes the smoothed gradient norm when using random, and the right one leveraging mixed-50% negatives. All the experiments were conducted on ZESHEL using 32 negatives.

Training & Evaluation setup Similarly as in related work (Logeswaran et al., 2019; Wu et al., 2020; Zhang and Stratos, 2021), the highest scoring candidate entities from the candidate retriever are passed to a ranker, which is a cross-encoder consisting of one BERT transformer. The cross-encoder Logeswaran et al. (2019) is used to select the best entity out of the candidate pool. It takes as input $\tau_{m,e}$, which is the concatenation of mention/context and entity representations τ_m and τ_e. We then obtain a dense vector representation for a mention-entity pair $\mathbf{y_{m,e}} = \mathrm{T}_{\mathrm{cross}}(\tau_{m,e})$, where $\mathrm{T}_{\mathrm{cross}}(\tau_{m,e})$ is the BERT transformer of the cross-encoder and $\mathrm{red}(\cdot)$ is a mean pooling function that takes the mean over input tokens embeddings. Entity candidates are scored by applying a linear layer $\mathrm{s}_{\mathrm{cross}}(m,e) = \mathbf{y_{m,e}W}$.

We pick the best performing retrieval model on recall@16 and use it to retrieve top 16 candidate entities for each mention. As the number of candidate entities is relatively low, we do not perform negative sampling and optimise the cross-encoder with the CE loss (Eq. 1). We report the micro-averaged unnormalized accuracy on the MedMentions dataset and macro-averaged unnormalized accuracy on the ZESHEL dataset in line with the prior work (Zhang and Stratos, 2021; Wu et al., 2020). The results are shown in the Table 5.

Results In Table 5 we can observe the downstream effect of having a candidate generator model with high recall@1 performance. On ZESHEL, We can see that a cross-encoder trained with the top 16 candidates from our best performing candidate generator achieved higher accuracy than Wu et al. (2020) who used the top 64 candidates. Moreover, similarly as with the candidate retrieval, FGSM boosts performance. For completeness, we have also included the state-of-the-art results from Zhang and Stratos (2021) who used 64 candidates and a larger BERT model in the cross-encoder. In our experiments we were limited to a single GPU with 16 GB memory which restricted us to a low number of maximum candidates, namely 16. We strongly believe that including more candidates than 16 would boost the performance of our method.

On MedMentions a cross-encoder trained with the top 16 candidates from our best performing candidate generator model achieved a competitive accuracy of 74%. The accuracy further increased to 74.6% when adding FGSM regularisation, coming close to the state-of-the-art performance of Varma et al. (2021), which includes additional post-processing.

C Training details

The hyperparameters used for conducting the experiments are visible in Table 6. We use a single NVIDIA V100 GPU with 16 GB of GPU memory for all model trainings.

D Biomedical Out of Knowledge Base dataset details

We constructed the OKB dataset by replacing the label of a set of mentions from the MedMentions corpus (Mohan and Li, 2019) with the NIL class. Namely we pick the mentions belonging to 11 types: *Mental Process, Health Care Related Organization, Element Ion or Isotope, Medical Device, Health Care Activity, Diagnostic Procedure, Professional or Occupational Group, Mental Process, Laboratory Procedure, Regulation or Law,*

Param	Bi-encoder	Cross-Encoder
Input sequence length	128	256
learning rate	1e-5	2e-5
warmup proportion	0.25	0.2
eps	1e-6	1e-6
gradient clipping value	1.0	1.0
effective batch size	32	4
epochs	7	5
learning rate scheduler	linear	linear
optimiser	AdamW	AdamW
α	32	-
δ	0.0	-
FGSM λ	1	1
FGSM ϵ	0.01	0.01

Table 6: Learning parameters for the bi-encoder and cross-encoder.

Organization, Professional Society. The final OKB subset includes approximately 24K mentions and 3K unique entities.

To ensure that the OKB dataset does not suffer from easy inferences and allows us to evaluate model performance. We ensured that the zero-shot distribution of the OKB mentions and types across the train/validation/test split was in line with the zero-shot distribution of mentions and types in the whole dataset. Additionally, we verified that there is no significant overlap between mention surface forms across the splits. Moreover, we looked at the length of entity descriptions which are used to create entity representations checking that the OKB mentions entity representations statistics are similar to the statistics computed using the whole dataset.

E Gradient norm analysis

	Train	Dev	Test
Mentions	14K	4.8K	4.7K
Entities	2.2K	1.1K	1.1K
% Entities seen	100	57.7	57.5

Table 7: Statistics of the OKB MedMentions subset.

Figure 3 shows the behaviour of the gradient $l2$ norm for both losses. We can see that for both random and mixed negatives, the norm of the Proxy-based loss has considerably lower variance. This is visible particularly when using the mixed negatives.

BERT for Long Documents: A Case Study of Automated ICD Coding

Arash Afkanpour **Shabir Adeel*** **Hansenclever Bassani***

Arkady Epshteyn* **Hongbo Fan*** **Isaac Jones*** **Mahan Malihi***

Adrian Nauth* **Raj Sinha*** **Sanjana Woonna*** **Shiva Zamani***

Elli Kanal† **Mikhail Fomitchev**† **Donny Cheung**†

Google

arashaf@google.com

Abstract

Transformer models have achieved great success across many NLP problems. However, previous studies in automated ICD coding concluded that these models fail to outperform some of the earlier solutions such as CNN-based models. In this paper we challenge this conclusion. We present a simple and scalable method to process long text with the existing transformer models such as BERT. We show that this method significantly improves the previous results reported for transformer models in ICD coding, and is able to outperform one of the prominent CNN-based methods.

1 Introduction

The International Classification of Diseases (ICD) codes provide a standard way of keeping track of diagnoses and procedures during a patient visit. These codes are used worldwide for epidemiological studies, billing and reimbursement, and research in health care. The codes are maintained by the World Health Organization (WHO) and are revised and updated periodically. As of 2022 the ICD codes are in the 11th revision.

Assigning ICD codes to a clinical note, such as a discharge summary, is done by professional medical coders. Human coders require extensive training, and the process of coding is often time-consuming, costly, and error-prone. Due to these challenges there is an incentive to automate the coding process. Therefore in recent years this problem has gained interest among machine learning researchers in health care (See, Mullenbach et al. (2018); Li and Yu (2020); Zhang et al. (2020) and references therein). On the surface, the problem can be considered as a multi-label document classification problem. However, there are aspects of the problem that make it particularly challenging. The primary challenge is that there are tens of thousands of classes. For instance, billable ICD-10-CM codes consist of approximately 73,000 codes. In addition, the distribution of the codes is not uniform. Many of the codes are related to rare conditions and are mentioned infrequently in text, which makes it difficult to train a reliable classifier for them.

Transformer-based language models developed based on self attention (Vaswani et al., 2017) have become the state-of-the-art across many NLP problems by outperforming previous solutions that were mostly based on recurrent neural networks (RNN) and convolutional neural networks (CNN). So one would expect that they perform well in ICD coding too. However, examining the literature of ICD coding methods reveals that transformer-based solutions fail to outperform CNN-based models. Many studies have applied the BERT language model (Devlin et al., 2018) to this task, for example Pascual et al. (2021); Singh et al. (2020); Biseda et al. (2020); Amin et al. (2019). More recently, Ji et al. (2021) performed a comprehensive quantitative study to compare BERT and some of its variants pre-trained on medical text against CNN-based models such as Mullenbach et al. (2018) and Cao et al. (2020) to answer the question of whether the magic of BERT (as observed across many NLP problems) also applies to automated ICD coding. They concluded that BERT cannot outperform CNN-based models in the full ICD code case.

Unlike RNN or CNN models, which in theory can process sequences of arbitrary length, transformers' computational complexity scales quadratically with sequence length. This means that most of these models can handle limited size sequences. For instance, BERT models usually are pre-trained and fine-tuned on sequences with at most 512 tokens. Clinical notes normally contain long snippets of text beyond the sequence limit of transformers. We hypothesize that this constraint could explain

* Equal contribution
† Technical leadership

Proceedings of the 13th International Workshop on Health Text Mining and Information Analysis (LOUHI), pages 100 - 107
December 7, 2022 ©2022 Association for Computational Linguistics

the poor performance of transformers in this task, and will present empirical evidence for that.

We emphasize that we do not claim to achieve state-of-the-art performance in ICD coding, or that our design is the most efficient transformer architecture for processing long text. For a review of efficient transformers see Tay et al. (2020) and references therein. Our goal is to provide new empirical evidence that shows even the standard transformer models can outperform some of the previous prominent methods and are a viable solution for ICD coding.

2 Related work

Medori and Fairon (2010) applied a rule-based method to extract important snippets of text and encode them with ICD codes. Perotte et al. (2014) proposed SVM classification with bag-of-words features. They experimented with both flat SVM (i.e. one classifier per code) and a hierarchical classifier.

With the success of deep learning in NLP tasks, many researchers focused on using RNN and CNN models for ICD coding. CNN models provide a convenient way to learn a contextual representation of text in NLP problems (Chen, 2015). For example, Mullenbach et al. (2018) proposed the *CAML* model: a convolutional layer on word2vec embedding vectors to learn a contextual representation for each word. The word representations are combined into a class-specific document representation using the attention mechanism. They also suggested a method to leverage code descriptions via a regularization term. Li and Yu (2020) proposed Multi-filter Residual CNN (MultiResCNN) that uses convolutional layers with different kernel sizes to capture patterns with different lengths. Additionally, they used residual blocks on top of the convolutional layer. Similar to Mullenbach et al. (2018) they employed a per-class attention mechanism to make the document representation attend to different parts of the input for each code.

Recurrent neural networks (RNN) are also studied extensively for ICD coding. Shi et al. (2017) applied LSTM at character and word level to encode both the clinical note and the code description. Baumel et al. (2018) employs a two-layer bidirectional Gated Recurrent Unit (GRU) model, where the first layer encodes individual sentences, and the second layer encodes the document.

With the success of transformer architectures

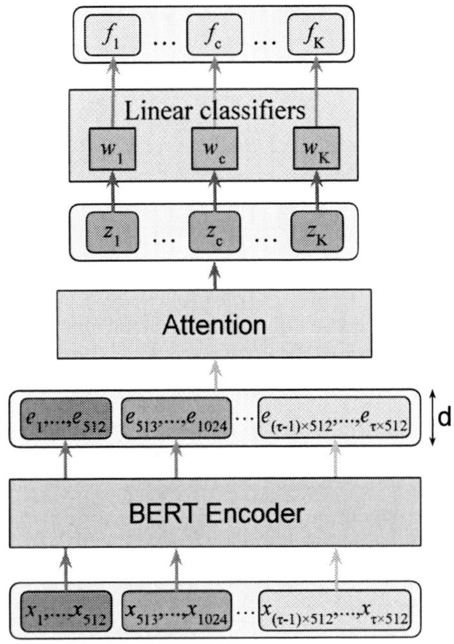

Figure 1: Model architecture proposed for handling long text inputs.

across many NLP tasks, researchers focused their attention to designing such models for ICD coding. BERT-XML (Zhang et al., 2020) with access to a large corpus of private data managed to pre-train the model with sequence length of 1024. Most of the work in this area, however, considered the standard BERT model and its variants pre-trained on medical text to encode the document (Pascual et al., 2021; Singh et al., 2020; Biseda et al., 2020; Amin et al., 2019). One observation with these models was that they were unable to outperform CNN-based models. Ji et al. (2021) performed a comprehensive study to answer a few research questions on the suitability of BERT models for ICD coding. They studied and compared different variants of BERT pre-training. They also proposed a hierarchical attention method so that long clinical notes can be processed with a BERT model with a limit of 512 tokens. Most importantly, they compared different BERT variants against traditional CNN-based models, and through extensive experiments showed that BERT-based models are not capable of outperforming CNN-based models in ICD coding. In the next sections we show that a simple method that enables processing of long text with transformers will attain results that contradict the findings of Ji et al. (2021).

3 Method

In this section we explain our method for building a model to predict medical codes. As illustrated in Figure 1, our model consists of an encoder that calculates token-level representation of the input text. This can be done in various ways, e.g. Mullenbach et al. (2018) used word2vec and a CNN layer to calculate word-level representations. We choose the BERT language model for this purpose. A class-specific representation of the document is then calculated using class-specific attention vectors, similar to Mullenbach et al. (2018). For d-dimensional token representations and K classes, this layer requires $d \times K$ parameters. Linear binary classifiers are built on top of the document representation to produce the probability that the document belongs to any of the K classes. This layer requires $(d+1) \times K$ parameters (one scalar for the offset).

Let $X = [x_1, \ldots, x_s]$ denote the tokenized input sequence with s tokens. Let $e_1(X), \ldots, e_s(X)$ denote the representation of tokens $1, \ldots, s$ obtained from an encoder. That is,

$$e_i(X) = \phi(x_i|X), \quad i \in \{1, \ldots, s\},$$

where ϕ is an encoder, such as BERT, that returns a context-dependent representation for each token. For each class c, token-level representations are combined into a single vector that represents the entire document using the attention mechanism:

$$z_c(X) = \sum_{i=1}^{s} \alpha_{c,i}(X) e_i(X),$$

where

$$\alpha_{c,i}(X) = \frac{\exp\left(\langle e_i(X), q_c \rangle\right)}{\sum_{j=1}^{s} \exp\left(\langle e_j(X), q_c \rangle\right)}, \quad (1)$$
$$i \in \{1, \ldots, s\},$$

are the normalized attention coefficients and $\langle \cdot, \cdot \rangle$ denotes inner product, and q_c is the d-dimensional attention vector for class c. The predicted probability of the model for class c is calculated by

$$f_c(X) = \sigma\left(\langle z_c(X), w_c \rangle + b_c\right),$$

where w_c is the weight vector for class c, b_c is the scalar offset for class c, and σ is the sigmoid function.

3.1 Handling long text

Language models such as BERT can handle input text up to a certain length. For example, BERT can take input of at most 512 tokens. While it is possible to pre-train the model on longer sequences (mostly to learn useful positional embedding vectors), memory requirement grows quadratically with input size. So pre-training a BERT model on longer text is not scalable.

There are transformer-based models that can handle long sequences, such as BigBird (Zaheer et al., 2020), ETC (Ainslie et al., 2020), Longformer (Beltagy et al., 2020), and LongT5 (Guo et al., 2021). There are a few factors that limit their usability in the medical coding task. For example, these models are usually designed to train on TPU, so training on GPU is often a slow process, if feasible, especially for longer sequences. Also, pre-trained checkpoints of these models are limited, unlike the BERT models that have many pre-trained variants including those pre-trained on medical text.

In this paper, we propose a simple idea, which enables us to use a vanilla BERT model on long sequences. Inspired by the local attention feature of CNN models, we propose to split the input text into (optionally overlapping) segments of 512 tokens. These segments are passed sequentially to a BERT model, and the token representations are concatenated to form $[e_1(X), ..., e_{512}(X), e_{513}(X), ..., e_{1024}(X), ...]$. One may argue that a limitation of this approach is that the token representations are calculated with a 512-token attention span. However, we have observed that in practice this method performs well. In fact, we conjecture that in many cases short snippets of text (as evidence) are sufficient for assigning the correct ICD codes to the input document. Algorithm 1 shows the training procedure.

4 Evaluation

We evaluate the accuracy of the proposed method with several sequence lengths and compare it against the CAML method (Mullenbach et al., 2018), which is one of the prominent CNN-based methods for ICD coding.

4.1 Data sets

For this task we chose the publicly-available MIMIC-III (Johnson et al., 2016) and MIMIC-IV

Algorithm 1 Training on a single example.

Input: tokenized input text of length s: $X = [x_1, \ldots, x_s]$, sparse binary label vector $Y = [y_1, \ldots, y_K]$ for K classes, where $y_c = 1$ if the example belongs to class c, and 0 otherwise.

Pad input text $X = [x_1, \ldots, x_s]$ to length $\tau \times 512$ to obtain $X' = [x_1, \ldots, x_s, \ldots, x_{\tau \times 512}]$, where $\tau = \lceil s/512 \rceil$.

Split X' into segments of 512 tokens: $S_1 = [x_1, \ldots, x_{512}], S_2 = [x_{513}, \ldots, x_{1024}], \ldots, S_\tau$.

Pass S_i's, $i \in \{1, \ldots, \tau\}$ sequentially to the BERT module and obtain the corresponding token representations.

Concatenate token representations from all sequences to obtain $[e_1, \ldots, e_s, \ldots, e_{\tau \times 512}]$.

Calculate class-specific document representations by $z_c(X) = \sum_{i=1}^{s} \alpha_{c,i}(X) e_i(X)$, with $\alpha_{c,i}$ from Eq. 2.

Calculate model predictions for all classes: $f_c(X) = \sigma\left(\langle z_c(X), w_c \rangle + b_c\right),\ c \in 1, \ldots, K$.

Calculate and apply gradient updates for loss function $\sum_{c=1}^{K} \ell(y_c, f_c(X))$, where ℓ is binary cross-entropy.

(Johnson et al., 2020) data sets. MIMIC-III is a large de-identified data set of over 40,000 patients admitted to intensive care units at the Beth Israel Deaconess Medical Center. The data set contains structured and unstructured data, including lab measurements, vital signs, medications, clinical notes, etc. Following previous studies, we focus on predicting ICD codes for discharge summaries where each note corresponds to a hospital stay event. MIMIC-IV is an update to MIMIC-III, which incorporates contemporary data. It is sourced from two in-hospital database systems: a custom hospital wide EHR and an ICU specific clinical information system.

Each discharge summary in MIMIC-III is manually coded by human coders with one or more ICD-9 codes that specify diagnoses and procedures of that particular stay. The data set contains 8,921 unique ICD-9 codes, including 6,918 diagnosis and 2,003 procedure codes. There are patients with multiple admissions and therefore multiple discharge summaries. To be consistent with the previous studies and to ensure that all of the notes of a patient are assigned to one of train/validation/test sets we use the data split provided by Mullenbach et al. (2018). This results in 47,724 discharge summaries for training, 1,632 summaries and 3,372 summaries for validation and test sets respectively.

The discharge summaries in MIMIC-IV are additionally labeled with ICD-10 codes. At the time of writing this paper the MIMIC-Note module, which contains the discharge summaries, is not yet publicly available. In our experiments we only consider the ICD-10 diagnosis set, which contains 72,748 codes in the data set.

For tokenizing text we used the standard BERT vocabulary and tokenizer (Devlin et al., 2018). Figure 2 shows the cumulative distribution function of the number of tokens per note for MIMIC-III and MIMIC-IV.

4.2 Models

Our classification model uses a BERT language model with the method described in Section 3. We dub this model *LongBERT* below. The BERT checkpoint we use in the experiments is a model with 2 transformer blocks and 256-dimensional embedding vectors. The checkpoint can be downloaded from TensorFlow Hub.[1]

The baseline model (*BERT-baseline*) was trained and evaluated on the first 512 tokens of input text. To measure the impact of sequence length we trained and evaluated similar models on the first s tokens of each note, with $s \in \{1024, 2048, 4096, 8192\}$. All BERT parameters and the additional attention and classification parameters were fine-tuned during training. We used the Adam optimizer (Kingma and Ba, 2014) with a learning rate of 2e-4. The batch size was set to 4 in all experiments, except for the models trained with the sequence length of 8192 which were trained with the batch size of 2 to avoid running out of memory. The models were trained for 1 million steps (each step is one batch). No hyper-parameter tuning was performed except for the number of training steps. The best model corresponds to the training step that achieves the highest validation micro F1 score.

[1] https://tfhub.dev/tensorflow/small_bert/bert_en_uncased_L-2_H-256_A-4/2

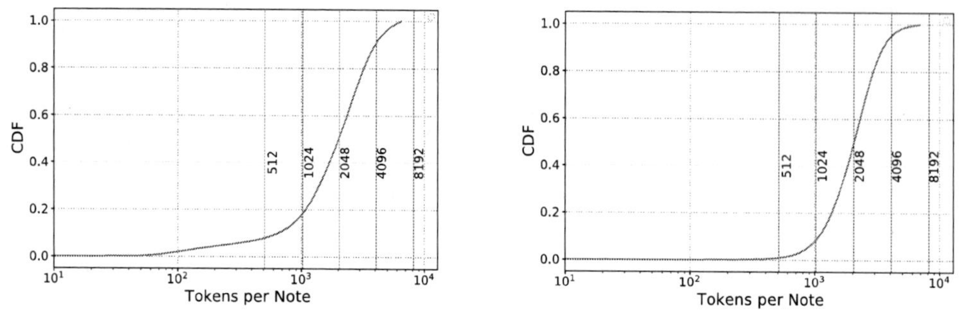

Figure 2: Cumulative distribution function (CDF) of the number of tokens per note for MIMIC-III (left) and MIMIC-IV (right) data sets.

We compare these models against a CAML model trained on sequences of 2500 words following Mullenbach et al. (2018). The hyperparameters were set according to the optimal values obtained in Mullenbach et al. (2018). Training was performed for 1 million steps, and the best model was selected according to validation micro F1 score.

Following previous work, in the MIMIC-III experiments, training and evaluation was performed on the full ICD-9 label set as well as the 50 most frequent codes. In the MIMIC-IV experiment, we consider only the ICD-10 diagnosis codes. Each ICD code has its own attention and classification weight vectors in the models. Table 1 breaks down the number of parameters of the models in the experiments.

4.3 Evaluation metrics

Our primary evaluation metric is micro-averaged F1 (micro F1 for short). Micro-averaged values are calculated by treating each code as a (binary) label for each note. That is, each (note, code) pair is counted as one instance for calculating the metrics. Let,

$$\text{micro precision} = \frac{\sum_{x,c} TP(x,c)}{\sum_{x,c} TP(x,c) + FP(x,c)},$$

$$\text{micro recall} = \frac{\sum_{x,c} TP(x,c)}{\sum_{x,c} TP(x,c) + FN(x,c)},$$

where $TP(x,c) = 1$ if class c is a true positive prediction for note x and 0 otherwise. $FP(x,c)$ (false positive) and $FN(x,c)$ (false negative) are defined analogously. Finally, micro F1 is the harmonic mean of micro precision and micro recall:

$$\text{micro F1} = 2\,\frac{\text{micro precision} \times \text{micro recall}}{\text{micro precision} + \text{micro recall}}.$$

The optimal threshold on model predictions, which is used to calculate $TP/FP/FN$ counts, is obtained by a grid search to maximize the validation set F1 score.

Additionally we report precision-recall AUC (PR-AUC), and ROC-AUC. In contrast to F1 score, these metrics are independent of a specific operating point and provide an aggregated view of model accuracy.

4.4 Results

Table 2 shows the results of the LongBERT and CAML models on the MIMIC-III full-code test set. Table 3 shows accuracy metrics obtained on the MIMIC-IV diagnosis code data set. Bold numbers represent the best value of each metric. A clear trend observed in both data sets is that as the sequence length of LongBERT increases, the accuracy of the model improves. These results demonstrate that the capability to process long text is critical in achieving high accuracy.

The LongBERT models with sequence lengths of 4096 and 8192 both outperform the CAML model. This finding contradicts the previous finding of Ji et al. (2021). While their hierarchical attention proposal and our method both handle long text by breaking it into segments of 512 tokens, one key difference is that they use the CLS token representation from each segment, whereas we use individual token representations. The best MIMIC-III full-code performance reported in Ji et al. (2021) was F1 = 0.47 with BioBERT full-text (Lee et al., 2020)

	MIMIC-III full	MIMIC-III top 50	MIMIC-IV diagnosis
Language model	9,591,040	9,591,040	9,591,040
Attention layer	2,283,776	12,800	18,623,488
Classification layer	2,292,697	12,850	18,696,236
Total	14,167,513	9,616,690	46,910,764

Table 1: Breakdown of the number of parameters of BERT-baseline and LongBERT with 2 transformer blocks and 256-dimensional embedding vectors. MIMIC-III full contains 8,921 classes, and MIMIC-IV diagnosis contains 72,748 classes.

checkpoint and hierarchical attention, while our small vanilla BERT model with sequence length of 8192 achieves F1 = 0.5680. These results show that with a proper modeling approach transformer-based models are indeed capable of outperforming CNN-based models in ICD coding.

MIMIC-III top 50. Following previous work, we also trained and evaluated the models on the MIMIC-III 50 most frequent codes. Table 4 shows the results. Similar to the full-code case we observe that processing longer segments results in higher accuracy.

In this case, however, there is no clear winner between LongBERT and CAML. While LongBERT achieves a higher micro F1 score, the CAML model has a higher PR-AUC. We conjecture that the smaller performance difference between the two models in this experiment compared to the full-code experiment is due to the amount of information in the data sets. By removing many of the labels in the top-50 experiment we essentially remove information. This information is more helpful to larger models (i.e. transformers) than smaller models, such as CAML. As a result, we observe a larger performance gap in the full-code experiment between LongBERT and CAML.

We also note that the accuracy numbers of the CAML model in this experiment are higher than those reported in Mullenbach et al. (2018). One difference here is that we do not discard notes that aren't assigned any of the top 50 codes as was done in the original paper. Such notes are used as negative examples for the top 50 codes. Therefore our data set contains more negative examples than the data set used in Mullenbach et al. (2018).

5 Discussion

Most of the existing BERT models pre-trained on generic or medical text can take input segments of up to 512 tokens. Clinical notes, however, are much longer than this limit. To deal with this limitation, much of the existing works in automated ICD coding that use BERT limit the input to the model by truncating the text or selecting specific spans of text. This results in loss of information and poor performance.

In this paper we proposed a simple method to apply BERT models to sequences longer than 512 tokens. Our method is simple and consists of two key components: (i) apply BERT sequentially to (optionally overlapping) segments of 512 tokens, and (ii) concatenate token-level representations from all segments, and combine them using a class-specific attention layer.

We demonstrated that processing long text sequences minimizes information loss and is critical for achieving high performance in automated ICD coding. We also showed that contrary to previous findings, this method with even a small vanilla BERT model outperforms CNN-based methods, and achieves competitive performance.

Future steps include evaluating medical variants of BERT, and exploring other transformer-based architectures that were designed to handle long sequences.

Limitations

While our method enables the processing of text longer than 512 tokens, one of the limitations of this approach is that context-dependent token representations are still calculated using a window of 512 tokens. Despite good performance of this method in practice, there could be cases where a context window of longer than 512 must be used to make accurate predictions.

Furthermore, while our method reduces computational complexity from quadratic (in sequence length) to linear, the memory requirement of the model could still be prohibitive in certain cases. For instance, for sequence length of 8192, and a

	Seq. length	Micro F1	Precision	Recall	PR-AUC	ROC-AUC
CAML	2500 (words)	0.5465	0.5973	0.5036	0.5361	**0.9831**
BioBERT full-text (Ji et al., 2021)	entire note	0.470	N/A	N/A	N/A	0.974
BERT-baseline	512	0.4149	0.4769	0.3672	0.3793	0.9745
LongBERT	1024	0.4697	0.5421	0.4144	0.4309	0.9766
LongBERT	2048	0.5036	0.5777	0.4463	0.4703	0.9794
LongBERT	4096	0.5514	0.6038	0.5074	0.5305	0.9820
LongBERT	8192	**0.5680**	**0.6148**	**0.5278**	**0.5402**	0.9827

Table 2: Accuracy metrics in the MIMIC-III full-code experiment.

	Seq. length	Micro F1	Precision	Recall	PR-AUC	ROC-AUC
CAML	2500 (words)	0.5439	0.5739	0.5169	0.5313	**0.9889**
BERT-baseline	512	0.4010	0.4298	0.3757	0.3580	0.9883
LongBERT	1024	0.4607	0.5094	0.4205	0.4254	0.9839
LongBERT	2048	0.4852	0.5268	0.4497	0.4559	0.9852
LongBERT	4096	0.5635	0.5925	0.5371	0.5450	0.9850
LongBERT	8192	**0.5703**	**0.6046**	**0.5397**	**0.5517**	0.9871

Table 3: Accuracy metrics in the MIMIC-IV diagnosis experiment.

	Seq. length	Micro F1	Precision	Recall	PR-AUC	ROC-AUC
CAML	2500 (words)	0.6390	**0.6506**	0.6278	**0.6410**	0.9102
BERT-baseline	512	0.5027	0.5367	0.4727	0.5117	0.8360
LongBERT	1024	0.5568	0.5923	0.5252	0.5406	0.8560
LongBERT	2048	0.5908	0.5987	0.5832	0.5604	0.8834
LongBERT	4096	0.6375	0.6157	0.6609	0.6229	0.9115
LongBERT	8192	**0.6522**	0.6417	**0.6629**	0.6303	**0.9181**

Table 4: Accuracy metrics in the MIMIC-III top-50 experiment.

small BERT checkpoint with only two transformer blocks we had to reduce batch size to 2 in order to train the models. Using a larger BERT checkpoint for long sequences requires more memory and multiple GPUs, which increases the cost of compute.

References

Joshua Ainslie, Santiago Ontanon, Chris Alberti, Vaclav Cvicek, Zachary Fisher, Philip Pham, Anirudh Ravula, Sumit Sanghai, Qifan Wang, and Li Yang. 2020. Etc: Encoding long and structured inputs in transformers. *arXiv preprint arXiv:2004.08483.*

Saadullah Amin, Günter Neumann, Katherine Dunfield, Anna Vechkaeva, Kathryn Annette Chapman, and Morgan Kelly Wixted. 2019. Mlt-dfki at clef ehealth 2019: Multi-label classification of icd-10 codes with bert. In *CLEF (Working Notes)*, pages 1–15.

Tal Baumel, Jumana Nassour-Kassis, Raphael Cohen, Michael Elhadad, and Noémie Elhadad. 2018. Multi-label classification of patient notes: case study on icd code assignment. In *Workshops at the thirty-second AAAI conference on artificial intelligence.*

Iz Beltagy, Matthew E Peters, and Arman Cohan. 2020. Longformer: The long-document transformer. *arXiv preprint arXiv:2004.05150.*

Brent Biseda, Gaurav Desai, Haifeng Lin, and Anish Philip. 2020. Prediction of icd codes with clinical bert embeddings and text augmentation with label balancing using mimic-iii. *arXiv preprint arXiv:2008.10492*.

Pengfei Cao, Yubo Chen, Kang Liu, Jun Zhao, Shengping Liu, and Weifeng Chong. 2020. Hypercore: Hyperbolic and co-graph representation for automatic icd coding. In *Proceedings of the 58th Annual Meeting of the Association for Computational Linguistics*, pages 3105–3114.

Yahui Chen. 2015. Convolutional neural network for sentence classification. Master's thesis, University of Waterloo.

Jacob Devlin, Ming-Wei Chang, Kenton Lee, and Kristina Toutanova. 2018. Bert: Pre-training of deep bidirectional transformers for language understanding. *arXiv preprint arXiv:1810.04805*.

Mandy Guo, Joshua Ainslie, David Uthus, Santiago Ontanon, Jianmo Ni, Yun-Hsuan Sung, and Yinfei Yang. 2021. Longt5: Efficient text-to-text transformer for long sequences. *arXiv preprint arXiv:2112.07916*.

Shaoxiong Ji, Matti Hölttä, and Pekka Marttinen. 2021. Does the magic of bert apply to medical code assignment? a quantitative study. *Computers in Biology and Medicine*, 139:104998.

Alistair Johnson, Lucas Bulgarelli, Tom Pollard, Steven Horng, Leo Anthony Celi, and Roger Mark. 2020. Mimic-iv. *version 0.4). PhysioNet. https://doi. org/10.13026/a3wn-hq05*.

Alistair EW Johnson, Tom J Pollard, Lu Shen, Liwei H Lehman, Mengling Feng, Mohammad Ghassemi, Benjamin Moody, Peter Szolovits, Leo Anthony Celi, and Roger G Mark. 2016. Mimic-iii, a freely accessible critical care database. *Scientific data*, 3(1):1–9.

Diederik P Kingma and Jimmy Ba. 2014. Adam: A method for stochastic optimization. *arXiv preprint arXiv:1412.6980*.

Jinhyuk Lee, Wonjin Yoon, Sungdong Kim, Donghyeon Kim, Sunkyu Kim, Chan Ho So, and Jaewoo Kang. 2020. Biobert: a pre-trained biomedical language representation model for biomedical text mining. *Bioinformatics*, 36(4):1234–1240.

Fei Li and Hong Yu. 2020. Icd coding from clinical text using multi-filter residual convolutional neural network. In *Proceedings of the AAAI Conference on Artificial Intelligence*, volume 34(05), pages 8180–8187.

Julia Medori and Cédrick Fairon. 2010. Machine learning and features selection for semi-automatic icd-9-cm encoding. In *Proceedings of the NAACL HLT 2010 Second Louhi Workshop on Text and Data Mining of Health Documents*, pages 84–89.

James Mullenbach, Sarah Wiegreffe, Jon Duke, Jimeng Sun, and Jacob Eisenstein. 2018. Explainable prediction of medical codes from clinical text. *arXiv preprint arXiv:1802.05695*.

Damian Pascual, Sandro Luck, and Roger Wattenhofer. 2021. Towards bert-based automatic icd coding: Limitations and opportunities. *arXiv preprint arXiv:2104.06709*.

Adler Perotte, Rimma Pivovarov, Karthik Natarajan, Nicole Weiskopf, Frank Wood, and Noémie Elhadad. 2014. Diagnosis code assignment: models and evaluation metrics. *Journal of the American Medical Informatics Association*, 21(2):231–237.

Haoran Shi, Pengtao Xie, Zhiting Hu, Ming Zhang, and Eric P Xing. 2017. Towards automated icd coding using deep learning. *arXiv preprint arXiv:1711.04075*.

AK Singh, Mounika Guntu, Ananth Reddy Bhimireddy, Judy W Gichoya, and Saptarshi Purkayastha. 2020. Multi-label natural language processing to identify diagnosis and procedure codes from mimic-iii inpatient notes. *arXiv preprint arXiv:2003.07507*.

Yi Tay, Mostafa Dehghani, Dara Bahri, and Donald Metzler. 2020. Efficient transformers: A survey. *ACM Computing Surveys (CSUR)*.

Ashish Vaswani, Noam Shazeer, Niki Parmar, Jakob Uszkoreit, Llion Jones, Aidan N Gomez, Lukasz Kaiser, and Illia Polosukhin. 2017. Attention is all you need. *Advances in neural information processing systems*, 30.

Manzil Zaheer, Guru Guruganesh, Kumar Avinava Dubey, Joshua Ainslie, Chris Alberti, Santiago Ontanon, Philip Pham, Anirudh Ravula, Qifan Wang, Li Yang, et al. 2020. Big bird: Transformers for longer sequences. *Advances in Neural Information Processing Systems*, 33:17283–17297.

Zachariah Zhang, Jingshu Liu, and Narges Razavian. 2020. Bert-xml: Large scale automated icd coding using bert pretraining. *arXiv preprint arXiv:2006.03685*.

Parameter Efficient Transfer Learning for Suicide Attempt and Ideation Detection

Bhanu Pratap Singh Rawat[1,*], **Hong Yu**[1,2,3,‡]
[1]CICS, UMass-Amherst, [2]U.S. Department of Veterans Affairs,
[3]Center of Biomedical and Health Research in Data Sciences
*brawat@umass.edu, ‡hong_yu@uml.edu

Abstract

Pre-trained language models (LMs) have been deployed as the state-of-the-art natural language processing (NLP) approaches for multiple clinical applications. Model generalisability is important in clinical domain due to the low available resources. In this study, we evaluated transfer learning techniques for an important clinical application: detecting suicide attempt (SA) and suicide ideation (SI) in electronic health records (EHRs). Using the annotation guideline provided by the authors of ScAN (Rawat et al., 2022), we annotated two EHR datasets from different hospitals. We then fine-tuned ScANER (Rawat et al., 2022), a publicly available SA and SI detection model, to evaluate *five* different parameter efficient transfer learning techniques, such as adapter-based learning and soft-prompt tuning, on the two datasets. Without any fine-tuning, ScANER achieve macro F1-scores of 0.85 and 0.87 for SA and SI evidence detection across the two datasets. We observed that by fine-tuning less than $\sim 2\%$ of ScANER's parameters, we were able to further improve the macro F1-score for SA-SI evidence detection by 3% and 5% for the two EHR datasets. Our results show that parameter-efficient transfer learning methods can help improve the performance of publicly available clinical models on new hospital datasets with few annotations.

1 Introduction

In the past decade, 90% of the US hospitals have adopted a certified electronic health record (EHR) system (IT, 2022). This has led to an enormous availability of EHRs with rich information about patients' health (Henry et al., 2016). With the advancement of natural language processing (NLP), there has been a significant improvement in the development of clinical models and systems to extract clinically relevant information from the EHRs for further downstream tasks (Uzuner et al., 2011; Rawat et al., 2022). Recent years have seen clinical datasets being publicly released for different NLP tasks such as named entity recognition, relation extraction, text de-identification and disease classification (Pampari et al., 2018; Sun et al., 2013; Henry et al., 2020). Medical Information Mart for Intensive Care - III (MIMIC) (Johnson et al., 2016) has enabled a large and continually growing set of de-identified EHR notes from an intensive care unit for developing other publicly available datasets such as emrQA (Pampari et al., 2018), ScAN (Rawat et al., 2022) and adverse drug reaction (ADR) extraction (Henry et al., 2020).

This increase in availability of the clinically annotated datasets has led to the improvement in performance of different NLP models. While this improvement is great, a key question is whether these improvements generalize to new datasets of the same task or not. This question is quite difficult to answer because it requires annotating multiple datasets or new datasets with the same guidelines when it is already difficult to annotate a single dataset (Laparra et al., 2021; Futoma et al., 2020). In this study, we evaluate different parameter efficient transfer learning techniques on the task of an important clinical application, namely suicide attempt (SA) and suicide ideation (SI) detection from EHRs.

Recently, a SA-SI detection dataset (ScAN) (Rawat et al., 2022) was publicly released in an effort to extract suicidal information from patients' EHRs. ScAN was released along with the annotation guidelines used by the experts and the baseline model to detect the suicidal evidences from EHR notes (ScANER). We followed the annotation guideline to annotate two new datasets: EHR notes from School of Medicine at University of Pittsburgh (hereby referred as ScAN_UP) and EHR notes from the US Veterans Health Administration (ScAN_VA). We used ScAN and ScANER as our base dataset and model for creating the two new datasets and evaluating different transfer learning

Proceedings of the 13th International Workshop on Health Text Mining and Information Analysis (LOUHI), pages 108 - 115
December 7, 2022 ©2022 Association for Computational Linguistics

techniques. In order to evaluate the transfer learning performance of ScANER, we kept the size of ScAN_UP and ScAN_VA relatively smaller than ScAN for further fine-tuning. These fine-tuned models could eventually help clinical professionals in making patient-aware clinical judgements for further treatments.

Pre-trained language models have significantly grown in size since the inception of BERT (Devlin et al., 2018) model. BERT was introduced with 110 million parameters but recent LMs such as generative pre-trained transformer (GPT-3) (Brown et al., 2020) and Open Pretrained Transformer (OPT) (Zhang et al., 2022) have $\sim$ 175 billion parameters. Given their unprecedented performance gains over different downstream tasks, the researchers in the clinical community have also adopted these models. But all hospitals or medical organizations do not have the resources to adapt these billion parameter models in their ecosystem. Hence it is important to evaluate parameter-efficient transfer learning techniques that keep most of the model parameters frozen during fine-tuning on a newer dataset for the same task. We decided to try five different techniques: fine-tuning the classification layer, BitFit (Zaken et al., 2021), adding adapter modules (Houlsby et al., 2019), soft-prompt fine-tuning (Lester et al., 2021) and tuning the last four layers (Lee et al., 2019). Most of these techniques require fine-tuning of less than 2% of ScANER's parameters except tuning the last four layers which requires tuning of $\sim$ 23% parameters.

In this study, we found that ScANER achieves $>$ 85% macro F1-score for SA-SI evidence detection on two new datasets without any fine-tuning. We were able to further improve the SA-SI evidence detection by 3% for ScAN_UP and 5% for ScAN_VA by fine-tuning less than $\sim$ 2% of ScANER's parameters. Both ScAN_UP and ScAN_VA contain less than 8% annotations when compared to the original ScAN dataset. This shows that parameter-efficient transfer learning methods can help in improving the performance of publicly available clinical models on new hospital datasets with few annotations.

2 Dataset

In order to evaluate different transfer learning techniques, we focused heavily on choosing a task that has a publicly available dataset along with the annotation guidelines and the baseline model. The annotation guidelines are very important because they would help us in keeping the annotation decisions across different datasets uniform. Hence, we chose the task of detecting suicide attempt and ideations events in EHRs because of the availability of ScAN dataset (Rawat et al., 2022). The annotations guidelines for creating ScAN are publicly available along with their proposed baseline model (ScANER).

2.1 ScAN: Suicide Attempt and Ideation Events Dataset

ScAN (Rawat et al., 2022) is a publicly available SA and SI events dataset which is a subset of the MIMIC-III (Johnson et al., 2016) dataset. The EHRs were filtered for the hospital stays that consisted of diagnostic codes associated with suicide and overdose. These EHRs were annotated at sentence-level for SA and SI events. Each hospital-stay consisting of multiple EHR notes, such as nursing note, physician note, and discharge summary, was also annotated for SA and SI. ScAN consists of 12,759 EHR notes with 19,960 unique evidence annotations for suicidal behavior. The publicly available annotation guidelines of ScAN allows the creation of new datasets for the same task with uniform annotations.

We decided to annotate two parallel datasets using the EHR notes of patients at School of Medicine, University of Pittsburgh and EHR notes of Veterans at Veteran Health Administration. For both datasets, we filtered the notes using the phrases related to suicidal behavior extracted from the ScAN dataset, such as *overdose*, *suicide attempt*, and *killing myself*. We were not able to map different EHRs from the same hospital-stay. Hence, we decided to focus only on extracting SA-SI evidence paragraphs from the EHRs using the *evidence retriever* module of ScANER. The *evidence retriever* module consists of a pre-trained LM (medRoBERTa) in a multi-task setting to extract all the evidence paragraphs from the EHR notes of the patients.

2.2 School of Medicine, University of Pittsburgh

There were 99,736 EHR notes available from the School of Medicine, University of Pittsburgh. After filtering notes with the help of the selected keywords for suicidal behavior we were able to find 220 unique EHR notes with a mention of SA or SI. The dataset was annotated by two expert annotators

	ScAN_UP (220 EHRs)			ScAN_VA (880 EHRs)		
Evidence	*Yes*	*No*		*Yes*	*No*	
Train	302	517		1171	2171	
Validation	72	108		233	467	
Test	258	491		968	1927	
SA	*Positive*	*Neg_Unsure*	*Neutral-SA*	*Positive*	*Neg_Unsure*	*Neutral-SA*
Train	199	35	585	419	35	2888
Validation	47	11	125	77	8	615
Test	149	42	558	340	44	2511
SI	*Positive*	*Negative*	*Neutral-SI*	*Positive*	*Negative*	*Neutral-SI*
Train	80	34	702	566	440	2316
Validation	13	15	151	98	91	506
Test	60	42	638	440	364	2066

Table 1: The distribution of evidences paragraphs in ScAN_UP and ScAN_VA for train, validation and test sets. A paragraph is considered an *evidence*, labeled as *Yes*, if it has at least one sentence annotated as SA or SI. A *No* evidence paragraph is *Neutral-SA* and *Neutral-SI*.

under the supervision of a senior physician. Following the annotation guidelines provided via ScAN (Rawat et al., 2022), we created four categories for SA: *positive*, *negative*, *unsure* and *neutral-SA*. A paragraph is marked *positive* for SA if it mentions a positive suicide attempt, such as 'tried to hang myself'. A *negative* SA annotation denotes an accidental self-inflicted harm which could be misinterpreted as a suicide attempt such as a clinically diagnosed 'accidental overdose'. An annotation is marked as *unsure* for SA if it is not clear from the text whether the suicide attempt is positive or negative. Any paragraph with none of the SA annotation would be considered as *neutral-SA*. For SI, we have three categories: *positive*, *negative* and *neutral-SI*. As per ScAN (Rawat et al., 2022), we also merged our two labels *negative* and *unsure* for suicide attempt to create one label: *neg_unsure*. Similar to the original dataset, ScAN_UP is also highly imbalanced consisting of only few instances of *neg_unsure* SA labeled paragraphs.

This resulted in 853 unique annotations at sentence level where 613 were for SA and 240 for SI. Similar to ScAN (Rawat et al., 2022), we also created paragraphs from the EHR notes using an overlapping window of 5 sentences. We divided the EHRs into train, validation and test set in the ratio of 50 : 10 : 40. This resulted in total 632 *evidence* paragraphs, where an evidence paragraph is any paragraph which contains at least one annotation related to SA or SI. The annotators achieved an agreement of 97.76% at paragraph-level and 100% on document-level. The distribution of the paragraphs for SA and SI is provided in Table 1.

2.3 Veterans Healthcare Administration (VHA)

In the VHA system, we found hundreds of thousands EHR notes with keywords related to suicidal behavior. We sampled 883 notes from all the available notes to keep the size of VHA dataset roughly 4 times bigger than ScAN_UP. The dataset was again annotated by two annotators under the guidance of a senior physician. The annotators achieved an agreement of 93.97% at paragraph-level and 100% agreement on document-level. There were total of 1371 unique annotations for suicide attempt and 2270 for suicide ideation. As a preventive measure by VHA, Veterans with any form of suicidal behavior are regularly screened for suicidal ideation resulting in an inflated number of negative SI annotations in ScAN_VA dataset. Similar to ScAN_UP, we created paragraphs from the EHR notes using an overlapping window of 5 sentences. We divided the EHRs into train, validation and test set in the ratio of 50 : 10 : 40. This resulted in a total of 2372 *evidence* paragraphs. The distribution of the paragraphs for SA and SI across train, validation and test set is provided in Table 1.

These two datasets are quite different from each other as the EHR notes used for ScAN_UP are written for civilians whereas the notes for ScAN_VA are written for Veterans and contain medical linguistics specific to veteran healthcare administration. As mentioned earlier, the *negative* SI annotations are frequently observed in ScAN_VA as compared to ScAN_UP. Thus the label distribution is also quite different amongst the two datasets. These two datasets would provide a good challenge to ScANER and it's further fine-tuned versions using different transfer learning techniques.

3 Methodology

ScANER (Rawat et al., 2022) consists of two submodules: (a) an *evidence retriever module* that extracts the evidence paragraphs related to SA and SI events and (b) a *predictor module* that predicts SA or SI label for a patient's hospital stay using all the EHR notes from the hospital admission. For our two datasets, ScAN_UP and ScAN_VA, we have sentence-level annotations in an EHR but do not have all the EHRs for patients' single admission. Hence, we only focus on the first module of ScANER which can be used to extract all the evidence paragraphs from an EHR. The *evidence retriever module* consists of a medRoBERTa model trained in a multi-task learning setting to identify the evidence paragraphs along with classifying the SA and SI event label for the paragraphs. We used the ScANER model trained on the original ScAN (Rawat et al., 2022) dataset for our experiments. We used *five* different transfer learning techniques with varying number of trainable parameters on ScAN_UP and ScAN_VA.

3.1 Fine-tuning the classifier layers

ScANER consists of three classification layers for predicting the evidence class label, SA label and SI label. We decided to only fine-tune these three final classification layers on our datasets while freezing the rest of the encoder parameters. This is the most parameter efficient transfer learning technique as it uses only $\sim$ 8 thousand parameters, out of the available 125 million, refer Table 2. This technique takes the least amount of resources for fine-tuning but provides very low capacity for the model to learn new information or patterns.

3.2 Soft prompt tuning

Soft prompt tuning (Lester et al., 2021) is a powerful technique for adapting pre-trained models for new downstream tasks. For prompt tuning, all the encoder parameters are frozen during fine-tuning except a few additional k tunable tokens for each downstream task. These tunable soft-prompts help the model in adapting to new tasks using the previously trained encoder parameters. The length of the soft-prompts (k) can be tuned as a hyper-parameter. These soft-prompts can be initialized randomly or using an existing embedding from the encoder's vocabulary (Lester et al., 2021) related to the downstream task at hand. We experimented with different length of soft prompts ranging from 10 to 40 and initializing the soft prompts with the embedding of the word 'the' and 'suicide'. This transfer learning technique uses only 0.02% of ScANER's parameters.

3.3 BitFit

BitFit (Zaken et al., 2021) is a sparse fine-tuning technique that modifies only the bias terms of the trained model. Zaken et al. (2021) showed that on small to medium sized training datasets, BitFit is competitive with fine-tuning the entire training model. BitFit is also a light fine-tuning method that only uses 0.2% of ScANER's parameters.

3.4 Adapters

Adapter modules (Houlsby et al., 2019) were proposed as another efficient transfer learning technique which requires adding a few trainable parameters for the downstream task while freezing all the original encoder parameters. Adapters require more additional tunable parameters as compared to soft prompt tuning because adapter modules are added in multiple transformer layers of the encoder. Though in comparison to training all the model parameters, it only adds $\sim 2\%$ parameters to the ScANER model.

3.5 Fine-tuning few last layers

Lee et al. (2019) studied the effect of freezing multiple early encoder layers and found that only a fourth of the final layers need to be fine-tuned to achieve 90% of the performance achieved via full model training. We experimented with fine-tuning last *two* to *five* layers for our new datasets. As compared to the earlier transfer learning methods, this technique requires the most number of parameters even with fine-tuning of only last 2 layers ($\sim 11\%$).

Evaluation Metrics As our main task is to classify a paragraph as an evidence or not, we looked at the accuracy, macro F1-score and weighted F1-score on the test sets of ScAN_UP and ScAN_VA. Since, the models are being fine-tuned in the multi-task setting we would also look at the auxiliary tasks of predicting the SA and SI labels for the paragraphs. Accuracy and weighted F1-score provides

ScAN_UP		Evidence			SA			SI		
Transfer Learning	# Tunable Params ↑	Acc	F1	Wt-F1	Acc	F1	Wt-F1	Acc	F1	Wt-F1
ScANER	-	0.88	0.87	0.88	0.81	0.57	0.82	0.89	0.58	0.88
Classifier	8 Thousand	0.88	0.87	0.88	0.85	0.54	0.82	0.89	0.56	0.88
Soft Prompt-tuning	23 Thousand	0.91	0.90	0.91	0.86	0.56	0.84	0.88	0.49	0.86
BitFit	130 Thousand	0.88	0.87	0.88	0.85	0.54	0.83	0.89	0.54	0.88
Adapter	2 Million	0.91	0.90	0.91	0.87	0.56	0.84	0.89	0.50	0.87
Last 4 layers	28 Million	0.89	0.88	0.89	0.85	0.54	0.83	0.89	0.52	0.87
All layers	125 Million	0.91	0.90	0.91	0.87	0.56	0.84	0.89	0.52	0.87
ScAN_VA		Evidence			SA			SI		
Transfer Learning	# Tunable Params ↑	Acc	F1	Wt-F1	Acc	F1	Wt-F1	Acc	F1	Wt-F1
ScANER	-	0.86	0.85	0.86	0.81	0.49	0.84	0.79	0.63	0.79
Classifier	8 Thousand	0.90	0.88	0.89	0.90	0.50	0.89	0.81	0.63	0.81
Prompt-tuning	23 Thousand	0.90	0.89	0.90	0.91	0.52	0.89	0.81	0.63	0.80
BitFit	130 Thousand	0.91	0.89	0.90	0.91	0.52	0.90	0.82	0.64	0.81
Adapter	2 Million	0.91	0.90	0.91	0.92	0.53	0.91	0.82	0.65	0.82
Last 4 layers	28 Million	0.90	0.89	0.90	0.90	0.51	0.89	0.82	0.64	0.81
All layers	125 Million	0.92	0.91	0.92	0.93	0.58	0.92	0.84	0.70	0.84

Acc: Accuracy, F1: Macro F1-score, Wt-F1: Weighted F1-score

Table 2: Evidence, SA and SI classification performance of all the transfer learning techniques on ScAN_UP and ScAN_VA datasets. The transfer learning techniques are *Classifier* layers tuning, Soft *prompt-tuning* (Lester et al., 2021), *BitFit* (Zaken et al., 2021), *Adapter* modules fine-tuning (Houlsby et al., 2019) and fine-tuning *last 4 layers* (Lee et al., 2019). *ScANER* (Rawat et al., 2022) refers to the original model without any fine-tuning on ScAN_UP and ScAN_UP and *all layers* refers to the fine-tuning of all the parameters of ScANER model.

overall model performance whereas the macro F1-scores provides class level model performance and is quite important in our cases as our dataset is highly imbalanced (refer Table1). All the final hyper-parameter settings for the transfer learning techniques are provided in Appendix A.

4 Results and Discussion

For evidence retrieval, even without fine-tuning the original ScANER model is able to achieve a macro-F1 score of 0.87 and 0.85 for ScAN_UP and ScAN_VA datasets respectively, refer Table 2. When all the parameters of ScANER are fine-tuned, the macro F1-score of the *evidence retriever* module improved by 3% and 6% for evidence retrieval for ScAN_UP and ScAN_VA. For ScAN_VA, the macro F1-score for SA and SI also improved by 9% and 7% respectively. But for ScAN_UP, the performance for both SA and SI classification dropped when all the layers of the encoder are fine-tuned. This is mainly because of the extreme imbalance for both SA and SI in the ScAN_UP dataset. The accuracy and weighted F1-score performance for SA classification improved by 6% and 2% respectively because the fine-tuned ScANER model performed well for the *positive* and *neutral-SA* class but performed poorly for the under-represented *neg_unsure* class. We tried multiple techniques to counter the imbalance, such as up-sampling and weighted log-loss learning as described in Rawat et al. (2022), but none of the techniques helped in improving the performance of fully-trained model on ScAN_UP. One thing to notice is that the performance for the main task of *evidence retrieval* improved for both datasets with transfer learning. We also observed that the performance improvement is not strictly correlated with the number of tunable parameters available for transfer learning.

ScAN_UP The adapter module and soft-prompt fine-tuning performed the best for the evidence retrieval task. Both techniques were able to achieve the same performance as the fully-trained *evidence retrieval* module while using less than 2% of the module parameters. They also achieved similar SA prediction performance in terms of F1-score but under-performed for the SI prediction task. Amongst the two, adapter modules based fine-tuning performed better for SI prediction by 1%. These results are encouraging as they suggest that with the help of only 132 annotated EHRs and fine-tuning of less than 2% of the parameters, we can significantly improve the performance of the *evidence retrieval* module. BitFit performed almost similar to only classifier fine-tuning even when it has 16 times more tunable parameters. For last few layers technique, we found that tuning last 4 layers yield the best results. It was also able to improve over the baseline ScANER performance but under-performed as compared to adapter and soft-prompt tuning.

For adapter modules, we found that 64 dimensional adapters work the best for our dataset. For soft-prompt fine-tuning, we tried initializing the soft prompt randomly, using the existing vocabulary embedding of the token 'the', and the vocabulary embedding of the token 'suicide'. For our dataset, the model with soft-prompt initialized using the embedding of the token 'suicide' performed the best. It outperformed the model with the soft-prompts using the emebeddings of the token 'the' by $\sim$ 1%. The results also show that even without any fine-tuning ScANER can retrieve evidences with a strong performance of macro F1 of 0.87.

ScAN_VA This dataset is almost twice the size of ScAN_UP which allows the ScANER model to improve even more. This is evident as the fully-trained *evidence retrieval* module outperformed the original ScANER module by 6%, 9% and 7% for evidence, SA and SI classification respectively. The adapter based model is able to achieve the best macro F1-score of 0.90 amongst all the transfer learning fine-tuning techniques. It outperformed all the other models for SA and SI classification as well while improving the performance of the original ScANER model by 5% for evidence retrieval, 4% for SA classification and 2% for SI classification. Even the classifier only fine-tuning technique is able to improve the performance of *ScANER* by 3% for evidence detection. The rest of the fine-tuning techniques improved the macro F1-score for evidence retrieval by atleast 4%.

Recommendations We observed that for both datasets, adapter based fine-tuning performed the best for evidence retrieval and SA classification. It also outperformed the other transfer learning techniques for SI classification on ScAN_VA but under-performed on ScAN_UP. As a result, for improving any publicly available clinical model using transfer learning we would recommend the use of adapter modules. If the availability of computational resources is still a problem, we would recommend using soft-prompt based fine-tuning as it uses $\sim$ 86 times lesser parameters as compared to adapter modules while consistently performing very well across both datasets. BitFit performed well on ScAN_VA dataset but under-performed when compared with most of the fine-tuning techniques on ScAN_UP dataset.

Overall, ScANER generalizes well on new datasets and achieved a macro F1-score of 0.87 and 0.85 on two new datasets without any further fine-tuning. With the help of parameter efficient transfer learning techniques, such as adapter and soft-prompt fine-tuning, we can significantly improve the performance of ScANER on new datasets. We observed that the SA-SI label distribution and the size of the dataset can also significantly affect the SA-SI classification performance of the fine-tuned models.

5 Related Works

Laparra et al. (2021) performed an extensive review to study the recent work on building more adaptable and generalizable NLP models for clinical domain using adaptive and transfer learning techniques. They reviewed the most recent relevant work to characterize different type of methods and tasks that are being used and studied in the clinical domain. They showed that most of the work is using pre-trained language models such as BioBERT (Lee et al., 2020) and clinicalBERT (Alsentzer et al., 2019). Laparra et al. (2021) also discussed work that uses multi-task learning, sequential transfer learning and cross-lingual adaptation but did not review any recently developed parameter efficient transfer learning techniques such as adapter modules (Houlsby et al., 2019), soft-prompt tuning (Lester et al., 2021) and BitFit (Zaken et al., 2021). They also mentioned that the high costs of creating and distributing new clinical datasets favor creating a new dataset for a new task rather than creating another dataset for an existing task. In order to mitigate such imbalance, we study the effectiveness of transfer learning techniques by creating *two* new datasets for an existing task with a publicly available dataset (ScAN) and evaluating newly introduced transfer learning techniques.

Narayanan et al. (2020) studied different transfer learning techniques for adverse drug event (ADE) and medication entity extraction. They mainly focused on evaluating different biomedical contextual embeddings and using these pretrained embeddings for improved performance on their tasks. Similarly, Sun and Yang (2019) also studied the effectiveness of multilingual BERT and BioBERT for a named entity recognition (NER) task of extracting chemical and protein entities from Spanish biomedical texts. Zhou et al. (2019) adapted a CRF trained on general medical domain for NER on nursing handover data to achieve improved performance. A participant at MediQA 2019 challenge (Abacha et al., 2019) combined multiple classification tasks such as sentence classification, pairwise text classification and relevance ranking for improved performance in the shared task of the challenge. All the studies, either used a pre-trained LM or multi-task learning to improve the performance of their model on a task. Whereas in our study, we use an openly available trained LM-based classification model and further fine-tune it using recently developed parameter efficient transfer learning techniques (Houlsby et al., 2019; Lee et al., 2019; Lester et al., 2021; Zaken et al., 2021) to improve it's performance on two new datasets of the same downstream tasks.

6 Conclusion

In this paper, we evaluated different parameter efficient transfer learning techniques on the task of suicide attempt (SA) and suicide ideation (SI) events detection in the EHR notes. According to the publicly available annotation guidelines of ScAN (Rawat et al., 2022) dataset, we created two new datasets: ScAN_UP and ScAN_VA. We tested the baseline model ScANER on these two datasets and achieved macro F1-scores of 0.87 and 0.85 for SA-SI evidence detection. We were able to further improve the performance of ScANER by at least 3% after fine-tuning only 2% of ScANER's parameters. We show that parameter efficient transfer learning can help improve the performance of publicly available clinical models on new hospital datasets with few annotations. We would recommend the use of adapter modules for further transfer learning of clinical models as they consistently performed well for SA-SI detection while tuning only 2% of the parameters. If the computational resources are still a constraint, we would recommend using soft-prompt tuning as they only tune 0.02% of the parameters while achieving a performance quite close to adapter module tuning.

References

Asma Ben Abacha, Chaitanya Shivade, and Dina Demner-Fushman. 2019. Overview of the mediqa 2019 shared task on textual inference, question entailment and question answering. In *Proceedings of the 18th BioNLP Workshop and Shared Task*, pages 370–379.

Emily Alsentzer, John R Murphy, Willie Boag, Wei-Hung Weng, Di Jin, Tristan Naumann, and Matthew McDermott. 2019. Publicly available clinical bert embeddings. *arXiv preprint arXiv:1904.03323*.

Tom Brown, Benjamin Mann, Nick Ryder, Melanie Subbiah, Jared D Kaplan, Prafulla Dhariwal, Arvind Neelakantan, Pranav Shyam, Girish Sastry, Amanda Askell, et al. 2020. Language models are few-shot learners. *Advances in neural information processing systems*, 33:1877–1901.

Jacob Devlin, Ming-Wei Chang, Kenton Lee, and Kristina Toutanova. 2018. Bert: Pre-training of deep bidirectional transformers for language understanding. *arXiv preprint arXiv:1810.04805*.

Joseph Futoma, Morgan Simons, Trishan Panch, Finale Doshi-Velez, and Leo Anthony Celi. 2020. The myth of generalisability in clinical research and machine learning in health care. *The Lancet Digital Health*, 2(9):e489–e492.

J Henry, Yuriy Pylypchuk, Talisha Searcy, and Vaishali Patel. 2016. Adoption of electronic health record systems among us non-federal acute care hospitals: 2008–2015. *ONC data brief*, 35(35):2008–2015.

Sam Henry, Kevin Buchan, Michele Filannino, Amber Stubbs, and Ozlem Uzuner. 2020. 2018 n2c2 shared task on adverse drug events and medication extraction in electronic health records. *Journal of the American Medical Informatics Association*, 27(1):3–12.

Neil Houlsby, Andrei Giurgiu, Stanislaw Jastrzebski, Bruna Morrone, Quentin De Laroussilhe, Andrea Gesmundo, Mona Attariyan, and Sylvain Gelly. 2019. Parameter-efficient transfer learning for nlp. In *International Conference on Machine Learning*, pages 2790–2799. PMLR.

Health IT. 2022. Non-federal acute care hospital electronic health record adoption.

Alistair EW Johnson, Tom J Pollard, Lu Shen, H Lehman Li-Wei, Mengling Feng, Mohammad Ghassemi, Benjamin Moody, Peter Szolovits, Leo Anthony Celi, and Roger G Mark. 2016. Mimic-iii, a freely accessible critical care database. *Scientific data*, 3(1):1–9.

Egoitz Laparra, Aurelie Mascio, Sumithra Velupillai, and Timothy Miller. 2021. A review of recent work in transfer learning and domain adaptation for natural language processing of electronic health records. *Yearbook of Medical Informatics*, 30(01):239–244.

Jaejun Lee, Raphael Tang, and Jimmy Lin. 2019. What would elsa do? freezing layers during transformer fine-tuning. *arXiv preprint arXiv:1911.03090*.

Jinhyuk Lee, Wonjin Yoon, Sungdong Kim, Donghyeon Kim, Sunkyu Kim, Chan Ho So, and Jaewoo Kang. 2020. Biobert: a pre-trained biomedical language representation model for biomedical text mining. *Bioinformatics*, 36(4):1234–1240.

Brian Lester, Rami Al-Rfou, and Noah Constant. 2021. The power of scale for parameter-efficient prompt tuning. *arXiv preprint arXiv:2104.08691*.

Sankaran Narayanan, Kaivalya Mannam, Sreeranga P Rajan, and P Venkat Rangan. 2020. Evaluation of transfer learning for adverse drug event (ade) and medication entity extraction. In *Proceedings of the 3rd Clinical Natural Language Processing Workshop*, pages 55–64.

Anusri Pampari, Preethi Raghavan, Jennifer Liang, and Jian Peng. 2018. emrQA: A large corpus for question answering on electronic medical records. *arXiv preprint arXiv:1809.00732*.

Bhanu Pratap Singh Rawat, Samuel Kovaly, Wilfred Pigeon, and Hong Yu. 2022. ScAN: Suicide attempt and ideation events dataset. In *Proceedings of the 2022 Conference of the North American Chapter of the Association for Computational Linguistics: Human Language Technologies*.

Cong Sun and Zhihao Yang. 2019. Transfer learning in biomedical named entity recognition: an evaluation of bert in the pharmaconer task. In *Proceedings of The 5th Workshop on BioNLP Open Shared Tasks*, pages 100–104.

Weiyi Sun, Anna Rumshisky, and Ozlem Uzuner. 2013. Annotating temporal information in clinical narratives. *Journal of biomedical informatics*, 46:S5–S12.

Özlem Uzuner, Brett R South, Shuying Shen, and Scott L DuVall. 2011. 2010 i2b2/va challenge on concepts, assertions, and relations in clinical text. *Journal of the American Medical Informatics Association*, 18(5):552–556.

Elad Ben Zaken, Shauli Ravfogel, and Yoav Goldberg. 2021. Bitfit: Simple parameter-efficient fine-tuning for transformer-based masked language-models. *arXiv preprint arXiv:2106.10199*.

Susan Zhang, Stephen Roller, Naman Goyal, Mikel Artetxe, Moya Chen, Shuohui Chen, Christopher Dewan, Mona Diab, Xian Li, Xi Victoria Lin, et al. 2022. Opt: Open pre-trained transformer language models. *arXiv preprint arXiv:2205.01068*.

Liyuan Zhou, Hanna Suominen, Tom Gedeon, et al. 2019. Adapting state-of-the-art deep language models to clinical information extraction systems: Potentials, challenges, and solutions. *JMIR medical informatics*, 7(2):e11499.

A Hyper-parameters for transfer learning techniques

Transfer Learning	# Prompts	LR	Epochs	Size
Classifier	-	1e-3	5	-
Soft prompts	20	1e-3	5	-
BitFit	-	1e-3	5	-
Adapter	-	1e-3	5	64
Last 4 layers	-	1e-4	5	-

Table 3: Best hyperparameters for classifier only, BitFit, soft promp, adapters, and last 4 layers fine-tuning

Automatic Patient Note Assessment without Strong Supervision

Jianing Zhou[1], Vyom Thakkar[1], Rachel Yudkowsky[2], Suma Bhat[1] and William F. Bond[3]
[1]University of Illinois at Urbana-Champaign
[2]University of Illinois at Chicago
[3]Jump Simulation, OSF Healthcare, University of Illinois College of Medicine at Peoria
[1]{zjn1746, spbhat2, vnt2}@illinois.edu
[2]rachely@uic.edu
[3]william.f.bond@jumpsimulation.org

Abstract

Training of physicians requires significant practice writing patient notes that document the patient's medical and health information and physician diagnostic reasoning. Assessment and feedback of the patient note requires experienced faculty, consumes significant amounts of time and delays feedback to learners. Grading patient notes is thus a tedious and expensive process for humans that could be improved with the addition of natural language processing. However, the large manual effort required to create labeled datasets increases the challenge, particularly when test cases change. Therefore, traditional supervised NLP methods relying on labelled datasets are impractical in such a low-resource scenario. In our work, we proposed an unsupervised framework as a simple baseline and a weakly supervised method utilizing transfer learning for automatic assessment of patient notes under a low-resource scenario. Experiments on our self-collected datasets show that our weakly-supervised methods could provide reliable assessment for patient notes with accuracy of 0.92.

1 Introduction

Sponsored by the Federation of State Medical Boards (FSMB) and the National Board of Medical Examiners (NBME), the United States Medical Licensing Examination (USMLE) is a "three-step examination for medical licensure in the U.S. that assesses a physician's ability to apply knowledge, concepts, and principles, and to demonstrate fundamental patient-centered skills, that are important in health and disease and that constitute the basis of safe and effective patient care."[1] Prior to 2020, the USMLE Step 2 exam included a second component, Step 2 Clinical Skills, that used a simulated clinical examination with standardized patients to assess various clinical competencies, including the ability to document relevant patient history and differential diagnoses in a written patient note. After the discontinuation of the USMLE Step 2 Clinical Skills examination, medical schools may have more motivation to include a clinical skills examination that requires patient note writing after observing standardized patients (Tsichlis et al., 2021). Patient notes, as one type of health documents, document clinical findings and reflect examinees' ability to gather information and communicate their findings to patients and colleagues. Therefore, in Step 2 Clinical Skills, examinees' written patient notes were assessed manually by experienced physician raters. More than 30,000 examinees took this examination each year, resulting in more than 330,000 patient notes that were graded by more than 100 raters (Sarker et al., 2019). The case-specific nature of the patient notes and large volume of exams make the human scoring process time-consuming and tedious. Additionally, it is well-documented that human judgement in general is prone to bias and errors (Engelhard Jr et al., 2018). Training of qualified physician graders also requires assessment and feedback from medical experts, costing significant amounts of time. The manual effort required in grading medical examinations makes this a challenging problem to be addressed with the addition of NLP techniques.

NLP has been applied to automatically process health documents, including assessing practical clinical content from patient notes (Latifi et al., 2016; Sarker et al., 2019). Specifically, patient notes after simulated patient encounters are required to contain specific information, which is specified by items in a checklist created through faculty consensus. Figure 1 shows an example of patient note and checklist items. The task of automatic patient note assessment aims to judge if the given checklist items are included in the patient notes by exactly same expressions or synonymous expressions. Equivalents may be true synonyms, acceptable abbreviations, or answer alternatives

[1]https://www.usmle.org/

Proceedings of the 13th International Workshop on Health Text Mining and Information Analysis (LOUHI), pages 116 - 126
December 7, 2022

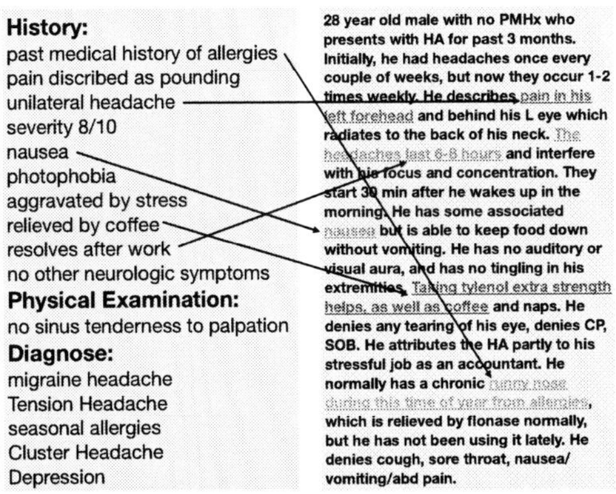

Figure 1: An example of patient note (right) and checklist (left). The challenges to NLP include the use of synonyms of checklist items, non-standard abbreviations, different expressions of negation and non-continuous occurrence of checklist items in patient notes.

deemed acceptable. Notes are further complicated by indications of body side (right or left), frequent negations, strings of positive or negative findings, and nonstandard abbreviations used by learners. Learners may use medical terms to describe findings (cholelithiasis) or lay terms (gall stones) and are typically judged the same if correct. Ideally, the NLP model would directly identify the phrases in patient notes correlated with the given checklist items for the most granular grading analysis and feedback to learners in formative settings. Therefore, we study automatic patient note assessment as two tasks: (i) directly judging if the given checklist items are entailed in the patient notes (a natural language inference task), and (ii) identifying the phrases in patient notes correlated with the given checklist items (a named entity recognition task).

Despite its importance, the task of automatic grading of patient notes remains under-explored with only a few works that have studied it (Yim et al., 2019; Sarker et al., 2019). Traditional supervised models have been utilized for this task (Latifi et al., 2016; Yim et al., 2019), but are limited in scope because they rely on large scale annotated datasets. The significant manual effort associated with labeled dataset creation makes these methods difficult and impractical. Besides, the traditional supervised models trained on data with prior clinical cases will be less effective for new clinical cases. Another challenge lies in the inconsistency between the checklist item and the corresponding phrase(s) in the patient note owing to their being non-exact matches occurring as, for instance, synonyms or abbreviations.

To overcome the limitations of previous works and the challenges of traditional supervised models for a low-resource scenario, we propose our method without strong supervision. First we propose a simple baseline unsupervised method with a pipeline framework which could be used in a zero-resource scenario. Then we propose our weakly supervised method utilizing multi-level transfer learning, including data-level and task-level. Data-level transfer learning refers to the ability of transferring knowledge learned from data in one domain to another domain. Task-level transfer learning refers to the ability of transferring knowledge learned from one task to another task. A BERT model (Devlin et al., 2019) pretrained on biomedical texts and a publicly available dataset[2] are used for data-level transfer learning. A key assumption is that judging if the checklist item is entailed in the patient note and identifying the corresponding phrases in the patient note are mutually related and thus we treat the automatic grading as a multi-task learning problem. With experiments on our self-collected datasets, we show that our weakly supervised method achieves a state-of-the-art performance.

Overall, the main contributions are as follows:

- We study an under-explored task of automatic patient note assessment and apply novel NLP methods to solve this task.
- We propose propose a weakly supervised method utilizing multi-level transfer learning at both data- and task-level. Furthermore, a multi-task learning mechanism is proposed for task-level transfer.
- Experimental results on case-specific datasets show that our weakly supervised method achieves SOTA performance. A unique contribution of our work not studied before and critically important for a low-resource scenario is understanding the effect of out-of-domain data. Our analyses show that our method has the ability of data-level transfer learning and task-level transfer learning even using instances that are not case-specific.

[2]The USMLE® Step 2 Clinical Skills Patient Note was made available for research purposes by NBME and can be requested at https://www.nbme.org/services/data-sharing. For more details about the corpus, see (Yaneva et al., 2022).

2 Related Work

Being a research task that is currently under-explored, there are very few works studying automatic patient note assessment. The most closely related task that past works focus on is automatic short answer grading (ASAG) for scientific topics (Liu et al., 2016; Hermet et al.; Mitchell et al., 2002; Sukkarieh and Pulman, 2005; Sukkarieh and Bolge, 2010; Dzikovska et al., 2012; D'Mello et al., 2008; Zhu et al., 2022; Haller et al., 2022), which is different from complex domain-specific answer assessment (e.g. medical domain in our work). ASAG aims to grade free text that answers to a prompt categorically or numerically. Produced by ETS, C-rater (Leacock and Chodorow, 2003) is one example system for ASAG focusing on grading school-level examinations based on the presence or absence of required answers. Text goes through a sequence of NLP modules for spelling correction, syntactic analysis, pronoun resolution, morphological analysis and synonym detection. Generated canonical representations are then fed into a maximum entropy model for classification. (Nehm et al., 2012) also focused on a similar task of awarding content points for specific items for college biology essays. Two text analytic platforms are utilized: SPSS Text Analysis 3.0 (SPSSTA) relying on hand-crafted vocabulary and rules and Summarization Integrated Development Environment (SIDE) using a classic bag-of-words representation and support vector machine. With the development of transformers, different transformers and large pre-trained models including BERT and RoBERTa have also been applied (Zhu et al., 2022).

While there are some works on ASAG for scientific topics, only three works studied automatic patient note assessment (Latifi et al., 2016; Sarker et al., 2019; Yim et al., 2019). Inspired by the works on ASAG, the first two (Latifi et al., 2016; Yim et al., 2019) studied two systems: a feature based system including an n-gram feature extraction followed by a SVM and a simple BERT based neural network. The third (Sarker et al., 2019) followed previous works on ASAG and leveraged the pipeline framework. Their system employs a sequence of modules including text normalization, lexicon-based matching, fuzzy matching and supervised concept detection all utilizing significant manual annotation and brute force exhaustive searches. Inspired by these works, we also proposed a pipeline model without supervision, which could be used under zero-resource scenario. A key departure from the prior pipeline efforts is our non-reliance on task-specific manual annotation.

Datasets	Headache	Abdominal Pain
Total num. of patient notes	510	570
Average Num. of Tokens in patient notes	132.35	97.05
Label Distribution	258/252	337/233
IAA	0.916	0.938
History Checklist	11	8
PEXAM Checklist	1	6
DDX Checklist	5	5
Total	17	19

Table 1: Statistics of our datasets. **History**, **PEXAM** and **DDX** represents the number of checklist items on History, Physical Examination and Diagnose. **Total** represents the number of all checklist items. **Label Distribution** is represented as the number of label 1 and the number of label 0. **IAA** refers to inter-annotator agreement evaluated by Cohen's kappa coefficient.

However, the methods proposed in previous works are insufficient for the task of automatic patient note assessment. N-gram features and SVMs are limited for extracting linguistic and semantic features, especially for complex domain-specific text. BERT based model requires a huge amount of annotated data for training, which is usually unavailable for our task, whereas, pipeline models have the obvious problem of error propagation. Therefore, we also proposed an end-to-end model, which utilizes multi-level transfer learning to alleviate the dependence on annotated data.

3 Datasets

We used two datasets in our study. The first is a self-collected dataset on two clinical cases— headache and abdominal pain—collectively referred to as **case-specific** datasets. Data from the same clinical case is referred to as in-domain data and data from a different clinical case is referred to as out-of-domain data. Collected data pertains to patient notes written by examinees, where each note covers three sections: (i) *history*, (ii) *physical exam* and (iii) *differential diagnosis*. Patient note in each section should pertain to items from the same domain in the checklist, which includes 17 checklist items for the headache case and 19 for the abdominal pain case. The checklist item may contain fine-grained medical concepts (e.g., 'headache') and general descriptions (e.g., 'pain started two weeks ago'). The medical concepts included in the checklist items for different cases may be similar or vastly different, depending on the clinical condition being portrayed by the patient. As part of the grading process two expert raters, typically physician faculty members, are asked to judge if the checklist items are stated in the patient notes. Inter-annotator agreements on both cases are reported in Table 1. For both cases, the inter-annotator agreements are above 0.9, which shows the reliability of our constructed dataset. Additionally, for the purpose of our experiments the raters were asked to identify the phrases in the patient notes that correspond to the checklist items when the checklist item was matched. The tokens in the highlighted phrases were labeled following the BIO convention (Ramshaw and Marcus, 1999). Due to the cost of physician faculty rater time, we only collected data from 30 examinees. Of these patient notes from 25 examinees were used for fine-tuning and those from the remaining 5 were set aside for testing.

A second dataset is the USMLE® Step 2 Clinical Skills Patient Note (Yaneva et al., 2022), which contains a total of 43,985 patient note history portions from 10 clinical cases, where 2,840 patient notes (284 notes per case) were annotated with concepts from the exam scoring rubrics. At the time of the writing of this paper the dataset was used for a Kaggle competition on automated scoring of clinical patient notes[3], and only a subset of 100 patient notes from the annotated data were made available to the public (the remaining 184 notes per case were used as a test set for the competition). Therefore, the study presented here has used a subset of 100 annotated patient notes per case, which was not large enough to be directly used for training or fine-tuning the model but was still considered as a diverse but related dataset. This dataset is referred to as **generic** dataset in the rest of the paper.

4 Baseline Method

In this section, we introduce our proposed unsupervised method used as a baseline model for written patient note assessment. This approach utilized Amazon Comprehend Medical[4] for the purpose of medical entity extraction. Amazon Comprehend Medical is an API that performs various types of text analysis for the medical domain, and is a service that is provided by Amazon Web Services (AWS). We made use of the medical entity detection feature of this API, that allowed extraction and detection of six different types of medical entities: anatomy, medical conditions, medications, protected health information, test treatment procedures as well as time expressions in the medical context (Bhatia et al., 2019).

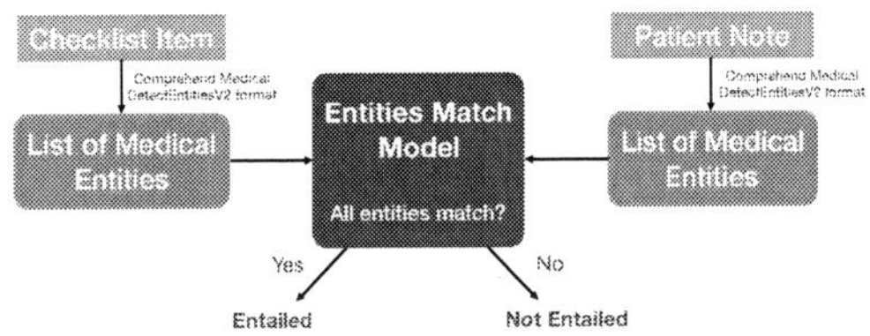

Figure 2: Unsupervised Method Model

4.1 Model Architecture

Here we describe the architecture of our unsupervised model presented in Figure 2. Taking the patient note $p = \{w_1^p, ..., w_n^p\}$ and checklist item $c = \{w_1^c, ..., w_m^c\}$ as input, the model aims to predict if the given checklist item is included in the given patient note.

Medical Entity Extraction. The first step converts the text of the checklists items and patient notes into the medical entities object format used by Amazon Comprehend Medical. The purpose is to then easily establish matching between medical entities extracted from checklist items and the medical entities extracted from patient notes.

Medical Entity Match. In the second step, we run a match-detection function based on each medical entity extracted from the checklist item and the medical entities extracted from the patient note. The match-detection function first filters the list of medical entities in the patient note by category. Comprehend Medical has six different entity categories, hence, if we are trying to find a match for a medical condition entity, then only medical condition entities are processed as potential candidate matches. This reduces the search space in the patient note based on medical entity categories. Once we obtain candidate matches by filtering based on these categories, we compare the similarity between the checklist item entity and the candidate entity from the patient note. If there is a surface level similarity (character-by-character equality), then we have found a match. If not, we compute a similarity score between the checklist item and the patient note medical entity using BioWordVec (Zhang et al., 2019). If the similarity score is beyond a certain threshold empirically chosen to be 0.8, only then we characterize the pair of entities

[3]https://www.kaggle.com/c/nbme-score-clinical-patient-notes/data

[4]https://aws.amazon.com/cn/comprehend/medical/

as a match. A checklist item is considered entailed by the patient note if all of the medical entities in the checklist item have a match in the patient note.

5 Weakly Supervised Method

In this section, we provide the details of our proposed weakly supervised method. In our work, the first task of judging checklist items' entailment by the patient note is formulated as a natural language inference task. The second task of identifying phrases that correspond to a checklist item can be treated as labeling the span of corresponding phrases, which is similar to named entity recognition. Therefore, we refer to it as the NER-related task. These two tasks are mutually beneficial in our setting; identification of corresponding phrases directly means the checklist item is entailed by the patient note and the entailment of checklist item indicates that the corresponding phrases are in the patient note. In order to harness this mutual benefit, we propose a multi-task transfer learning setting with a mutual feedback mechanism. Using this method, data from different clinical cases could help the model to learn the basic concepts of our tasks and build appropriate representation for underlying medical concepts. Therefore, we also utilize data from different clinical cases for transferring common medical and task knowledge. Finally, we propose a multi-level transfer learning method including task-level and data-level transfer learning which removes the need for large-scale annotated corpora and is thus weakly supervised.

5.1 Model Architecture

Here we describe the architecture of our model, which is related to the task-level transfer learning. Figure 3 shows the architecture of our multi-level transfer learning model. Taking the patient note $p = \{w_1^p, ..., w_n^p\}$ and the checklist item $c = \{w_1^c, ..., w_m^c\}$ as input, the model aims to predict if the given checklist item is entailed by the given patient note and also identifies the span of the expressions corresponding to the given checklist item. BIO labels are used to label the span of the target phrases. In our model, the lower encoder layers are used for extracting the hidden representations of the input text and are shared across all tasks and data while the top task-specific layers with a mutual feedback mechanism are used for different tasks. The mutual feedback mechanism is used for sharing knowledge across different tasks via outputs of different task-specific layers. The architecture details are as follows:

Encoder Layers. The encoder layers are used to extract contextual embeddings for input text. We use BERT model as our encoder shared across different tasks. For BERT model, [CLS] is used at the start of the input and [SEP] is used to separate patient note and checklist item. Therefore, the final input to the encoder is $\{[\text{CLS}], w_1^p, ..., w_n^p, [\text{SEP}], w_1^c, ..., w_m^c, [\text{SEP}]\}$. The output contextual embeddings would be $X = \{x_{[\text{CLS}]}, x_1^p, ..., x_n^p, x_{[\text{SEP}]}, x_1^c, ..., x_m^c, x_{[\text{SEP}]}\}$.

Task-Specific Layers. For task-specific layers, different layers take different outputs of encoder layers as input. For NLI task, the contextual embedding $x_{[\text{CLS}]}$ is used as input because the whole sequence information are encoded into this embedding (Devlin et al., 2019). For NER task, the contextual embeddings of each token in the patient note $\{x_1^p, ..., x_n^p\}$ are used:

$$[p_s, p_{ns}] = \text{NLI}(x_{[\text{CLS}]})$$
$$[p_B^i, p_I^i, p_O^i] = \text{NER}(x_i^p)$$

where $\text{NLI}(\cdot)$ represents the NLI task layer and $[p_s, p_{ns}]$ is the output distribution with p_s as the probability of checklist item is stated and p_{ns} as the probability of checklist item is not stated. $\text{NER}(\cdot)$ represents the NER task layer and $[p_B^i, p_I^i, p_O^i]$ is the output distribution with p_B^i as the probability of token i is predicted as the beginning of the target phrase, p_I^i as the probability of token i is predicted as inside the target phrase and p_O^i as the probability of token i is predicted as outside the target phrase.

Mutual Feedback Mechanism As stated before, the NLI task and NER task can actually benefit each other. Therefore, the output of one task could be used to enhance the input of another task. Then the enhanced inputs would be fed into two new task-specific layers for these two tasks. For the NLI task, the output from the previous NER task layer is used to enhance the input as follows:

$$[p_B^{ave}, p_I^{ave}, p_O^{ave}] = [\frac{1}{n}\sum_{i=1}^{n} p_B^i, \frac{1}{n}\sum_{i=1}^{n} p_I^i, \frac{1}{n}\sum_{i=1}^{n} p_O^i]$$
$$\hat{x}_{[\text{CLS}]} = \text{cat}(x_{[\text{CLS}]}, [p_B^{ave}, p_I^{ave}, p_O^{ave}])$$
$$[\hat{p}_s, \hat{p}_{ns}] = \text{NLI}_{new}(\hat{x}_{[\text{CLS}]})$$

where $[\hat{p}_s, \hat{p}_{ns}]$ is the final output distribution for the whole sequence. The average of the output distribution over all the tokens in the patient note is used to enhance the input. Therefore, if the

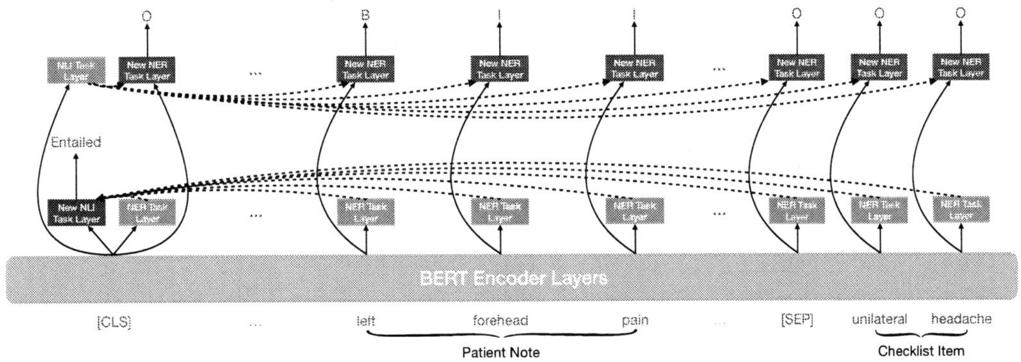

Figure 3: Architecture of our weakly supervised method. Dashed arrows represent outputs from NLI and NER task layers that are used to enhance the input to the new task layers. New NLI and NER task layers take outputs from NLI and NER task layers and outputs from the BERT encoder layers as input and generate the final outputs.

target phrase corresponding to the checklist item is identified, p_B^{ave} and p_I^{ave} would be non-zero, which could be used to guide the new NLI layer. Similarly, for the NER task, the output from previous NLI task layer is used to enhance the input:

$$\hat{x}_i^p = \text{cat}(x_i^p, [p_s, p_{ns}])$$
$$[\hat{p}_B^i, \hat{p}_I^i, \hat{p}_O^i] = \text{NER}_{new}(\hat{x}_i^p)$$

where $[\hat{p}_B^i, \hat{p}_I^i, \hat{p}_O^i]$ is the final output distribution for token i. The output distribution from previous NLI layer is used to enhance the input. If the checklist item is found to be stated in the patient note, p_s would be much larger than p_{ns} which could be used as a guidance for new NER layer. Finally, the enhanced input $\hat{x}_i^p$ is fed into the new NER layer.

5.2 Training Protocol

We use a simple joint training objective for our model, which is the sum of the sequence classification loss and the token classification loss, each of which is given by the corresponding cross-entropy loss. This training allows the task-level transfer as shown in Section 8.2.

The model is first trained with the generic dataset to learn the basic concept pertaining to the two tasks and the common medical/clinical knowledge. Then for new clinical cases with a few annotated instances, the model is fine-tuned with the case-specific data. Our hypothesis is that with the knowledge of the two tasks and the common medical knowledge learned during training, the model should be able to transfer to new clinical cases without the need for a large scale annotated dataset. In addition, for new clinical cases without any annotated data, our model can still be used because the knowledge of the two tasks and the common medical knowledge learned during training can be transferred to new clinical cases. The ability of data-level transfer is presented in Section 8.1.

6 Experiments

6.1 Baselines

Due to the fact that related prior works did not release their codes and did not provide enough details for reproduction, we only test one baseline model for comparison with our proposed unsupervised and weakly supervised methods on the NLI-related task. Besides, we also use one baseline model for comparison on the NER-related task.

- **NLI model**: For the NLI-related task of judging if the given checklist item is stated in the patient note, a simple BERT sentence pair classification model is used as the baseline with only the sequence classification loss as the training objective.
- **NER model**: For the NER-related task of identifying corresponding phrases in the patient note given checklist items, a simple BERT token classification model is used as baseline, which generates BIO labels to label the span of the target phrases. For this NER model, only the token classification loss is used as the training objective.

For both the baselines, the experimental settings and the parameters are set to be the same as those in our weakly supervised method. In addition, the baseline models and our weakly supervised model are trained on the same data but with different la-

Methods	Headache				Abdominal Pain			
	History	PEXAM	DDX	Total	History	PEXAM	DDX	Total
Unsupervised	0.72	0.30	0.87	0.63	0.68	0.58	0.93	0.73
NLI baseline	0.83	0.88	0.89	0.87	0.88	0.70	0.89	0.82
Weakly Supervised	0.91	0.94	0.94	0.93	0.91	0.90	0.93	0.91

Table 2: Performance of different methods on NLI-related task. Accuracy is used for evaluation. **History**, **PEXAM** and **DDX** represents the accuracy averaged on History, Physical Examination and Diagnose checklist items respectively. **Total** represents the accuracy averaged on all checklist items

Methods	Headache				Abdominal Pain			
	History	PEXAM	DDX	Total	History	PEXAM	DDX	Total
NER baseline	0.56	0.54	0.49	0.53	0.57	0.55	0.53	0.55
Weakly Supervised	0.60	0.59	0.63	0.61	0.61	0.62	0.63	0.62

Table 3: Performance of different methods on NER-related task. F1 score is used for evaluation.

bels. That is, for the NLI model, only the sequence-level labels indicating if the given checklist item is entailed or not are used. For the NER model, only the token-level labels indicating if each token belongs to the target phrase are used for training.

6.2 Evaluation Metrics

For the different tasks, different evaluation metrics are used. Accuracy defined as $\frac{\text{Num of Correct Predictions}}{\text{Num of All Predictions}}$ is used for NLI-related task, whereas the NER-related task, we use the F1 score, which is widely used for NER.

Number	Headache Case		Abdominal Pain	
	NLI	NER	NLI	NER
0	0.89	0.32	0.82	0.42
1	0.90	0.41	0.87	0.50
5	0.92	0.54	0.89	0.56
10	0.93	0.60	0.90	0.60
15	0.93	0.61	0.91	0.62
20	0.93	0.60	0.90	0.62
25	0.93	0.61	0.91	0.62

Table 4: Performance of weakly supervised method. The weakly supervised method is fine-tuned on different number of in-domain data. **NLI** refers to NLI-related task that is evaluated by accuracy. **NER** represents NER-related task that is evaluated by F1 score[5].

7 Results

The performances of the different models on the NLI task of judging if the checklist item is entailed by the patient note are summarized in Table 2. We find that our proposed unsupervised framework achieves an average accuracy of 0.63 across all the checklist items on the headache dataset and an average accuracy of 0.73 on the abdominal pain dataset. Compared with the unsupervised method, our weakly supervised method achieves a much better performance showing an average accuracy of 0.93 on the headache dataset and 0.91 on the abdominal pain dataset. As shown in Table 2, our proposed weakly supervised method outperforms the baseline NLI model and the unsupervised method across all the sections (checklist items averaged by section—history, physical exam and diagnosis) by a large margin. Looking at the accuracy values averaged across each section, we notice that our proposed weakly supervised method performs consistently well on all the checklist types whereas the baseline NLI model and the unsupervised method only perform well on specific sections.

For the NER task of identifying the corresponding phrases, our weakly supervised method achieves an F1 score of 0.61 on the headache dataset and 0.62 on abdominal pain dataset, which is better than the performance of the baseline NER model as shown in Table 3. This demonstrates that our proposed weakly supervised method utilizing multi-level transfer learning achieves the SOTA performance in both tasks when compared to all the baselines and our unsupervised method.

8 Analysis

In this section, we provide some ablation studies to analyze the contribution of data and the different modules used in our weakly supervised method.

[5]Due to the space limitation, the analysis on the number of in-domain data is provided in the appendix.

Methods	Headache				Abdominal Pain			
	History	PEXAM	DDX	Total	History	PEXAM	DDX	Total
Unsupervised	0.72	0.30	0.87	0.63	0.68	0.58	0.93	0.73
Weakly Supervised wo Fine-tuning	0.83	0.88	0.89	0.87	0.88	0.70	0.89	0.82
Weakly Supervised + Out-of-domain Data	0.83	0.94	0.89	0.89	0.87	0.81	0.90	0.86
Weakly Supervised + In-domain Data	0.91	0.94	0.94	0.93	0.91	0.90	0.93	0.91

Table 5: Performance of our methods on the NLI-related task. Our weakly supervised method is fine-tuned using different data sizes to show the ability of data-level transfer learning. Accuracy is used as the evaluation metric.

Tasks	Methods	Headache			Abdominal Pain		
		No Fine-tune	Out-of-domain	In-domain	No Fine-tune	Out-of-domain	In-domain
NLI	**NLI baseline**	0.82	0.84	0.87	0.77	0.80	0.82
	Ours	0.87	0.89	0.93	0.82	0.86	0.91
NER	**NER baseline**	0.08	0.36	0.53	0.08	0.40	0.55
	Ours	0.32	0.47	0.61	0.42	0.51	0.62

Table 6: Comparison between baseline models with single-task training and our weakly supervised model with multi-task training. For the NLI task, accuracy is used for evaluation, and for the NER task, F1 score is used.

8.1 Data-Level Transfer

Here we explore our method's data-level transfer learning ability, which is reflected in the performance of the model with the out-of-domain data. Two settings are used for our experiments. In the first setting we train our weakly supervised model with the generic dataset and then directly test the model on the case-specific dataset with no fine-tuning. In the second setting, we train our weakly supervised model with the generic dataset and then fine-tune it on the case-specific dataset on one of the two clinical cases. After training and fine-tuning, we test our model on the case-specific dataset related to another clinical case (e.g., fine-tuning on abdominal pain and testing on headache).

From the results in Table 5, for the first setting, we see that our model outperforms the other models even when trained on the generic dataset alone. For the second setting, when fine-tuned on the abdominal pain dataset and tested on the headache case-specific dataset, our model's performance improved from 0.87 to 0.89. Similarly, when fine-tuned on the headache dataset and tested on the abdominal pain dataset, our model's performance improved from 0.82 to 0.86. This shows that even the out-of-domain data can aid the performance of our weakly supervised method, suggesting that our method can transfer knowledge learned from out-of-domain data to new cases.

In addition, we also provide an analysis on the influence of training data amount on the performance in the Appendix. It is concluded that our weakly supervised method only requires a small number of in-domain data for fine-tuning to achieve a satisfactory performance for both tasks.

8.2 Task-Level Transfer

In this part we show our weakly supervised method's ability of task-level transfer learning via comparison between our model with multi-task training and the baseline models with single-task training. Our model and the baseline models are trained on the same datasets for a fair comparison.

Results are presented in Table 6. When only trained on the generic data and tested on the NLI-related task (corresponding to **No Fine-tune** columns), our model with multi-task training has an averaged accuracy of 0.87 on the headache case and 0.82 on the abdominal pain case whereas the NLI baseline model has an accuracy of only 0.82 on headache case and 0.77 on abdominal pain case. For the NER-related task, when only trained on the data from kaggle dataset, our model has an averaged F1 score of 0.32 on the headache case and 0.42 on the abdominal pain case whereas the NLI baseline model has an averaged F1 score of only 0.08 on headache and 0.08 on abdominal pain.

When in-domain data is used for fine-tuning, our model with multi-task training still outperforms the NLI and NER baseline models on both tasks. When trained using the generic data, fine-tuned on the in-domain data and tested on the NLI-related task (corresponding to **In-domain** columns), our model with multi-task training has an averaged accuracy of 0.93 on the headache case and 0.91 on abdominal pain case whereas the NLI baseline model has an averaged accuracy of only 0.87 on headache case and 0.82 on abdominal pain case. For the NER-related task, our model has an averaged F1 score of 0.61 on headache case and 0.62 on abdominal pain case whereas the NLI baseline model has an averaged F1 score of 0.53 on headache case and 0.55 on abdominal pain case.

Similarly, when out-of-domain data is used for fine-tuning (corresponding to **Out-of-domain** columns), our model with multi-task training also outperforms the NLI and NER baseline models on both tasks. Therefore, as shown by the higher performance compared with the NLI and NER baseline models, our weakly supervised model benefits from the multi-task training and shows a strong ability of task-level transfer learning.

9 Practical Impact

Our system is currently undergoing pilot testing by learners and faculty to assess the perceived impact on providing more immediate and automated feedback. The immediacy is important so that the case is fresh, and it will likely impact debriefing of simulation cases, potentially making debriefing more focused on areas of learner struggles identified in the notes. With our system grading all learners, automated feedback could be provided to the users and learners in time, which can be used to help their study of patient notes writing.

10 Conclusion and Future Work

In this paper, we study the problem of automatic written medical examination assessment. The complexity and huge manual effort required make data resources for this task very limited. Therefore, traditional NLP systems relying on large annotated corpora are impractical. With these factors in mind, we proposed a weakly supervised method. Our weakly supervised method utilizes multi-level transfer learning including data-level transfer learning and task-level transfer learning to simultaneously judge if the checklist item is stated in the patient note and also identify spans of relevant phrases. Experiments on two self-collected datasets show that our weakly supervised method is able to achieve the SOTA performance on both tasks. Therefore, our weakly supervised method can correctly judge if the checklist item is stated in the given patient note and can also find the relevant phrases most of the time. Our future work involves developing more effective transfer learning mechanisms to improve the performance on identifying the relevant phrases.

Acknowledgement

This project was funded (in part) by a National Board of Medical Examiners (NBME) Edward J. Stemmler, MD Medical Education Research Fund grant. The project and the views expressed in this publication do not necessarily reflect the position or policy of NBME, and NBME support provides no official endorsement.

The authors wish to acknowledge: Dr. Yoon Soo Park from the University of Illinois College of Medicine at Chicago for contributions to the Stemmler Fund project; Internal Medicine clerkship directors Drs. Saurabh Bansal and Manajyoti Yadav from the University of Illinois College of Medicine at Peoria for grading assistance. We also thank our research staff, including Rebecca Ebert-Allen, Rebecca Ruger, Alyse Dutcher, Ryan Klute for their assistance during the project.

References

Parminder Bhatia, Busra Celikkaya, Mohammed Khalilia, and Selvan Senthivel. 2019. Comprehend medical: a named entity recognition and relationship extraction web service. In *2019 18th IEEE International Conference On Machine Learning And Applications (ICMLA)*, pages 1844–1851. IEEE.

Jacob Devlin, Ming-Wei Chang, Kenton Lee, and Kristina Toutanova. 2019. BERT: Pre-training of deep bidirectional transformers for language understanding. In *Proceedings of the 2019 Conference of the North American Chapter of the Association for Computational Linguistics: Human Language Technologies, Volume 1 (Long and Short Papers)*, pages 4171–4186, Minneapolis, Minnesota. Association for Computational Linguistics.

Myroslava O Dzikovska, Rodney Nielsen, and Chris Brew. 2012. Towards effective tutorial feedback for explanation questions: A dataset and baselines. In *Proceedings of the 2012 Conference of the North American Chapter of the Association for Computational Linguistics: Human Language Technologies*, pages 200–210.

Sidney D'Mello, Tanner Jackson, Scotty Craig, Brent Morgan, P Chipman, Holly White, Natalie Person, Barry Kort, R El Kaliouby, Rosalind Picard, et al. 2008. Autotutor detects and responds to learners affective and cognitive states. In *Workshop on emotional and cognitive issues at the international conference on intelligent tutoring systems*, pages 306–308.

George Engelhard Jr, Jue Wang, and Stefanie A Wind. 2018. A tale of two models: Psychometric and cognitive perspectives on rater-mediated assessments using accuracy ratings. *Psychological Test and Assessment Modeling*, 60(1):33–52.

Stefan Haller, Adina Aldea, Christin Seifert, and Nicola Strisciuglio. 2022. Survey on automated short answer grading with deep learning: from word embeddings to transformers. *arXiv preprint arXiv:2204.03503*.

Matthieu Hermet, Stan Szpakowicz, and Lise Duquette. Automated analysis of students' free-text answers for computer-assisted assessment1.

Syed Latifi, Mark J Gierl, André-Philippe Boulais, and André F De Champlain. 2016. Using automated scoring to evaluate written responses in english and french on a high-stakes clinical competency examination. *Evaluation & the health professions*, 39(1):100–113.

Claudia Leacock and Martin Chodorow. 2003. C-rater: Automated scoring of short-answer questions. *Computers and the Humanities*, 37(4):389–405.

Chia-Wei Liu, Ryan Lowe, Iulian Serban, Mike Noseworthy, Laurent Charlin, and Joelle Pineau. 2016. How NOT to evaluate your dialogue system: An empirical study of unsupervised evaluation metrics for dialogue response generation. In *Proceedings of the 2016 Conference on Empirical Methods in Natural Language Processing*, pages 2122–2132, Austin, Texas. Association for Computational Linguistics.

Tom Mitchell, Terry Russell, Peter Broomhead, and Nicola Aldridge. 2002. Towards robust computerised marking of free-text responses. In *Proceedings of the 6th International Computer Assisted Assessment (CAA) Conference*.

Ross H Nehm, Minsu Ha, and Elijah Mayfield. 2012. Transforming biology assessment with machine learning: automated scoring of written evolutionary explanations. *Journal of Science Education and Technology*, 21(1):183–196.

Lance A Ramshaw and Mitchell P Marcus. 1999. Text chunking using transformation-based learning. In *Natural language processing using very large corpora*, pages 157–176. Springer.

Abeed Sarker, Ari Z Klein, Janet Mee, Polina Harik, and Graciela Gonzalez-Hernandez. 2019. An interpretable natural language processing system for written medical examination assessment. *Journal of biomedical informatics*, 98:103268.

Jana Sukkarieh and Eleanor Bolge. 2010. Building a textual entailment suite for the evaluation of automatic content scoring technologies. In *Proceedings of the Seventh International Conference on Language Resources and Evaluation (LREC'10)*.

Jana Z Sukkarieh and Stephen G Pulman. 2005. Information extraction and machine learning: Automarking short free text responses to science questions. In *Proceedings of the 2005 conference on artificial intelligence in education: Supporting learning through intelligent and socially informed technology*, pages 629–637.

Jason T Tsichlis, Andrew M Del Re, and J Bryan Carmody. 2021. The past, present, and future of the united states medical licensing examination step 2 clinical skills examination. *Cureus*, 13(8).

Victoria Yaneva, Janet Mee, Le Ha, Polina Harik, Michael Jodoin, and Alex Mechaber. 2022. The usmle® step 2 clinical skills patient note corpus. In *Proceedings of the 2022 Conference of the North American Chapter of the Association for Computational Linguistics: Human Language Technologies*, pages 2880–2886.

Wen-wai Yim, Ashley Mills, Harold Chun, Teresa Hashiguchi, Justin Yew, and Bryan Lu. 2019. Automatic rubric-based content grading for clinical notes. In *Proceedings of the Tenth International Workshop on Health Text Mining and Information Analysis (LOUHI 2019)*, pages 126–135.

Yijia Zhang, Qingyu Chen, Zhihao Yang, Hongfei Lin, and Zhiyong Lu. 2019. Biowordvec, improving biomedical word embeddings with subword information and mesh. *Scientific data*, 6(1):1–9.

Xin Hua Zhu, Han Wu, and Lanfang Zhang. 2022. Automatic short answer grading via bert-based deep neural networks. *IEEE Transactions on Learning Technologies*, pages 1–1.

A Appendix

A.1 Training Data Amount

First, we analyze the influence of training data amount on the performance. We use different amounts of in-domain data to fine-tune our weakly supervised method. As shown in Table 4, we notice that when the data used for fine-tuning is less than 10 patient notes the amount of training data has a big influence on the performance, with the performance improving with the increase of training data. When the training data is increased from 0 to 10 patient notes, the averaged accuracy increased from 0.89 to 0.93 and the F1 score increased from 0.32 to 0.6 on headache case. case of 0 training instances corresponds to the out-of-the-box performance of the weakly supervised model. On the abdominal pain case, the averaged accuracy increased from 0.85 to 0.9 and the F1 score increased from 0.42 to 0.6. However, when the data used for fine-tuning is more than 10, the performance did not change significantly with the increase of training data. Based on this, we note that our weakly supervised method only requires a small number of in-domain data for fine-tuning to achieve a satisfactory performance for both tasks.

A.2 Limitations

Although our weakly supervised model shows a satisfactory performance on NLI-related task after fine-tuning on in-domain data, the performance on NER-related task is still limited. Therefore, our weakly supervised model is limited on relating phrases in the patient notes with the given checklist item. Besides, without fine-tuning on in-domain data, the performance on NLI-related task is not good enough, which means that our weakly supervised model still relies on annotated in-domain data. In addition, it is obvious that our unsupervised model has a much worse performance compared with our weakly supervised model. Therefore, one important limitation lies on the relying of in-domain data given that unsupervised model's performance is unsatisfactory and weakly supervised model need in-domain data for fine-tuning.

Another important limitation lies in the data used for testing. In our experiments, we only use patient notes from 5 examinees for testing, which is not a large test set. Therefore, future studies should consider validating these results with larger samples and wider variety of cases.

DDI-MuG: Multi-aspect Graphs for Drug-Drug Interaction Extraction

Jie Yang[1], Yihao Ding [1], Siqu Long[1], Josiah Poon[1], Soyeon Caren Han[1, 2*]
[1]The University of Sydney, [2] The University of Western Australia
{jyan4704, ydin0771, slon6753}@uni.sydney.edu.au,
{josiah.poon, caren.han}@sydney.edu.au,
caren.han@uwa.edu.au

Abstract

Drug-drug interaction (DDI) may lead to adverse reactions in patients, thus it is important to extract such knowledge from biomedical texts. However, previously proposed approaches typically focus on capturing sentence-aspect information while ignoring valuable knowledge concerning the whole corpus. In this paper, we propose a Multi-aspect Graph-based DDI extraction model, named DDI-MuG. We first employ a bio-specific pre-trained language model to obtain the token contextualized representations. Then we use two graphs to get syntactic information from input instance and word co-occurrence information within the entire corpus, respectively. Finally, we combine the representations of drug entities and verb tokens for the final classification. It is encouraging to see that the proposed model outperforms all baseline models on two benchmark datasets. To the best of our knowledge, this is the first model that explores multi-aspect graphs to the DDI extraction task, and we hope it can establish a foundation for more robust multi-aspect works in the future.

1 Introduction

According to statistics from the U.S. Centers of Disease Control and Prevention, from 2015 to 2018, 48.6 % of Americans used at least one prescription drug in 30 days[1]. More seriously, 20% of the elderly took more than 10 drugs simultaneously (Zhang et al., 2020). However, drug-drug interaction (DDI) may occur when patients take multiple drugs, resulting in reduced drug effectiveness or even, possibly, adverse drug reactions (ADRs) (Zhu et al., 2020). Therefore, the study of DDI extraction can be considerably important to patients' healthcare, as well as clinical research. Currently, a number of drug databases, such as DailyMed (Barrière and Gagnon, 2011), TWOSIDES (Tatonetti et al., 2012) and DrugBank (Wishart et al., 2017) can be used for retrieving DDI knowledge directly. However, with the exponential growth in biomedical literature, huge amounts of the most current and valuable knowledge remain hidden in biomedical literature (Zhang et al., 2020). Thus, the development of an automatic tool to extract DDI is an urgent need.

During the past few years, various deep learning-based approaches, such as (Liu et al., 2016; Zhang et al., 2018; Li and Ji, 2019; Ren et al., 2019; Mondal, 2020; Asada et al., 2020; Fatehifar and Karshenas, 2021; Shi et al., 2022) have been proposed to extract DDI knowledge. It is worth noting that compared with Convolutional Neural Networks (CNNs) and Long Short-Term Memory (LSTM), which are sequential-based architectures, Graph Neural Networks (GNNs) can better deal with complex structural knowledge. Based on this, Li and Ji (2019) combined a Bio-specific BERT (Devlin et al., 2019) and Graph Convolutional Network (GCN) (Kipf and Welling, 2017) to capture contextualized representation together with syntactic knowledge. Shi et al. (2022) adopted the Graph Attention Network (GAT) (Veličković et al., 2018) on an enhanced dependency graph to obtain higher-level drug representations for DDI extraction. However, as examples in Table 1, all the previous models only pay attention to the sentence-aspect features, and do not even exploit the corpus knowledge, which could cause essential clues to be overlooked.

To alleviate the issues mentioned above, in this work, we propose a multi-aspect graphs-based DDI extraction model, DDI-MuG, which can make use of the information in both sentence and corpus aspects. First, we use PubMedBERT to obtain sentence semantic representation. We then apply a GCN with an average pooling layer to capture syntactic features from the input instance, and another GCN with average pooling is employed to model

*Corresponding Author (caren.han@sydney.edu.au)
[1]https://www.cdc.gov/nchs/data/hus/2019/039-508.pdf

Proceedings of the 13th International Workshop on Health Text Mining and Information Analysis (LOUHI), pages 127 - 137
December 7, 2022 ©2022 Association for Computational Linguistics

Table 1: Summary of previous neural network-based models and our proposed model

Model	Sentence (semantic)	Sentence (syntactic)	Corpus
AB-LSTM (Sahu and Anand, 2018)	GloVe (Pennington et al., 2014)	No	No
DCNN(Liu et al., 2016)	Order embedding(Lai et al., 2016)	No	No
ASDP-LSTM (Zhang et al., 2018)	Word2Vec(Mikolov et al., 2013)	Dependency parse	No
RHCNN (Sun et al., 2019)	Bio-word emb.(Pyysalo et al., 2013)	Dependency parse	No
GCNN-DDI (Xiong et al., 2019)	Bio-word emb.(Pyysalo et al., 2013)	Dependency parse	No
BERTChem-DDI(Mondal, 2020)	BioBERT(Jinhyuk et al., 2019)	No	No
BERTDesc-DDI(Asada et al., 2020)	SciBERT(Beltagy et al., 2019)	No	No
DDI-MuG (Ours)	PubMedBERT(Gu et al., 2021)	Dependency parse	PMI

the word co-occurrence in the corpus level simultaneously. After that, an attentive pooling is used to integrate and obtain the optimal feature from the output of PubMedBERT and both sentence-aspect and corpus-aspect graphs. Finally, we employ a fully connected neural network in the output layer for the classification. Our proposed model is evaluated on two benchmark datasets: DDIExtraction-2013 (Herrero-Zazo et al., 2013) and TAC 2018 corpora (Demner-Fushman et al., 2018). Experimental results show that our proposed model improves the performance of DDI extraction effectively.

To recap, the main contributions of our work can be summarized as follows:

- We propose a novel neural model, named DDI-MuG, to exploit information from sentence-aspect and corpus-aspect graph. As far as we know, this is the first model that utilizes multi-aspect graphs for the DDI extraction task.
- We explore the effectiveness of different components in DDI-MuG. Experimental results indicate that knowledge from multi-aspect graphs are complementary, and their effective combination can largely improve the performance.
- We evaluate the proposed model on two benchmark datasets, and achieve new state-of-the-art performance on both of them.

The rest of the paper is organized as follows. First, we introduce the background in Section 1. Then, several related works are introduced in Section 2. Next, in Section 3, we explain the framework in the proposed model in detail. We then describe the two benchmark datasets, evaluation metrics, and parameters setting in Section 4. Section 5 presents the experimental results and discussion, and finally, we conclude this work in Section 6.

2 Related Works

Knowledge in many applications is exceedingly complex for a single-aspect network to learn robust representations. Multi-aspect networks have thus emerged naturally in different fields. Khan and Blumenstock (2019) developed a multi-aspect GCNs model to consider different aspects of phone networks for poverty research. They employed subspace analysis and a manifold ranking procedure in order to merge multiple views and prune the graph, respectively. Liu et al. (2020) first constructed semantic-based, syntactic-based, and sequential-based text graphs, and then utilized an inter-graph propagation to coordinate heterogeneous information among graphs. In order to exploit richer sources of graph edge information, Gong and Cheng (2019) resorted to multi-dimensional edge weights to encode edge directions. Similarly, Huang et al. (2020) used multi-dimensional edge weights to exploit multiple attributes, adapting the edge weights before entering into the next layer.

3 Methods

The architecture of the proposed model is illustrated in Figure 1. First, we obtain the contextual semantic representation of the input instances by PubMedBERT. Then, a sentence-aspect graph is constructed to encode the syntactic feature from the dependency path, while a corpus-aspect graph is used to explore word co-occurrence within the entire corpus. Based on the vocabulary and instances analysis, we find that the part-of-speech (POS) tag of words, especially words corresponding to verbs, might be helpful for the final representation. Therefore, we subsequently feed the representations of verbs and drug entities from PubMedBERT, together with the two graphs, into an

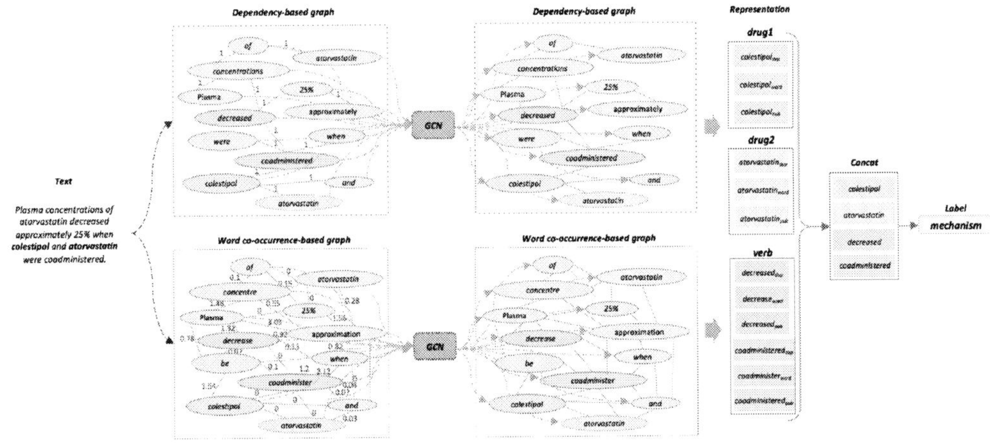

Figure 1: The proposed model architecture. This example is selected from DDIExtraction-2013 dataset. Two drugs are labeled in bold. As the space is limited, only part of the edges are shown in the word co-occurrence-based graph.

attentive pooling layer, to distinguish important features from all representations. Finally, a fully connected layer with softmax is employed to perform the classification. The process is described in the following subsections in detail.

3.1 Encoding sentences with PubMedBERT

PubMedBERT was pre-trained on 14 million biomedical abstracts with 3.2 billion words from scratch. Given an input sentence $S = [w_1, w_2, ..., w_n, ..., w_t]$ with drug entities d_1 and d_2, we convert each word w_i into word pieces and then feed them into PubMedBERT. After the PubMedBERT calculation, we employ average pooling to aggregate vectorial representations of word pieces as the word representations. We denote the two drugs and verbs representations by $drug1_{pub}$, $drug2_{pub}$, and $verbs_{pub}$ respectively.

3.2 Graph construction

Considering a graph with *n* nodes, the node *i* at the *l*-th layer is updated based on the representation of all neighborhood nodes in the *(l-1)*-th layer as follows:

$$H^l = \sigma(\hat{A}H^{l-1}W^l) \quad (1)$$

Here, $\hat{A} = \widetilde{D}^{-\frac{1}{2}}\widetilde{A}\widetilde{D}^{\frac{1}{2}}$ represents the normalized adjacency matrix, and $\widetilde{A} = A + I$ is the adjacency matrix with added self-connections. $\widetilde{D}$ is the diagonal node degree matrix with $\widetilde{D}(i,i) = \sum_j \widetilde{A}(i,j)$. $H^l \in R^{n*d_l}$ is the node embedding matrix at the *l*-th layer, *n* is the number of nodes, d_l indicates the dimension of the node features. Finally, $W^l \in R^{d_l*d_{l+1}}$ denotes a layer-specific trainable weight matrix, and σ is a nonlinear function.

For each input instance, we encode a dependency graph from the current instance and a word co-occurrence over the entire corpus.

3.2.1 Sentence-aspect dependency graph

Dependency parser is widely used in relation classification tasks with the aim of exploring syntactic information of sentence. We apply the Stanford dependency parser (Chen and Manning, 2014) to extract dependency syntactic information. Figure 2 shows the dependency relation of the input text in Figure 1. The connection from *coadministered* to *colestipol* means that *coadministered* is the head word of *colestipol*, and *"nsubjpass"* denotes the *"passive nominal subject"* dependency relation between the two words. We use the word embedding from PubMedBERT as the initial node representations, and set edge weights as 0 or 1 to indicate if two nodes are connected in the dependency path.

Let the node representations in *l*-th layer of the dependency graph be M^l. We apply two graph convolutional layers to update each node, thus the updated M^2 is expressed as follows:

$$M^2 = \sigma(\hat{A}M^1W^2) \quad (2)$$

Then, an average pooling layer is applied to get the syntactic-based sentence embedding. Let $d_1, d_2, ..., d_n, ..., d_t$ be the updated node representations obtained from graph convolutional layers, the output of dependency graph, G_{Dep}, is shown as:

$$G_{Dep} = \underset{1 \leq i \leq t}{avg}[d_i] \tag{3}$$

We denote the outputs of drug and verbs representations as $drug1_{dep}$, $drug2_{dep}$, and $verbs_{dep}$, respectively.

3.2.2 Corpus-aspect word co-occurrence graph

Information on the co-occurrence of words indicates the connection between them, such as whether they form as a common phrase or provide clues for classification tasks. Firstly, we first lemmatize each word with Natural Language Toolkit (NLTK) [2]. Then we connect all word pairs in graph, and employ point-wise mutual information (PMI) (Turney, 2001), a word associations measure, to store the word correlation information as an edge weight as follows:

$$A_{ij} = \begin{cases} 1, & i = j \\ PMI(i,j), & i \neq j, PMI(i,j) > 0 \\ 0, & i \neq j, PMI(i,j) \leq 0 \end{cases} \tag{4}$$

The PMI between any two words is calculated as:

$$PMI(i,j) = log\frac{p(i,j)}{p(i)p(j)}, \tag{5}$$

$$p(i,j) = \frac{\#W(i,j)}{\#W}, p(i) = \frac{\#W(i)}{\#W}. \tag{6}$$

where *i, j* are words, *#W(i,j)* is the number of examples in a fixed sliding window that contains both words, *#W(i)* is the number of instances in the sliding window that contain word *i*, and *#W* is the total number of sliding windows. It is worth noting that the entire input sentence is set as the sliding window. Suppose there are 31,738 instances in the corpus, and the word of *"decrease"* and *"coadminister"* appear 1,821 and 953 times respectively, and that they occur 27 times together in the whole corpus. Based on Formula 5 to 6, the PMI between this two words is -4.8. A positive PMI value corresponds to a high correlation between two words, while a negative value means that the two words have a small probability or no probability of occurrence. When two words have a negative PMI value, we view them as non-co-occurring and set their edge weight as 0.

Suppose the node representations in *l*-th layer is N^l. Similar to the dependency graph, the updated N^2 is shown as:

$$N^2 = \sigma(\hat{A}N^1W^2) \tag{7}$$

After an average pooling layer was utilized to get the word co-occurrence-based embedding, the G_{Word} graph is expressed as:

$$G_{Word} = \underset{1 \leq i \leq t}{avg}[w_i] \tag{8}$$

where w_i is the updated *l*-th node representation from graph convolutional layers.

Drug and verbs representations, denotes by $drug1_{word}$, $drug2_{word}$, and $verbs_{word}$, are extracted from G_{Word} and used as input for the next layer.

3.3 Attentive pooling layer

So far, given two drug entities and verbs, we have obtained rich feature representations from PubMedBERT and two graphs. As each instance has a different number of verbs, we apply an attentive pooling to get a fixed-length representation for verbs. In detail, this pooling mechanism computes the weights of feature vectors by using an attention mechanism, allowing it to learn the most significant feature effectively. Let A_{drug1} and A_{drug2} be the combined representation of drug entities from PubMedBERT and the two graphs, and A_{verbs} be the corresponding verbs representation:

$$A_{drug1} = [drug1_{pub}; drug1_{dep}; drug1_{word}] \tag{9}$$

$$A_{drug2} = [drug2_{pub}; drug2_{dep}; drug2_{word}] \tag{10}$$

$$A_{verbs} = [verbs_{pub}; verbs_{dep}; verbs_{word}] \tag{11}$$

where [;] denotes concatenation. These three representations are fed into the attentive pooling layer separately as follows:

$$H_{drug1} = tanh(A_{drug1}) \tag{12}$$

$$\alpha = Softmax(w^a H_{drug1}) \tag{13}$$

$$z_{drug1} = \alpha A_{drug1} \tag{14}$$

where w^a is the learning parameter, α is the attention weights. z_{drug1}, z_{drug2} and z_{verbs} are the representation of the two drugs and verbs, as the output of the attentive pooling layer.

[2]https://www.nltk.org/

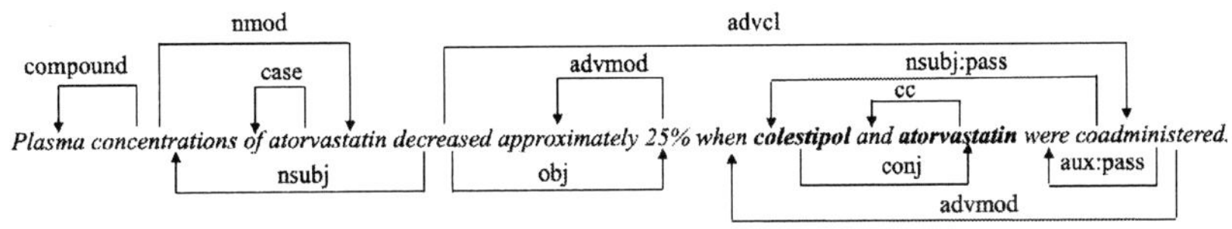

Figure 2: An example of dependency relation. Two drugs are labeled in bold.

3.4 Fully connected and softmax layer

In this layer, the updated representation of two drugs and verbs are concatenated as z_{total}, and a nonlinear activation functions *tanh* is then applied over z_{total} into a fully connected layer. Finally, we deploy a softmax with a dropout layer to get the probability score for each class. The process is expressed as follows:

$$z_{total'} = tanh(z_{total}) \quad (15)$$

$$p(y|x) = Softmax(W^s z_{total'} + b^s) \quad (16)$$

where $z_{total'}$ is the output of the fully connected layer, W^s and b^s are the softmax matrix and the bias parameter, respectively.

4 Experiments

In our experiments, two public DDI extraction corpora, i.e., DDIExtraction-2013 and TAC 2018, were used to evaluate the proposed model. This section introduces the two corpora in detail and then presents the evaluation metrics and parameters setting.

4.1 DDIExtraction-2013 dataset

We obtained the corpus from the challenge SemEval-2013 Task 9 (Segura-Bedmar et al., 2013). This corpus is the major dataset that can be used to evaluate and compare the performance of DDI extraction models. It contains manually annotated sentences from 175 abstracts in MedLine[3], and 730 abstracts in DrugBank[4]. There are four kinds of positive interaction types: *Advice, Effect, Mechanism, Int.* If the two drugs are unrelated, their relations are labeled as *Negative*. The definitions of the five types are as follows:

- **Advice**: a recommendation or advice regarding the simultaneous use of two drugs is described between two drugs.
- **Effect**: an effect or a pharmacodynamic mechanism is described between two drugs.
- **Mechanism**: a pharmacokinetic mechanism is described between two drugs.
- **Int**: a DDI occurs between two drugs, but no additional information is provided.
- **Negative**: there is no interaction between two drugs.

The original corpus suffers from a serious data imbalance problem. For example, the ratio of *Int* to *Negative* instances in the training set is 1:123.7, which heightens the difficulty of classifying drug pairs that hold *Int* relations, and continually affect the overall performance. To alleviate this data imbalance issue, many negative examples are filtered out in earlier studies, e.g., (Kim et al., 2015; Liu et al., 2016; Zhao et al., 2016; Wang et al., 2017; Sahu and Anand, 2018; Zhu et al., 2020). To ensure that the experiment results can be compared fairly with other baseline models, we adopted three rules in (Liu et al., 2016) to remove negative instances:

- If both drugs have the same name, remove the corresponding instances. The assumption is that drug will not interact with itself.
- If one drug is a particular case or an abbreviation of the other, filter out the corresponding instances. Several patterns, such as *"DRUG-A (DRUG-B)"* and *"DRUG-A such as DRUG-B"*, are used to identify such cases.
- If both drugs appear in the same coordinate structure, filter out the corresponding instances. Also, we use some pre-defined patterns, like *"DRUG-A,* $(DRUG - N)^+$*, DRUG-B"*, to filter out such instances.

Table 2 summarizes the statistics and divisions of this corpora.

4.2 TAC 2018 corpus

One of the tasks in "Drug-Drug Interaction Extraction from Drug Labels" track of the Text Analysis

[3]https://www.nlm.nih.gov/bsd/medline.html

[4]https://go.drugbank.com/

Table 2: The statistics of DDIExtraction-2013 corpus.

		Training		Test	
		Original	Filtered	Original	Filtered
Positive	Advice	826	824	221	221
	Effect	1,687	1,676	360	358
	Mechanism	1,319	1,309	302	301
	Int	188	187	96	96
Negative		23,772	19,342	4,737	3,896
Overall		27,792	23,338	5,716	4,872

Conference (TAC) 2018[5] was to detect and extract DDIs from structured product labelings (SPLs). The organizers provided a set of 22 SPLs for training (Training-22). Two other datasets containing 57 and 66 SPLs were provided as test sets. The organizers also provided an additional 180 SPLs (NLM-180) to supplement the training set. Interactions in this corpus are classified into one of the following three types:

- **Pharmacokinetic**: This type includes phrases that demonstrate changes in physiological functions (Demner-Fushman et al., 2018), such as *decrease exposure, increased bioavailability.*
- **Pharmacodynamic**: This type includes phrases that describe the effects of the drugs, e.g., *blood pressure lowering.*
- **Unspecified**: This type corresponds to caution phrases, e.g., *avoid use.*

As the original corpus is in .XML format, we use the dataset in the KLncLSTMsentClf model (Baruah and Kolla, 2018) to train and evaluate our proposed model. In total, we obtain 6,436 training sentences by merging the training-22 and NLM-180 corpora. The two test sets contain 8,205 and 4,256 sentences, respectively.

4.3 Evaluation metrics

precision(P), recall(R) and *F-score(F)* are the major evaluation metrics in the DDI extraction task. In this paper, we adopt the standard micro-average *precision, recall* and *F-score* to evaluate the performance and the formulas are listed as follows:

$$Precision = \frac{TP}{(TP+FP)}, \tag{17}$$

$$Recall = \frac{TP}{(TP+FN)}, \tag{18}$$

[5]https://tac.nist.gov/2018/

$$F-score = \frac{2*P*R}{(P+R)}. \tag{19}$$

TP(true positive) represents the number of correctly classified positive instances, FP(false positive) denotes the number of negative instances that are misclassified as positive instances, and FN(false negative) is the number of positive instances that are misclassified as negative ones.

4.4 Parameters setting

In our experiment, PyTorch library (Paszke et al., 2019) is used as the computational framework. As there is no development or validation set in the original corpus, we randomly select 20% of the training dataset as the validation set to adjust the model parameters, and the remaining 80% as the training set. The parameters used are shown as follow:

- Maximal length n = 128.
- Embedding size of PubMedBERT m_1 = 768.
- Hidden layer dimension of dependency and co-occurrence graph m_2 & m_3 = 200.
- Mini-batch size = 32.
- Dropout rate p = 0.1.
- Learning rage lr = 0.0001.
- Number of epoch = 10.

5 Results and Discussion

5.1 Results on DDIExtraction-2013

5.1.1 Comparison with baseline methods

We compare the performance of our DDI-MuG with 11 baseline methods. The comparison results of different models are showed in Table 3. The highest value is labeled in bold, and the second highest value is marked underline. In general, deep neural network-based approaches achieve better performance than statistical ML-based methods. It demonstrates the capability and potential of utilizing neural network in DDI extraction task. A notable exception is that the F1-score of SVM-DDI (Kim et al., 2015) is slightly higher than the AB-LSTM model (Sahu and Anand, 2018). This might be due to SVM-DDI (Kim et al., 2015) benefiting from rich and complex lexical and syntactic handcraft features. It can be seen that our DDI-MuG obtains the best overall performances in view of precision and F1-score. In terms of the performances for all four types, DDI-MuG performs best

Table 3: Performance Comparisons on DDIExtraction-2013 Corpus. The highest value is labeled in bold, and the second highest value is marked underline.

Methods	Breakdown F1				Overall performance		
	Advice	Effect	Mechanism	Int	Precision	Recall	F1
Statistical ML-based methods							
UTurKu(Björne et al., 2013)	0.630	0.600	0.582	0.507	0.732	0.499	0.594
WBI(Thomas et al., 2013)	0.632	0.610	0.618	0.510	0.642	0.579	0.609
FBK-irst(Chowdhury and Lavelli, 2013)	0.692	0.628	0.679	0.547	0.646	0.656	0.651
SVM-DDI(Kim et al., 2015)	0.725	0.662	0.693	0.483	-	-	0.670
Deep neural network-based methods							
AB-LSTM(Sahu and Anand, 2018)	0.697	0.683	0.681	0.542	0.678	0.659	0.669
DCNN(Liu et al., 2016)	0.777	0.693	0.702	0.464	0.757	0.647	0.698
Joint AB-LSTM(Sahu and Anand, 2018)	0.794	0.676	0.763	0.431	0.734	0.697	0.715
ASDP-LSTM (Zhang et al., 2018)	0.803	0.718	0.740	0.543	0.741	0.718	0.729
RHCNN (Sun et al., 2019)	0.805	0.734	0.782	0.589	0.773	0.737	0.754
GCNN-DDI (Xiong et al., 2019)	0.835	0.758	0.794	0.514	0.801	0.740	0.770
DREAM(Shi et al., 2022)	0.848	0.761	0.816	0.551	0.823	0.747	0.783
Our methods							
DDI-MuG(with word. graph)	0.893	0.812	<u>0.871</u>	<u>0.599</u>	<u>0.868</u>	0.805	0.835
DDI-MuG(with dep. graph)	<u>0.900</u>	**0.826**	0.865	0.583	0.842	**0.835**	<u>0.839</u>
DDI-MuG	**0.907**	<u>0.823</u>	**0.893**	**0.606**	**0.870**	<u>0.824</u>	**0.847**

on *Advice*, *Mechanism* and *Int*, and obtain the second best performance on *Effect*. It is worth noting that all methods achieve relatively low performance on *Int*. This discrepancy might be caused by the insufficient training samples of *Int*, which leads to these models to be underfitting.

Then, we find out the contributions of multi-aspect graphs to the proposed model. By removing in turn the sentence-aspect dependency graph and corpus-aspect word co-occurrence graph, our method reduces to DDI-MuG(with word. graph) and DDI-MuG(with dep. graph), respectively. From Table 3, we can see that the F1-score of DDI-MuG(with dep. graph) is higher than the F1-score of DDI-MuG(with word. graph), which proves that the syntactic features are indeed valuable for identifying the interaction relation between two drugs. Overall, it can be seen that the F1-score of DDI-MuG surpass the DDI-MuG(with word. graph) and DDI-MuG(with dep. graph) by 0.012 and 0.008, seperately. This indicates that multi-aspect graphs are complementary to each other, and together can serve as an appropriate supplement to contextual information.

5.1.2 Impact of pre-trained embedding

To evaluate the efficiency of the pre-trained language model, we conduct the experiments of replacing PubMedBERT with other similar models. As shown in Table 4, the four bio-specific models, i.e., BioBERT, SciBERT, ouBioBERT(Wada et al., 2020), and PubMedBERT, leading to improvement over standard BERT. DDI-MuG by PubMedBERT achieves the best result for the reason that it was pre-trained on biomedical texts from scratch.

5.1.3 Error analysis

In addition to present the above achievements, it is necessary to discuss the limitations of our approach. One common type of error is that the four kinds of positive instances are often misclassified as negative instances. This is due to the imbalanced data that small instances categories being misclassified as large instance categories. There is another notable error that 34.4% of *Int* type instances are misclassified as *Effect* type. This is because that some *Int* instances have similar semantics to *Effect* instances. For example, in the following two instances:

- *"**arbiturates** may decrease the effectiveness of oral contraceptives, certain antibiotics, quinidine, **theophylline**, corticosteroids, anticoagulants, and beta blockers."*
- *"**sulfoxone** may increase the effects of **barbiturates**, tolbutamide, and uricosurics."*

The words *decrease* and *increase* are the clues for identifying interactions in the two semantically close sentences. However, the first instance belongs to the *Int* type, while the second belongs to *Effect*. The number of *Int* instances is far smaller than the number of *Effect* instances, which also leads to the occurrence of this kind of mistake.

Table 4: The effect of pre-trained embedding. The highest value is labeled in bold.

Pre-trained embedding	P	R	F1
DDI-MuG(by BERT)	0.801	0.790	0.795
DDI-MuG(by BioBERT)	0.843	0.816	0.829
DDI-MuG(by SciBERT)	0.839	0.825	0.832
DDI-MuG(by ouBioBERT)	0.850	**0.826**	0.838
DDI-MuG(by PubMedBERT)	**0.870**	0.824	**0.847**

5.1.4 Are verb representations really helpful?

In our previous vocabulary and instances analysis, we found that in the DDIExtraction-2013 corpus, when instances contain the words *inhibit, increased, decreased*, there is a great possibility that the drug pair has the *Mechanism* relation. On the other hand, when instances contain *avoided, recommended* or *administered*, the drug pair is likely to have the *Advice* relation.

Thus, to further investigate how the verbs are important for the final classification, we studied the effect of extracting DDI only from the drug information, without using the verbs knowledge. Table 5 shows the comparison of the performance with and without the verbs information. This result indicates verbs representation can serve as a supplement to improve the model performance.

Table 5: The comparison of with or without verbs information. The highest value is labeled in bold.

	Precision	Recall	F-score
DDI-MuG(drug-only)	0.863	0.823	0.843
DDI-MuG(all)	**0.870**	**0.824**	**0.847**

5.2 Results on TAC 2018

5.2.1 Comparison with baseline model

Since we use the same dataset as KLncLSTMsentClf (Baruah and Kolla, 2018), we view it as the baseline model. From Table 6, we can see that our proposed model achieves better results in both two test sets, which indicates the transferability of our proposed model.

6 Conclusions

In this paper, we propose DDI-MuG, a novel multi-aspect graphs framework for DDI extraction task. Concretely, a bio-specific pre-trained language model, PubMedBERT, is firstly employed to encode the context information of each word from the aspect of sentence semantic information. Then, two graphs are utilized to explore sentence syntactic and corpus word co-occurrence information, respectively. After that, attentive pooling mechanism is employed to update the representations of drug entities and verbs. Finally, by feeding the concatenated representation of the two drugs and verbs into a fully connected and softmax classifier, the interaction between two drugs is obtained. Extensive comparison experiments with baseline models on two public datasets verify the effectiveness of utilizing multi-aspect graphs in the DDI extraction task.

For the future work, there are at least two directions could be considered. Firstly, the performance on categories with small training samples, like *Int* in the DDIExtraction-2013 corpora, is unsatisfactory. The solution of contrastive learning can be explored. Secondly, drug knowledge from external databases could be integrated in the architecture for richer drug representations.

Table 6: Comparison with baseline models on the TAC 2018 corpus. The highest value is labeled in bold.

Dataset	Model	P	R	F1
Test1	KLncLSTMsentClf	0.470	0.620	0.530
Test1	DDI-MuG(with word. graph)	0.717	0.712	0.715
Test1	DDI-MuG(with dep. graph)	0.688	0.718	0.703
Test1	DDI-MuG(all)	**0.721**	**0.728**	**0.723**
Test2	KLncLSTMsentClf	0.490	0.670	0.567
Test2	DDI-MuG(with word. graph)	0.710	0.726	0.718
Test2	DDI-MuG(with dep. graph)	0.713	0.730	0.721
Test2	DDI-MuG(all)	**0.717**	**0.743**	**0.729**

References

Masaki Asada, Makoto Miwa, and Yutaka Sasaki. 2020. Using drug descriptions and molecular structures for drug–drug interaction extraction from literature. *Bioinformatics*, 37(12):1739–1746.

Caroline Barrière and Michel Gagnon. 2011. Drugs and disorders: From specialized resources to web data. In *Workshop on Web Scale Knowledge Extraction, 10th International Semantic Web Conference*.

Gaurav Baruah and Maheedhar Kolla. 2018. Klicklabs at the TAC 2018 drug-drug interaction extraction from drug labels track. In *Proceedings of the 2018 Text Analysis Conference, TAC 2018, Gaithersburg, Maryland, USA, November 13-14, 2018*. NIST.

Iz Beltagy, Kyle Lo, and Arman Cohan. 2019. Scibert: Pretrained language model for scientific text. In *Proceedings of the 2019 Conference on Empirical Methods in Natural Language Processing and the 9th International Joint Conference on Natural Language Processing*, pages 3615–3620, Hong Kong, China. Association for Computational Linguistics.

Jari Björne, Suwisa Kaewphan, and Tapio Salakoski. 2013. UTurku: Drug named entity recognition and drug-drug interaction extraction using SVM classification and domain knowledge. In *Second Joint Conference on Lexical and Computational Semantics (*SEM), Volume 2: Proceedings of the Seventh International Workshop on Semantic Evaluation (SemEval 2013)*, pages 651–659, Atlanta, Georgia, USA. Association for Computational Linguistics.

Danqi Chen and Christopher Manning. 2014. A fast and accurate dependency parser using neural networks. In *Proceedings of the 2014 Conference on Empirical Methods in Natural Language Processing (EMNLP)*, pages 740–750, Doha, Qatar. Association for Computational Linguistics.

Md. Faisal Mahbub Chowdhury and Alberto Lavelli. 2013. FBK-irst : A multi-phase kernel based approach for drug-drug interaction detection and classification that exploits linguistic information. In *Second Joint Conference on Lexical and Computational Semantics (*SEM), Volume 2: Proceedings of the Seventh International Workshop on Semantic Evaluation (SemEval 2013)*, pages 351–355, Atlanta, Georgia, USA. Association for Computational Linguistics.

Dina Demner-Fushman, Kin Wah Fung, Phong Do, Richard David Boyce, and Travis R. Goodwin. 2018. Overview of the tac 2018 drug-drug interaction extraction from drug labels track. *Theory and Applications of Categories*.

Jacob Devlin, Ming-Wei Chang, Kenton Lee, and Kristina Toutanova. 2019. BERT: Pre-training of deep bidirectional transformers for language understanding. In *Proceedings of the 2019 Conference of the North American Chapter of the Association for Computational Linguistics: Human Language Technologies, Volume 1 (Long and Short Papers)*, pages 4171–4186, Minneapolis, Minnesota. Association for Computational Linguistics.

Mohsen Fatehifar and Hossein Karshenas. 2021. Drug-drug interaction extraction using a position and similarity fusion-based attention mechanism. *Journal of Biomedical Informatics*, 115:103707.

Liyu Gong and Qiang Cheng. 2019. Exploiting edge features for graph neural networks. In *2019 IEEE/CVF Conference on Computer Vision and Pattern Recognition (CVPR)*, pages 9203–9211.

Yu Gu, Robert Tinn, Hao Cheng, Michael Lucas, Naoto Usuyama, Xiaodong Liu, Tristan Naumann, Jianfeng Gao, and Hoifung Poon. 2021. Domain-specific language model pretraining for biomedical natural language processing. *ACM Trans. Comput. Healthcare*, 3(1).

María Herrero-Zazo, Isabel Segura-Bedmar, Paloma Martínez, and Thierry Declerck. 2013. The ddi corpus: An annotated corpus with pharmacological substances and drug–drug interactions. *Journal of Biomedical Informatics*, 46(5):914 – 920.

Zhichao Huang, Xutao Li, Yunming Ye, and Michael K. Ng. 2020. Mr-gcn: Multi-relational graph convolutional networks based on generalized tensor product. In *Proceedings of the Twenty-Ninth International Joint Conference on Artificial Intelligence, IJCAI-20*, pages 1258–1264. International Joint Conferences on Artificial Intelligence Organization.

Lee Jinhyuk, Yoon Wonjin, and Kim. 2019. Biobert: a pre-trained biomedical language representation model for biomedical text mining. *Bioinformatics*, 36(4):1234–1240.

Muhammad Raza Khan and Joshua E. Blumenstock. 2019. Multi-gcn: Graph convolutional networks for multi-view networks, with applications to global poverty. *Proceedings of the AAAI Conference on Artificial Intelligence*, 33(01):606–613.

Sun Kim, Haibin Liu, Lana Yeganova, and W. John Wilbur. 2015. Extracting drug–drug interactions from literature using a rich feature-based linear kernel approach. *Journal of Biomedical Informatics*, 55:23–30.

Thomas N. Kipf and Max Welling. 2017. Semi-supervised classification with graph convolutional networks. In *5th International Conference on Learning Representations*, Toulon, France. OpenReview.net.

Siwei Lai, Kang Liu, Shizhu He, and Jun Zhao. 2016. How to generate a good word embedding. *IEEE Intelligent Systems*, 31(6):5–14.

Diya Li and Heng Ji. 2019. Syntax-aware multi-task graph convolutional networks for biomedical relation extraction. In *Proceedings of the Tenth International Workshop on Health Text Mining and Information Analysis (LOUHI 2019)*, pages 28–33, Hong Kong. Association for Computational Linguistics.

Shengyu Liu, Buzhou Tang, Qingcai Chen, and Xiaolong Wang. 2016. Drug-drug interaction extraction via convolutional neural networks. *Computational and Mathematical Methods in Medicine*, 2016.

Xien Liu, Xinxin You, Xiao Zhang, Ji Wu, and Ping Lv. 2020. Tensor graph convolutional networks for text classification. *Proceedings of the AAAI Conference on Artificial Intelligence*, 34(05):8409–8416.

Tomas Mikolov, Wen-tau Yih, and Geoffrey Zweig. 2013. Linguistic regularities in continuous space word representations. In *Proceedings of the 2013 Conference of the North American Chapter of the Association for Computational Linguistics: Human Language Technologies*, pages 746–751, Atlanta, Georgia. Association for Computational Linguistics.

Ishani Mondal. 2020. BERTChem-DDI : Improved drug-drug interaction prediction from text using chemical structure information. In *Proceedings of Knowledgeable NLP: the First Workshop on Integrating Structured Knowledge and Neural Networks for NLP*, pages 27–32, Suzhou, China. Association for Computational Linguistics.

Adam Paszke, Sam Gross, Francisco Massa, Adam Lerer, James Bradbury, Gregory Chanan, Trevor Killeen, Zeming Lin, Natalia Gimelshein, Luca Antiga, Alban Desmaison, Andreas Köpf, Edward Yang, Zach DeVito, Martin Raison, Alykhan Tejani, Sasank Chilamkurthy, Benoit Steiner, Lu Fang, Junjie Bai, and Soumith Chintala. 2019. *PyTorch: An Imperative Style, High-Performance Deep Learning Library*. Curran Associates Inc., Red Hook, NY, USA.

Jeffrey Pennington, Richard Socher, and Christopher Manning. 2014. GloVe: Global vectors for word representation. In *Proceedings of the 2014 Conference on Empirical Methods in Natural Language Processing (EMNLP)*, pages 1532–1543, Doha, Qatar. Association for Computational Linguistics.

Sampo Pyysalo, F Ginter, Hans Moen, T Salakoski, and Sophia Ananiadou. 2013. Distributional semantics resources for biomedical text processing. *Proceedings of Languages in Biology and Medicine*, pages 39–44.

Yafeng Ren, Hao Fei, and Donghong Ji. 2019. Drug-drug interaction extraction using a span-based neural network model. In *2019 IEEE International Conference on Bioinformatics and Biomedicine (BIBM)*, pages 1237–1239.

Sunil Kumar Sahu and Ashish Anand. 2018. Drug-drug interaction extraction from biomedical texts using long short-term memory network. *Journal of Biomedical Informatics*, 86:15–24.

Isabel Segura-Bedmar, Paloma Martínez, and María Herrero-Zazo. 2013. SemEval-2013 task 9 : Extraction of drug-drug interactions from biomedical texts (DDIExtraction 2013). In *Second Joint Conference on Lexical and Computational Semantics (*SEM), Volume 2: Proceedings of the Seventh International Workshop on Semantic Evaluation (SemEval 2013)*, pages 341–350, Atlanta, Georgia, USA. Association for Computational Linguistics.

Yong Shi, Pei Quan, Tianlin Zhang, and Lingfeng Niu. 2022. Dream: Drug-drug interaction extraction with enhanced dependency graph and attention mechanism. *Methods*, 203:152–159.

Xia Sun, Ke Dong, Long Ma, Richard Sutcliffe, Feijuan He, Sushing Chen, and Jun Feng. 2019. Drug-drug interaction extraction via recurrent hybrid convolutional neural networks with an improved focal loss. *Entropy*, 21(1).

Nicholas P. Tatonetti, Patrick P. Ye, Roxana Daneshjou, and Russ B. Altman. 2012. Data-driven prediction of drug effects and interactions. *Science translational medicine*, 4(125).

Philippe Thomas, Mariana Neves, Tim Rocktäschel, and Ulf Leser. 2013. WBI-DDI: Drug-drug interaction extraction using majority voting. In *Second Joint Conference on Lexical and Computational Semantics (*SEM), Volume 2: Proceedings of the Seventh International Workshop on Semantic Evaluation (SemEval 2013)*, pages 628–635, Atlanta, Georgia, USA. Association for Computational Linguistics.

Peter Turney. 2001. Mining the web for synonyms: Pmi-ir versus lsa on toefl. *Proceedings of the Twelfth European Conference on Machine Learning*, pages 491–502.

Petar Veličković, Guillem Cucurull, Arantxa Casanova, Adriana Romero, Pietro Liò, and Yoshua Bengio. 2018. Graph attention networks. In *International Conference on Learning Representations*.

Shoya Wada, Toshihiro Takeda, Shiro Manabe, Shozo Konishi, Jun Kamohara, and Yasushi Matsumura. 2020. A pre-training technique to localize medical bert and to enhance biomedical bert.

Wei Wang, Xi Yang, Canqun Yang, Xiaowei Guo, Xiang Zhang, and Chengkun Wu. 2017. Dependency-based long short term memory network for drug-drug interaction extraction. *BMC bioinformatics*, 18(Suppl 16):578–578.

David S Wishart, Yannick D Feunang, An C Guo, Elvis J Lo, Ana Marcu, Jason R Grant, Tanvir Sajed, Daniel Johnson, Carin Li, Zinat Sayeeda, Nazanin Assempour, Ithayavani Iynkkaran, Yifeng Liu, Adam Maciejewski, Nicola Gale, Alex Wilson, Lucy Chin, Ryan Cummings, Diana Le, Allison Pon, Craig Knox, and Michael Wilson. 2017. DrugBank 5.0: a major update to the DrugBank database for 2018. *Nucleic Acids Research*, 46(D1):D1074–D1082.

W. Xiong, F. Li, H. Yu, and D. Ji. 2019. Extracting drug-drug interactions with a dependency-based graph convolution neural network. In *2019 IEEE International Conference on Bioinformatics and Biomedicine (BIBM)*, pages 755–759, Los Alamitos, CA, USA. IEEE Computer Society.

Tianlin Zhang, Jiaxu Leng, and Ying Liu. 2020. Deep learning for drug–drug interaction extraction from the literature: a review. *Briefings in bioinformatics*, 21(5):1609–1627.

Yijia Zhang, Wei Zheng, Hongfei Lin, Jian Wang, Zhihao Yang, and Michel Dumontier. 2018. Drug–drug interaction extraction via hierarchical RNNs on sequence and shortest dependency paths. *Bioinformatics*, 34(5):828–835.

Zhehuan Zhao, Zhihao Yang, Ling Luo, Hongfei Lin, and Jian Wang. 2016. Drug drug interaction extraction from biomedical literature using syntax convolutional neural network. *Bioinformatics (Oxford, England)*, 32(22):3444–3453.

Yu Zhu, Lishuang Li, Hongbin Lu, Anqiao Zhou, and Xueyang Qin. 2020. Extracting drug-drug interactions from texts with biobert and multiple entity-aware attentions. *Journal of Biomedical Informatics*, 106:103451.

Divide and Conquer: An Extreme Multi-Label Classification Approach for Coding Diseases and Procedures in Spanish

Jose Barros
DCC, University of Chile
jose.barros.s@ug.uchile.cl

Matías Rojas
CMM, University of Chile
matias.rojas.g@ug.uchile.cl

Andrés Abeliuk
DCC, University of Chile
aabeliuk@dcc.uchile.cl

Jocelyn Dunstan
CMM, University of Chile
jdunstan@uchile.cl

Abstract

Clinical coding is the task of transforming medical documents into structured codes following a standard ontology. Since these terminologies are composed of hundreds of codes, this problem can be considered an Extreme Multi-label Classification task. This paper proposes a novel neural network-based architecture for clinical coding. First, we take full advantage of the hierarchical nature of ontologies to create clusters based on semantic relations. Then, we use a *Matcher* module to assign the probability of documents belonging to each cluster. Finally, the *Ranker* calculates the probability of each code considering only the documents in the cluster. This division allows a fine-grained differentiation within the cluster, which cannot be addressed using a single classifier. In addition, since most of the previous work has focused on solving this task in English, we conducted our experiments on three clinical coding corpora in Spanish. The experimental results demonstrate the effectiveness of our model, achieving state-of-the-art results on two of the three datasets. Specifically, we outperformed previous models on two subtasks of the CodiEsp shared task: CodiEsp-D (diseases) and CodiEsp-P (procedures). Automatic coding can profoundly impact healthcare by structuring critical information written in free text in electronic health records.

1 Introduction

The International Classification of Diseases (ICD) is a medical glossary published by the World Health Organization, which establishes specific coding rules for healthcare procedures and diseases. Mapping electronic health records into alphanumeric codes allows rapid summarization of information, which is necessary to calculate costs, support clinical decisions, and conduct detailed epidemiological studies. However, manual coding is time-consuming, resource-intensive, and error-prone, even for specialists. For this reason, developing tools to support this task is precious.

Extreme multi-label classification is a subset of the multi-label classification task in which the objective is to learn feature architectures and classifiers that can automatically tag a data point with the most relevant labels from a huge label set (Bhatia et al., 2016). The term extreme is justified in this case since the space of possible labels is generally very large and can exceed the number of documents in a given corpus.

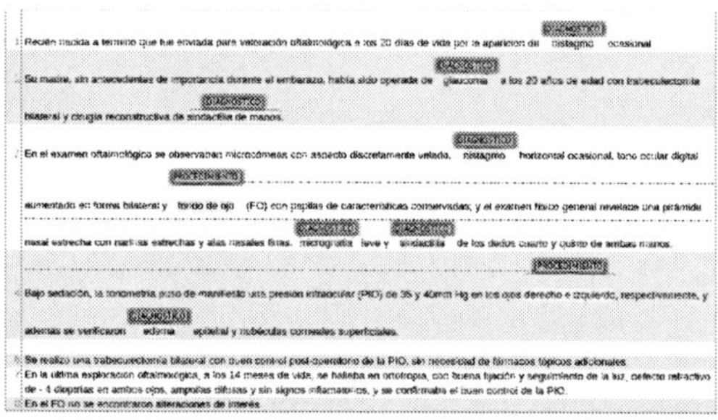

Figure 1: Example of a CodiEsp Electronic Health Record annotated, every diagnostic and procedure mention has a unique code. Every code from this Electronic Health Record is aggregated, and the document is labeled with all the codes present in the document. Each entity mention and its span is later used in the different data augmentation techniques explained in 3.6. Figure extracted from (Miranda-Escalada et al., 2020b).

Clinical coding is an important Natural Language Processing (NLP) task that seeks to automatically assign codes to medical documents following a standard terminology, such as the ICD glossary. Since each document can be labeled with more than one code from an extensive list, this task can be considered an Extreme Multi-label Classification task (Liu et al., 2017). An example of a Codiesp-D medical document is shown in Figure 1. Despite the availability of clinical coding datasets in Spanish, such as CANTEMIST (Miranda-Escalada

Proceedings of the 13th International Workshop on Health Text Mining and Information Analysis (LOUHI), pages 138 - 147
December 7, 2022

et al., 2020a) or CodiEsp (Miranda-Escalada et al., 2020b), the size of these resources is not yet comparable to that for the English language. For example, Codiesp has 1,000 clinical case reports, while the MIMIC-III dataset (Johnson et al., 2016) has 52.726 discharge summaries. This scarcity of data forces models in other languages to have different architectures than those trained on English datasets.

To fill this gap, we introduce a novel architecture for solving clinical coding on three Spanish clinical datasets. This architecture is composed of two modules: the Ranker and the Matcher. The first module calculates the probability of a document belonging to a cluster, while the second performs the classification of documents into codes. Each cluster is created previously by analyzing the ontology of each dataset. Our experimental results show the effectiveness of our model, achieving state-of-the-art performance on CodiEsp-D (diseases) and CodiEsp-P (procedures) according to the Mean Average Precision (MAP) and F_1 score.

2 Related Work

In recent years, there has been a growing interest from the NLP research community in studying the clinical coding task. Early work focuses on creating machine learning-based classifiers with heavy feature engineering (Larkey and Croft, 1995; Goldstein et al., 2007). However, as mentioned in Kaur et al. (2021) and Teng et al. (2022), recent deep learning advancements have greatly improved clinical coding models' performance for all languages.

Regarding extreme multi-label architectures, there is one work that heavily inspired this work, which is X-Transformers (Chang et al., 2020). It proposes creating a clusterization of labels using the distance between the label descriptions encoded using contextualized embeddings retrieved from transformers' language models. Then they predict the clusters using a transformers classifier, and finally, they predict the labels over the subset of predicted clusters using one-vs-rest linear classifiers. The design of this architecture was thought to handle corpora much larger than the ones we have studied in this work, thus prioritizing time efficiency much more.

One of the most popular datasets used for clinical coding for the Spanish language is CodiEsp (Miranda-Escalada et al., 2020b). Most of the work proposed formulated the problem as text classification. In López-Garcıa et al. (2020), they used a transformer-based model to classify the sentences of the documents. Then, the whole document set of codes is the union of the sentence-level codes. Other approaches focused on solving the problem as a Named Entity Recognition (NER) task. In Cossin and Jouhet (2020), they created a dictionary based on entity mentions and code definitions. Then, they matched spans of documents with the code definitions in the dictionary using a tree-based algorithm. Finally, other ensemble-based models combined text classification and NER tasks to solve the clinical coding problem. For example, Blanco et al. (2020) implemented a model that used string-matching encoders and one-vs-rest document classification. This model obtained the best results in the competition.

Another important task of clinical coding in Spanish is Cantemist (Miranda-Escalada et al., 2020a), which aims to identify codes present in cancer diagnoses. This task had two winner systems obtaining the same MAP score. The first model proposed by García-Pablos et al. (2020) used different transformer-based models to predict specific parts of a code. These models were ensembled using a novel voting system. The second winner was López-García et al. (2020), who reused their approach proposed in CodiEsp but further pre-trained a language model with a private oncology corpus.

Recent work by López-García et al. (2021) outperformed previous models in CodiEsp and CANTEMIST by a wide margin. First, they trained three multilingual language models using private oncology datasets and then fine-tuned these models for classifying documents into codes. To improve the performance of their models, they ensembled the results from five different instances of each trained transformer.

3 Model

Our proposed architecture comprises two main modules: the Matcher and the Ranker. The first module calculates the probability that a document belongs to some cluster, while the second one calculates the probability of codes in the document. The results of both modules are used to perform the final prediction of codes. This process is carried out by multiplying the probability of codes obtained from the Ranker for each document with the code cluster probability obtained from the Matcher module. We refer to this approach as the Divide and

emc

Code partitioning	Assigned Cluster
From a00 to b99	Some diseases caused by infections and parasites
From c00 to d49	Tumors and neoplasia
From d50 to d89	Diseases of the blood and hematopoietic organs
From h00 to h59	Diseases of the eye and its adnexa
From h60 to h95	Diseases of the ear and mastoid process

Table 1: Example of five clusters defined for CodiEsp Diagnostics.

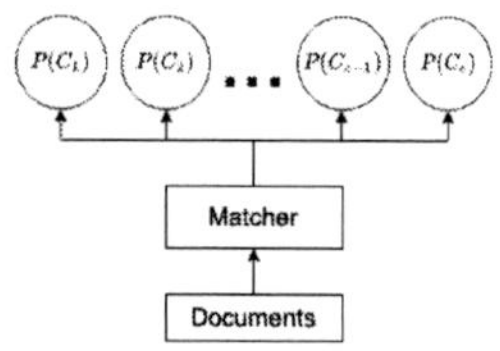

Figure 2: Overview of the Matcher module. $P(C_i)$: probability of document having a code in cluster i.

Conquer (D&C) model since dividing the original task into two simpler text classification subtasks allows us to improve the results considerably.

3.1 Clusters

As a preliminary step before training our model, we create partitions of semantically related codes based on the ontologies hierarchy. We will refer to these groups as clusters. In Table 1, we show an example of clusters defined in the CodiEsp Diagnostics corpus. Here, we used the first three letters of a code that, in the ontology, are related to a disease category.

For Codiesp-D, we created 21 clusters; for Codiesp-P 17 clusters; and for CANTEMIST 51 clusters. The clusters were defined using the categories systematized by the ontology's creators leveraging extensive work from clinicians worldwide to group semantically related codes, which gives us confidence about the quality of the selected clusters.

3.2 Matcher

As shown in Figure 2, the Matcher module assigns the probability of each document belonging to each cluster. Each document is categorized with the clusters mapped to the document labels, where each label belongs to a single cluster. This task can be formulated as multi-label text classification. Notably, the number of clusters is significantly lower than the number of labels on the corpus. For example, in the Codiesp-D subtask, the amount of different labels is 2.557, and the number of clusters is 21. This simplifies the task charged to the

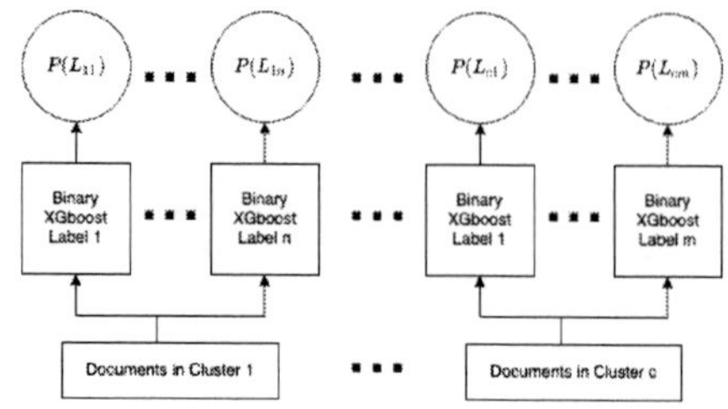

Figure 3: Overview of the Ranker module. $P(L_{ij})$: probability of document having a mention of code i in cluster j.

Matcher, classifying in fewer classes using significantly more documents per class.

To perform this classification, we decided to fine-tune a transformer-based architecture, as these models have boosted the performance of NLP architectures in several NLP tasks, including text classification. Transformers models are based entirely on attention, replacing the recurrent layers most commonly used in encoder-decoder architectures with multi-headed self-attention (Vaswani et al., 2017). This aims to draw global dependencies between input and output without the need for sequential computation of Recurrent Neural Networks (RNN) (Rumelhart et al., 1986) or Long short-term memory (LSTM) (Hochreiter and Schmidhuber, 1997).

3.3 Ranker

The Ranker module calculates, for each possible code, the probability of belonging to the document. This process is carried out by training a single binary model per code, following a one-vs-rest approach. Each model is trained only with documents with codes belonging to the cluster, which allows for a fine-grained differentiation between similar codes. This way, the gold labels of this task are the codes in the document.

Since each document can contain many codes, this problem, like the Matcher, can be formulated as a multi-label text classification task. However, this subtask is considered extreme since possible codes are much larger than the number of possible clusters in the other task. The Ranker module is based on the one-vs-rest approach, where the input documents are binary classifiers encoded using the TF-IDF method, and the output is fed into an Extreme Gradient Boosting (XGBoost) model (Chen et al., 2015).

We decided to use the Gradient Boosting Trees algorithm considering the computational cost of training one model per label and also the quality of the model's predictions. Although, as previously discussed, neural networks are the go-to choice when solving an NLP task, it is not feasible to train one neural network (specifically a Transformer or LSTM) per label due to the computational costs of training in an extreme multi-label environment. Because each cluster has fewer examples than the entire corpus, even training one neural network model per cluster yields worse results because of the data scarcity issue.

3.4 Combining output of Matcher and Ranker

Having trained both the Matcher and the Ranker, the issue of how to combine the results is left. To handle this task, we implemented two approaches; one that outputs probabilities of all labels and another that predicts the labels of the document.

First, to get the probabilities of all labels, it is important to note that the output probabilities of the Ranker are not precisely the probabilities of the label because it was trained only with documents in the cluster. More rigorously, these values can be defined as the conditional probabilities of the label given that it belongs to the cluster. Therefore, to compute the probabilities of the label, we can use the Bayes Theorem,

$$P(L) = P_{Matcher}(C) * P_{Ranker}(L|C), \quad L \in C. \tag{1}$$

Where $P_{Matcher}(C)$ is the probability that the Matcher module assigns a document to cluster C, and $P_{Ranker}(L|C)$ is the conditional probability that the Ranker module assigns a label L to a document given that it belongs to cluster C.

3.5 Ensemble

Using an ensemble of strong learners only leverages different runs of the same computationally expensive training process and thus confounds the advances obtained by creating better architectures. However, since most of the previous work proposed ensemble models, we implemented an ensemble strategy to perform a fair comparison with that systems.

The ranking of all the labels is done by summing the probabilities of the ensembled models for each label, where the prediction of the final labels is a union of the predicted labels in all the ensembled models.

3.6 Data Augmentation

In addition, to improve the performance of our models, we implemented four data augmentation techniques. Three methods are based on entity mentions associated with the codes, while the other uses code descriptions. Specifically, we added new documents as follows:

- **NE Mentions**: Each entity mentioned is a new document.
- **NE Sentences**: Each sentence in the original corpus is considered as a new document.
- **NE Stripped**: New documents only with words that participate in some entity mention.
- **Definition codes**: Each definition of a code is a new document.

The first three techniques need to have a corpus in which the different labels are associated with a span of the document (Named Entity), which is widespread because most corpora are created to solve Named Entity Recognition tasks and Text Classification tasks. Not all of these data augmentation techniques are used by the Matcher and the Ranker. In fact, the Matcher uses a transformer architecture that is trained using sentences and needs semantic context, so for the Matcher, NE Stripped would only make the results worse and is not used.

4 Experiments

4.1 Datasets

We conducted our experiments with three clinical coding corpora in Spanish. Table 2 shows a summary of descriptive statistics for each corpus: CodiEsp-D, CodiEsp-p, and CANTEMIST.

- **CodiEsp**[1] (Miranda-Escalada et al., 2020b): Corpus composed of 1,000 clinical cases manually annotated using the guidelines ICD10-Clinical Modification and ICD10-Procedure. This dataset was used for two shared tasks: CodiEsp Diagnostics (CodiEsp-D) and CodiEsp Procedures (CodiEsp-P).
- **CANTEMIST**[2] (Miranda-Escalada et al., 2020a): Corpus composed of 1,301 oncologic clinical case reports annotated using the ICD-O-3 codes.

[1] https://zenodo.org/record/3837305
[2] https://zenodo.org/record/3978041

	CodiEsp-D			CodiEsp-P			CANTEMIST		
	Train	Dev	Test	Train	Dev	Test	Train	Dev	Test
Documents	500	250	250	500	250	250	501	500	300
Avg document length	410	411	414	410	411	414	894	804	812
Avg codes per document	14.4	13.7	14.7	3.9	4.2	4.4	12.8	12	12.1
Avg clusters per document	4.9	4.9	4.8	1.9	2.0	2.0	2.8	2.8	2.8
Number of different codes	2557			870			850		
Number of clusters	21			17			51		

Table 2: Descriptive statistics of the datasets.

4.2 Settings

For ease of reading, we explain the different hyperparameters and strategies used for the Matcher and the Ranker training. We used the same data splits as in previous work (Miranda-Escalada et al., 2020b,a) to guarantee a fair comparison.

Regarding the Matcher module, we fine-tuned a Biomedical version of RoBERTa in Spanish, leaving only the last layer trainable. We trained our model during 15 epochs using the Adam with weight decay optimizer (Loshchilov and Hutter, 2017), which is an improved version of Adam (Kingma and Ba, 2014) using a batch size of 25 documents. To handle overfitting, we employed a linear scheduler with warmup, which linearly increases the learning from 0 to the max learning rate during warmup and then decreased the learning rate to 0. This module was implemented using the Flair framework (Akbik et al., 2019).

To choose the optimizer, we used the defaults of the Flair framework. The number of epochs was chosen after training on the train split and evaluating on the Codiesp-D validation split. The loss reduction stagnated at epoch 10. Given that we used a Linear Scheduler that decreases the learning rate for each epoch, we used 15 epochs to ensure that we reached the best performance.

Each one-vs-rest model of the Ranker was created using the XGBoost implementation provided by the Sklearn library (Pedregosa et al., 2011). Regarding the hyperparameters, we used the exact tree method, the ratio of the negative class to the positive class as the scaling weight, the Dart enhancer, and 60% of subsample and column subsample.

To ensure the reproducibility of our results, we released an open-source library[3] with the code of our experiments. This framework allows extending the model to other datasets by simply implementing the preprocessing data functions. Likewise, data augmentation techniques can be extended to other corpora by implementing a preprocessing function that obtains the span, the document, and the mention of a code. All the experiments were performed using a GPU Nvidia DGX A100.

[3] https://github.com/plncmm/dac-divide-and-conquer

4.3 Metrics

To evaluate the performance of our model, we compute the metrics used in previous work on CodiEsp and CANTEMIST. First, we calculate the MAP, which is a widely used evaluation score for ranking problems (Miranda-Escalada et al., 2020b) and has shown good discrimination and stability (Schütze et al., 2008). MAP is defined as the mean of the average precision (AP) of all documents:

$$AP = \frac{\sum P(k) * rel(k)}{\text{number of relevant labels}},$$

where $P(k)$ is the precision at position k, and $rel(k)$ is an indicator function equaling 1 if the item at rank k is a relevant label, zero otherwise. Second, we calculated the micro average F_1 score, corresponding to the harmonic mean of the precision and recall.

These metrics were evaluated on the test set provided by the shared tasks, so comparability to other models is assured. To correctly determine whether the differences between the performance of our model and the other models are reliable or due to statistical chance, we have done five different evaluation rounds, each with a different seed, ensuring different results. The results reported are the mean of these five evaluation rounds, and the standard deviation is also reported.

Regarding the performance of the ensemble models, the report of different evaluation rounds is unfeasible due to the high computational time cost. However, the ensemble interiorizes the statistical chance because it uses 15 different instances of the architecture.

To provide a more comprehensive analysis of the architecture, we computed metrics for each one of the modules. These metrics help us gain insights into which part of the architecture levels are acceptable and allow us to know when high scores for the architecture as a whole can be expected.

Regarding the Matcher module, we report the MAP and the F_1 score when the gold labels are the clusters. In the case of the Ranker module, we had to approach the issue of creating metrics that could evaluate its performance independently from the Matcher step, which is not straightforward. To overcome this issue, we have defined a weighted version of the metrics in which, for each cluster, we calculate the MAP and the F_1 score for that cluster's sentences. Then, the cluster metrics are aggregated and weighted by the number of sentences in each cluster. This metric indicates how well the Ranker is labeling the documents. This metric can be interpreted as what the metric would be if the Matcher had a perfect performance and thus acts as a ceiling for the DAC model's final performance.

5 Results

Table 3 shows the overall results of our model. We reported two different results for the DAC architecture: the average of five different evaluation rounds using the original approach and other results from a version that ensembled 15 different model instances.

Our base model achieves state-of-the-art results in both CodiEsp tasks, surpassing the best base model (Clinical Coding Transformers - Best (López-García et al., 2021)) by 8% in CodiEsp diagnostics and by 6% in CodiEsp procedures. Even in comparison with an ensemble of strong learners, which obtains a similar performance (Clinical Coding Transformers - Ensemble (López-García et al., 2021)), our base model surpasses their results by a small margin of 0.5% in CodiEsp Diagnostics and 0.2% in CodiEsp Procedures. Their results correspond to 15 different runs of 3 strong learners, where each language model was trained with a private oncology corpus. Unlike the mentioned work, we used only publicly available resources and a simpler architecture regarding computational cost.

Most notably, our ensemble-based version of 15 different instances of our model outperformed previous results in the CodiEsp tasks by a wide margin, outperforming state-of-the-art methods, including ensembles of strong learners, in CodiEsp-D and CodiEsp-P by 3.0% and 3.3%, respectively.

We hypothesize that the high performance of our model is explained since the original text classification task is reduced to two subtasks, where the number of possible labels is smaller. First, the Matcher module performs a text classification in which the number of labels equals the number of clusters. Second, the Ranker is trained only with documents belonging to a cluster, which allows for a fine-grained differentiation between similar codes.

We believe that the incapability to obtain state-of-the-art for CANTEMIST is because it is a simpler task than Codiesp-D and Codiesp-P. This is noticeable by looking at the performance of every model in each task. Our architecture is built to thrive under challenging tasks where a straightforward fine-tuning of a transformer is not the best approach. Nonetheless, our architecture is the third best evaluated for CANTEMIST, considering that ICB-UMA (López-García et al., 2020) and Clinical Coding Transformers (López-García et al., 2021) are from the same authors and used the same approach. Therefore, we obtained competitive results compared to state-of-the-art models and surpassed the performance of most of the systems (Miranda-Escalada et al., 2020a).

6 Module Analysis

In Table 4, we report the mean results using different language models and compare the performance with the ensemble for each corpus. We performed experiments for 15 instances of the architecture with different seeds, five using BioClinical RoBERTa, five using BioMedical RoBERTa, and five using BETO.

Notably, the MAP and F_1 scores for the Matcher are high in all experiments. This is required for the architecture to be competitive; otherwise, the error propagation leads to a low-quality final model. Another interesting finding is that we can see no significant difference between the domain-specific language models (BioMedical RoBERTa and Bio-Clinical) across all our experiments. However, the general-domain language model we have tested (BETO) has significantly lower performance on all tasks. Finally, it is worth mentioning that the ensemble-based architecture significantly outperforms all base models at hand, at least in the MAP metric. According to the F_1 metric, it surpasses the models in the CodiEsp tasks and fails in the Cantemist corpus. This adds room for improvement in how the class prediction is combined to calculate

Model	CodiEsp-D		CodiEsp-P		CANTEMIST	
	MAP	F_1	MAP	F_1	MAP	F_1
IXA-AAA (Blanco et al., 2020)	0.593	0.009	0.425	0.008	-	-
IAM (Cossin and Jouhet, 2020)	0.521	0.687	0.493	0.522	-	-
FLE (García-Santa et al., 2020)	0.519	0.679	0.443	0.514	-	-
The Mental Strokers (Costa et al., 2020)	0.517	0.591	0.445	0.488	-	-
Vicomtech (García-Pablos et al., 2020)	-	-	-	-	0.847	**0.855**
ICB-UMA (López-García et al., 2020)	0.482	0.009	-	-	0.847	0.013
Clinical Transformers - Best (López-García et al., 2021)	0.616	-	0.514	-	0.862	-
Clinical Transformers - Ensemble (López-García et al., 2021)	0.662	-	0.544	-	**0.884**	-
Divide and Conquer (DAC)	**0.665**	**0.746**	**0.545**	**0.553**	0.788	0.712
Divide and Conquer - Ensemble (DAC-E)	**0.682**	**0.744**	**0.562**	**0.560**	0.804	0.695

Table 3: Overall results on three clinical coding datasets. Results of the Clinical Transformers are taken from the author's paper, all the other results are obtained from the competitions overview. Some results are missing because those approaches were not implemented for the corresponding tasks.

Codiesp-D	Matcher		Ranker		DaC	
	MAP	F_1	MAP	F_1	MAP	F_1
BioClinical RoBERTa - Mean	0.930	0.852	0.729	**0.726**	**0.665**	0.727
BioMedical RoBERTa - Mean	**0.938**	**0.865**	**0.730**	**0.726**	**0.665**	**0.729**
BETO - Mean	0.916	0.824	0.728	0.728	0.653	0.713
Ensemble	-	-	-	-	**0.682**	**0.744**
Codiesp-P	Matcher		Ranker		DaC	
	MAP	F_1	MAP	F_1	MAP	F_1
BioClinical RoBERTa - Mean	0.941	**0.879**	0.614	0.584	0.545	**0.536**
BioMedical RoBERTa - Mean	**0.947**	0.867	**0.617**	**0.587**	**0.546**	0.531
BETO - Mean	0.936	0.853	0.612	**0.587**	0.533	0.525
Ensemble	-	-	-	-	**0.562**	**0.560**
CANTEMIST	Matcher		Ranker		DaC	
	MAP	F_1	MAP	F_1	MAP	F_1
BioClinical RoBERTa - Mean	**0.953**	**0.900**	0.821	0.711	**0.788**	0.706
BioMedical RoBERTa - Mean	0.948	0.898	0.819	**0.713**	0.784	**0.708**
BETO - Mean	0.915	0.857	**0.822**	0.712	0.763	0.692
Ensemble	-	-	-	-	**0.804**	0.695

Table 4: Report of metrics for each module and model trained in CodiEsp-D, CodiEsp-P, and CANTEMIST. The F_1 scores of both the DaC model and the Ranker use only the first three characters of the code as the label in Codiesp-D and the first four characters of the code in Codiesp-P. We used only the first characters following the procedures of evaluating the models created by the competition. The bolded results indicate the best metric score for each module, and the underline marks the worst performance.

the F_1 metric.

7 Conclusions and Future Work

This paper proposes a novel model for clinical coding in Spanish, outperforming previous results in two datasets; CodiEsp-D and CodiEsp-P. Our method uses a Divide and Conquer approach that creates semantic groups of codes to build an architecture composed of two specialized modules: the Matcher and the Ranker.

The clinical coding task is separated into two simpler tasks solved with the modules mentioned above. First, the Matcher predicts the clusters of each document, and then the Ranker predicts the codes of each document given a cluster. This division allows us to use state-of-the-art transformers to solve the task of cluster prediction and permits a fine-grained differentiation between similar codes in a cluster using XGBoost.

Although our base model achieves better results than previous ensemble-based models, we included the results of an ensemble strategy to perform a fair comparison with previous work. Our experimental results demonstrate that ensembling models yield better results than our base model. Furthermore, our DaC approach allows us to identify where future research can have a greater impact on improving accuracy. The results of each module indicate that there is more potential for improvement focusing on the Ranker module.

Future directions include implementing and testing the Divide and Conquer model on other multi-label text classification corpora. First, we expect to test the DaC architecture on clinical corpora in other languages, including languages with more resources, such as English. Second, we expect to test the architecture on other extreme multi-label classification corpora. This poses a challenge since the number of labels we have processed thus far, although very vast, falls into the category of small extreme multi-label classification datasets (Bhatia et al., 2016). We expect to encounter issues with the training time required to process other large corpora, forcing us to modify the library to optimize the speed.

In terms of improving the performance using this architecture, we identify opportunities to optimize the number of layers that we left fine-tuneable in the Matcher module, given that we have seen research that shows that fine-tuning more layers provides better results (Lee et al., 2019). Also, for the Ranker, we know that XGBoost can be trained with a ranking objective function, thus providing an alternative to the one-vs-rest approach. Implementing the Ranker using this approach would be faster to train and may provide similar or better results. In addition, we would like to improve data augmentation techniques by improving NER models. This can be achieved by using contextualized embeddings at the character level, which has been shown to improve the performance of models on various NLP tasks (Rojas et al., 2022a,b).

Finally, the DaC architecture is a black box when defining which labels to assign for each document. Recently, explainability features of the different architectures are gaining more relevance. It is paramount that the model's predictions are understood to help the user make appropriate choices (Duque et al., 2021). We expect to develop explainability to the labels predicted by providing textual queues of what features the model used to choose each label. The textual queues that support label assignment can be provided by the Ranker leveraging the explainability features of tree ensembles (Petkovic et al., 2018), and the textual queues that support the cluster choice can be obtained using the attention weights of the transformer model (Liu et al., 2021).

Limitations

Although our approach achieved excellent results across all the corpora in this research, they have clear limitations. The main drawback is that to apply this approach, it is necessary to have codes that can be clusterized. In fact, only a thorough categorization of similar codes into groups yields accurate results. Another major drawback is that the architecture predicts codes at the document level, thus having information that is not as complete as an entity-level prediction.

Finally, one limitation of the Matcher module is that it has a maximum document size of 512 tokens since it uses pre-trained transformers, which can contribute to losing important information on the cluster prediction process.

Acknowledgements

This work was funded by ANID Chile: Basal Funds for Center of Excellence FB210005 (CMM) and FB210017 (CENIA); Millennium Science Initiative Program ICN17_002 (IMFD) and ICN2021_004 (iHealth), and Fondecyt grant 11201250. Regarding hardware, the research was partially supported by the supercomputing infrastructure of the NLHPC (ECM-02) and the Patagón supercomputer of Universidad Austral de Chile (FONDEQUIP EQM180042).

References

Alan Akbik, Tanja Bergmann, Duncan Blythe, Kashif Rasul, Stefan Schweter, and Roland Vollgraf. 2019. Flair: An easy-to-use framework for state-of-the-art nlp. In *NAACL 2019, 2019 Annual Conference of the North American Chapter of the Association for Computational Linguistics (Demonstrations)*, pages 54–59.

Kush Bhatia, Kunal Dahiya, Himanshu Jain, Purushottam Kar, Anshul Mittal, Yashoteja Prabhu, and Manik Varma. 2016. The extreme classification repository: Multi-label datasets and code.

Alberto Blanco, Alicia Pérez, and Arantza Casillas. 2020. IXA-AAA at CLEF eHealth 2020 CodiEsp. Automatic Classification of Medical Records with Multi-label Classifiers and Similarity Match Coders. In *CLEF (Working Notes)*.

Wei-Cheng Chang, Hsiang-Fu Yu, Kai Zhong, Yiming Yang, and Inderjit S Dhillon. 2020. Taming pretrained transformers for extreme multi-label text classification. In *Proceedings of the 26th ACM SIGKDD international conference on knowledge discovery & data mining*, pages 3163–3171.

Tianqi Chen, Tong He, Michael Benesty, Vadim Khotilovich, Yuan Tang, Hyunsu Cho, Kailong Chen, et al. 2015. Xgboost: extreme gradient boosting. *R package version 0.4-2*, 1(4):1–4.

Sébastien Cossin and Vianney Jouhet. 2020. IAM at CLEF eHealth 2020: Concept Annotation in Spanish Electronic Health Records. In *CLEF (Working Notes)*.

Joao Costa, Inês Lopes, André V Carreiro, David Ribeiro, and Carlos Soares. 2020. Fraunhofer aicos at clef ehealth 2020 task 1: Clinical code extraction from textual data using fine-tuned bert models. In *CLEF (Working Notes)*.

Andres Duque, Hermenegildo Fabregat, Lourdes Araujo, and Juan Martinez-Romo. 2021. A keyphrase-based approach for interpretable ICD-10 code classification of Spanish medical reports. *Artificial Intelligence in Medicine*, 121:102177.

Aitor García-Pablos, Naiara Perez, and Montse Cuadros. 2020. Vicomtech at cantemist 2020. In *Proceedings of the Iberian Languages Evaluation Forum (IberLEF 2020), CEUR Workshop Proceedings*.

Nuria García-Santa, Kendrick Cetina, L Cappellato, C Eickhoff, N Ferro, and A Nevéol. 2020. FLE at CLEF eHealth 2020: Text Mining and Semantic Knowledge for Automated Clinical Encoding. In *CLEF (Working Notes)*.

Ira Goldstein, Anna Arzumtsyan, and Özlem Uzuner. 2007. Three approaches to automatic assignment of ICD-9-CM codes to radiology reports. In *AMIA Annual Symposium Proceedings*, volume 2007, page 279. American Medical Informatics Association.

Sepp Hochreiter and Jürgen Schmidhuber. 1997. Long short-term memory. *Neural computation*, 9(8):1735–1780.

Alistair EW Johnson, Tom J Pollard, Lu Shen, Li-wei H Lehman, Mengling Feng, Mohammad Ghassemi, Benjamin Moody, Peter Szolovits, Leo Anthony Celi, and Roger G Mark. 2016. MIMIC-III, a freely accessible critical care database. *Scientific data*, 3(1):1–9.

Rajvir Kaur, Jeewani Anupama Ginige, and Oliver Obst. 2021. A systematic literature review of automated ICD coding and classification systems using discharge summaries. *arXiv preprint arXiv:2107.10652*.

Diederik P Kingma and Jimmy Ba. 2014. Adam: A method for stochastic optimization. *arXiv preprint arXiv:1412.6980*.

Leah S Larkey and W Bruce Croft. 1995. Automatic assignment of ICD-9 codes to discharge summaries. Technical report, Technical report, University of Massachusetts at Amherst, Amherst, MA.

Jaejun Lee, Raphael Tang, and Jimmy Lin. 2019. What would elsa do? freezing layers during transformer fine-tuning. *arXiv preprint arXiv:1911.03090*.

Jingzhou Liu, Wei-Cheng Chang, Yuexin Wu, and Yiming Yang. 2017. Deep learning for extreme multi-label text classification. In *Proceedings of the 40th International ACM SIGIR Conference on Research and Development in Information Retrieval*, SIGIR '17, page 115–124, New York, NY, USA. Association for Computing Machinery.

Shengzhong Liu, Franck Le, Supriyo Chakraborty, and Tarek Abdelzaher. 2021. On exploring attention-based explanation for transformer models in text classification. In *2021 IEEE International Conference on Big Data*, pages 1193–1203. IEEE.

Guillermo López-García, José María Jerez, and Francisco José Veredas. 2020. ICB-UMA at CANTEMIST 2020: Automatic ICD-O Coding in Spanish with BERT. In *IberLEF@ SEPLN*, pages 468–476.

Guillermo López-Garcıa, José María Jerez, and Francisco José Veredas. 2020. ICB-UMA at CLEF e-health 2020 Task 1: Automatic ICD-10 coding in Spanish with BERT. In *Proc. Work. Notes CLEF, Conf. Labs Eval. Forum, CEUR Workshop*, pages 1–15.

Guillermo López-García, José María Jerez, Nuria Ribelles, Emilio Alba, and Francisco J Veredas. 2021. Transformers for clinical coding in spanish. *IEEE Access*, 9:72387–72397.

Ilya Loshchilov and Frank Hutter. 2017. Decoupled weight decay regularization. *arXiv preprint arXiv:1711.05101*.

Antonio Miranda-Escalada, Eulalia Farré, and Martin Krallinger. 2020a. Named entity recognition, concept normalization and clinical coding: Overview of the cantemist track for cancer text mining in spanish, corpus, guidelines, methods and results. In *Proceedings of the Iberian Languages Evaluation Forum (IberLEF 2020), CEUR Workshop Proceedings*.

Antonio Miranda-Escalada, Aitor Gonzalez-Agirre, Jordi Armengol-Estapé, and Martin Krallinger. 2020b. Overview of automatic clinical coding: Annotations, guidelines, and solutions for non-english clinical cases at codiesp track of clef ehealth 2020. In *CLEF (Working Notes)*.

Fabian Pedregosa, Gaël Varoquaux, Alexandre Gramfort, Vincent Michel, Bertrand Thirion, Olivier Grisel, Mathieu Blondel, Peter Prettenhofer, Ron Weiss, Vincent Dubourg, et al. 2011. Scikit-learn: Machine learning in python. *Journal of Machine Learning Research*, 12:2825–2830.

Dragutin Petkovic, Russ Altman, Mike Wong, and Arthur Vigil. 2018. Improving the explainability of random forest classifier–user centered approach. In *Proceedings of the Pacific Symposium on Biocomputing*, pages 204–215. World Scientific.

Matias Rojas, Felipe Bravo-Marquez, and Jocelyn Dunstan. 2022a. Simple yet powerful: An overlooked architecture for nested named entity recognition. In *Proceedings of the 29th International Conference on Computational Linguistics*, pages 2108–2117, Gyeongju, Republic of Korea. International Committee on Computational Linguistics.

Matías Rojas, Jocelyn Dunstan, and Fabián Villena. 2022b. Clinical flair: A pre-trained language model for Spanish clinical natural language processing. In *Proceedings of the 4th Clinical Natural Language Processing Workshop*, pages 87–92, Seattle, WA. Association for Computational Linguistics.

David E Rumelhart, Geoffrey E Hinton, and Ronald J Williams. 1986. Learning representations by back-propagating errors. *Nature*, 323(6088):533–536.

Hinrich Schütze, Christopher D Manning, and Prabhakar Raghavan. 2008. *Introduction to Information Retrieval*, volume 39. Cambridge University Press.

Fei Teng, Yiming Liu, Tianrui Li, Yi Zhang, Shuangqing Li, and Yue Zhao. 2022. A review on deep neural networks for icd coding. *IEEE Transactions on Knowledge and Data Engineering*, pages 1–1.

Ashish Vaswani, Noam Shazeer, Niki Parmar, Jakob Uszkoreit, Llion Jones, Aidan N Gomez, Łukasz Kaiser, and Illia Polosukhin. 2017. Attention is all you need. *Advances in neural information processing systems*, 30.

Curriculum-guided Abstractive Summarization for Mental Health Online Posts

Sajad Sotudeh [1*], **Nazli Goharian**[1], **Hanieh Deilamsalehy**[2], and **Franck Dernoncourt**[2]

[1]IRLab, Georgetown University
{sajad,nazli}@ir.cs.georgetown.edu
[2]Adobe Research
{deilamsa,franck.dernoncourt}@adobe.com

Abstract

Automatically generating short summaries from users' online mental health posts could save counselors' reading time and reduce their fatigue so that they can provide timely responses to those seeking help for improving their mental state. Recent Transformers-based summarization models have presented a promising approach to abstractive summarization. They go beyond sentence selection and extractive strategies to deal with more complicated tasks such as novel word generation and sentence paraphrasing. Nonetheless, these models have a prominent shortcoming; their training strategy is not quite efficient, which restricts the model's performance. In this paper, we include a curriculum learning approach to reweigh the training samples, bringing about an efficient learning procedure. We apply our model on *extreme* summarization dataset of MENTSUM posts —a dataset of mental health related posts from Reddit social media. Compared to the state-of-the-art model, our proposed method makes substantial gains in terms of ROUGE and BERTSCORE evaluation metrics, yielding 3.5% (ROUGE-1), 10.4% (ROUGE-2), and 4.7% (ROUGE-L), 1.5% (BERTSCORE) relative improvements.

1 Introduction

Summarization of mental health online posts is an emerging task that aims to summarize users' posts who are seeking help to enhance their mental state in online networks such as Reddit [1] and Reachout [2]. The post might address several issues of the user's concerns or simply be an elaboration on the user's mental and emotional situation. With user preference, each user-written post can be accompanied by a succinct summary (known as TL;DR [3]), condensing major points of the user post. This TL;DR summary is deemed to urge the counselors for a faster read of the user's posted content before reading the post in its entirety; hence, counsellors can provide responses promptly. Herein, we aim to improve state-of-the-art results reported in (Sotudeh, Goharian, and Young, 2022) for this task.

Large-scale deep neural models are often hard to train, leaning on intricate heuristic set-ups, which can be time-consuming and expensive to tune (Gong et al., 2019; Chen et al., 2021). This is especially the case for the Transformers-based summarizers, which have been shown to consistently outperform the RNN networks when rigorously tuned (Popel and Bojar, 2018), but also require heuristics such as specialized learning rates and large-batch training (Platanios et al., 2019). In this paper, we attempt to overcome the mentioned problem on BART (Lewis et al., 2020) Transformers-based summarizer by introducing a *Curriculum Learning (CL)* strategy (Bengio et al., 2009) for training the summarization model, leading to improved convergence time, and performance.

Inspired by humans' teaching style, *curriculum learning* suggests moving the teaching process from easier samples to more difficult ones and dates back to the nineties (Elman, 1993). The driving idea behind this approach is that networks can accomplish better task learning when the training instances are exposed to the network in a specific and certain order, from easier samples to more difficult ones (Chang et al., 2021). In the context of neural networks, this process can be thought of as a technique that makes the network robust to getting stuck at local optima, which is more likely in the early stages of the training process. Given the mentioned challenge of summarization networks, we utilize the SUPERLOSS (Castells et al., 2020) function that falls into the family of confidence-aware curriculum learning techniques, introducing a new parameter called confidence (i.e., σ) to the network.

* Work partially done during the internship at Adobe Research.

[1]https://www.reddit.com/

[2]https://au.reachout.com/

[3]TL;DR is the abbreviation of "Too Long, Didn't Read". We use "TL;DR" and "summary" exchangeably in this paper.

Proceedings of the 13th International Workshop on Health Text Mining and Information Analysis (LOUHI), pages 148 - 153
December 7, 2022

We validate our model on MENTSUM (Sotudeh, Goharian, and Young, 2022) dataset, containing over 24k instances mined from 43 mental health related communities on Reddit social media. Our experimental results show the efficacy of applying curriculum learning objectives on BART summarizer, achieving a new state-of-the-art performance.

2 Related Work

While majority of works in mental health research have focused on studying users' behavioral patterns through classification and prediction tasks (Choudhury et al., 2013; Resnik et al., 2013; Coppersmith et al., 2014; Yates et al., 2017; Cohan et al., 2017, 2018; MacAvaney et al., 2018), summarization of online mental health posts has been recently made viable. Sotudeh, Goharian, and Young (2022) were the first to step on this direction via introducing MENTSUM dataset of online mental health posts, pinpointing the baseline results. Curriculum Learning (CL) has gained growing interest from the research communities during the last decade (Tay et al., 2019; MacAvaney et al., 2020; Xu et al., 2020). Bengio et al. (2015) were the first to apply this strategy in the context of sequence prediction through *scheduled sampling* approach, which gently changes the training process from ground truth tokens to model-generated ones during decoding. Sample's *difficulty* is a key concept in this scheme as it is used to distinguish easy examples from difficult ones. Researchers have used many textual features as the "difficulty measure" including n-gram frequency (Haffari et al., 2009), word rarity and sentence length (Platanios et al., 2019). Recent works (Saxena et al., 2019; Cachola et al., 2020) have made use of confidence-aware approaches that learn the difficulty of training samples and dynamically reweight samples in the training process.

3 Our Approach

In this section, we describe the details of our proposed model, where a curriculum learning architecture is added to the BART's Transformers-based framework, upweighting easier training samples; hence, increasing their contribution in learning stage.

Curricular Learner for BART. Considering the applicability of curriculum learning in training large-scale networks, we aim to use it in our summarization framework. Before incorporating the curriculum learning strategy into our model's training stage, we first need to define the *difficulty* metric to distinguish the hardness of samples. In practice, estimating a prior difficulty for each sample is considered a complex task, so we propose to discriminate the samples with progressive signals —such as the respective sample loss at each training iteration— in the training process. In this context, CL is achieved by predicting the difficulty of each sample at the training iterations in the form of a weight, such that difficult samples receive lower weights during the early stages of training and vice versa. To model the curriculum, we propose to use SUPERLOSS (Castells et al., 2020) which is a generic loss criterion built upon the task loss function as shown in Figure 1.

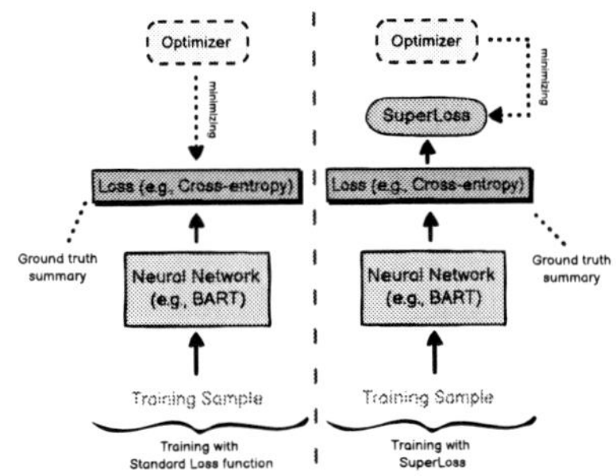

Figure 1: Training with standard loss function (left-hand side) and SuperLoss criteria (right-hand side)

More specifically, SUPERLOSS is a task-agnostic confidence-aware loss function that takes in two parameters: (1) the task loss $\mathcal{L}_i = \ell(y_i, \widehat{y_i})$, where y_i is neural network's (i.e., BART's generated summary) output and $\widehat{y_i}$ is the gold label (i.e., ground-truth summary); and (2) σ_i as the confidence parameter of the ith sample. SUPERLOSS is framed as $\mathrm{L}_\lambda(\mathcal{L}_i, \sigma_i)$ and computed as follows,

$$\mathrm{L}_\lambda(\mathcal{L}_i, \sigma_i) = (\mathcal{L}_i - \tau)\sigma_i + \lambda(\log \sigma_i)^2 \quad (1)$$

in which λ is the regularization parameter, and τ is the running or static average of task loss (i.e., $\mathcal{L}$) during the training. While SUPERLOSS provides a well-defined approach to curriculum learning strategy, learning σ parameter is not tractable for tasks with abundant training instances such as text summarization. To circumvent this issue and hinder imposing new learnable parameters, SUPERLOSS suggests using the converged value of σ_i at the limit,

$$\sigma^*_\lambda(\ell_i) = \arg\min_{\sigma_i} \mathrm{L}_\lambda(\ell_i, \sigma_i)$$
$$\mathrm{SL}_\lambda(\ell_i) = \mathrm{L}_\lambda(\ell_i, \sigma^*_\lambda(\ell_i, \sigma_i)) = \min_{\sigma_i} \mathrm{L}_\lambda(\ell_i, \sigma_i), \quad (2)$$

Using this technique, the confidence parameters are not required to be learned during the training. Castells et al. (2020) found out that $\sigma^*_\lambda(\ell_i)$ has a closed-form solution, computed as follows,

$$\sigma^*_\lambda(\ell_i) = e^{-W(\frac{1}{2}\max(-\frac{2}{e},\beta))}, \beta = \frac{\ell_i - \tau}{\lambda} \quad (3)$$

in which W is the Lambert W function. With this in mind, SUPERLOSS upweights easier samples dynamically during the training, providing a curriculum learning approach to summarization. We call this model CURRSUM in our experiments.

4 Experimental Setup

4.1 Dataset

We use the MENTSUM dataset in our experiments. This dataset contains over 24k post-TL;DR pairs, making up 21,695 (train), 1209 (validation), and 1215 (test) instances, and is gathered by mining 43 mental health subreddit communities on Reddit with rigorous filtering rules. We refer the readers to the main paper for more details on this dataset (Sotudeh, Goharian, and Young, 2022).

4.2 Comparison

We compare our model against the BART (Lewis et al., 2020) baseline, which does not utilize the curriculum learning objective. BART is an abstractive model that uses a pre-trained encoder-decoder architecture, unlike BERT which only utilizes a pre-trained encoder. As shown in (Sotudeh, Goharian, and Young, 2022), BART is the strongest baseline; thus, we apply CL on it to evaluate its impact on summarization. We refer the reader to the original paper for more extractive and abstractive baselines.

4.3 Implementation details

We use the Huggingface's Transformers library (Wolf et al., 2020) [4] to implement our models. We train all of our models for 8 epochs, performing evaluation step in intervals of 0.5 epochs, and use the checkpoint that achieves the best ROUGE-L score in the validation for the inference. AdamW optimizer (Loshchilov and Hutter, 2019) initialized with learning rate of $3e-5$, $(\beta_1, \beta_2) = (0.9, 0.98)$, and weight decay of 0.01 is used for all of our summarization models, as well as for BART. Cross-entropy loss is used for all models. To keep track of the learning process, we use Weights & Biases (Biewald, 2020) toolkit [5].

Model	R-1	R-2	R-L	BS
ORACLEEXT	35.98	11.59	23.21	82.72
BART (2020)	29.13	7.98	20.27	85.01
CURRSUM (Ours)	**30.16**	**8.82**	**21.24**	**86.32**

Table 1: ROUGE and BERTSCORE results on test set of MENTSUM dataset. As BART was the most performant baseline provided in (Sotudeh, Goharian, and Young, 2022), we evaluate the effectiveness of Curriculum on BART in this work. For other baselines, we refer the reader to the main paper.

5 Ethics and privacy

While we use publicly available Reddit data to train and validate our model, we recognize that special care should be taken when dealing with this type of data due to its sensitivity and users' privacy (Benton et al., 2017; Nicholas et al., 2020). Thus, we have made no attempts to identify, contact the users, or explore user relations to trace their social media account. It has to be mentioned that the MENTSUM dataset is distributed through Data Usage Agreement (DUA) [6] document to further ensure that users' identity would never be disclosed.

6 Results

Automatic evaluation. Table 1 reports the performance of the baseline model along with our model's in terms of ROUGE score variants (Lin, 2004) and BERTSCORE (Zhang et al., 2020) over MENTSUM dataset. As indicated, the best model is our CURRSUM that uses SUPERLOSS curriculum directly on top of the BART model and is a clear improvement over it across all metrics, achieving the current state-of-the-art performance. Specifically, our model outperforms its ground baseline that has no curriculum (i.e., BART) by improvement gains of 3.5%, 10.4%, 4.7%, 1.5% for ROUGE-1, ROUGE-2, ROUGE-L, BERTSCORE, respectively. Having looked at the ORACLEEXT scores which shows

[4] https://github.com/huggingface/transformers

[5] https://wandb.ai/

[6] https://ir.cs.georgetown.edu/resources/mentsum.html

Human-written	BART-generated	CURRSUM-generated
... having kids ... my ADHD ... opinions on parenthood...	*I think it's better to not have a child now than never having a child in the future.*	*I'm **thinking of having a child** and I **don't want to be a shi**y dad** for my **ADHD**. Any **parents with ADHD experience?***
... symptoms of depression ... go to get help ...	*I think I **am clinically depressed**, and I have pretty much **all the symptoms of depression**, so I am asking for help*	*I think I am **clinically depressed** xxx. I have pretty much **every symptom on this list**.*
... free of anxiety for a decade.. symptoms come ...	*I've been **free of anxiety** for less than a week and I'm afraid that it's only a temporary solution. I don't know what to do.*	*I've been **free of anxiety for a decade** and I'm afraid that if the **symptoms start to come back**, I'll be able to **slip back into it** again.*
... ask her out ... barely talked to her... mess up my chance...	*I like a girl at work and I want to **ask her out**, but I don't want to **ruin that chance** in the process.*	*I like a girl at work, but I don't know how to ask her out.*

Table 2: Four samples of the the human-written, BART-generated, and CURRSUM-generated TL;DR summaries. The human-written samples are partially shown to preserve users' privacy. That is, we have only shown the important human-written phrases to trace them within the generated summaries. The text that is unfaithful to the post (i.e., not supported by the user post) is in Gray and the salient information that is picked up by the summarization systems is shown in **Bold**.

the upper bound performance of an ideal extractive summarizer, it seems that there is room for improvement on the extractive setting to achieve state-of-the-art performance. More sophisticated models can invest in extractive or hybrid summarization models such as those done in (Gehrmann et al., 2018; MacAvaney et al., 2019; Sotudeh et al., 2020).

Case study and analysis. While our proposed model significantly improves upon the BART baseline, we recognize the limitations of ROUGE metric in evaluation of summarization systems (Cohan and Goharian, 2016). In order to explore the qualities and limitations of our work, we analyze the human-written TL;DRs along with the generated results by BART and our model, comparing them against each other. Table 2 shows four samples of the human and systems generated TL;DRs. As seen, our model can improve the faithfulness of the summary [7] in the first, second, and third samples. Having looked at other cases in our study, it appears that curriculum learning positively mitigates faithfulness errors. This might be attributed to the fact that the summarizer can achieve an improved *understanding* of the source document when the contribution of each sample is controlled in each iteration of the learning process. Looking at the second sample, it turns out that our model can improve the conciseness of the summaries; that is, briefly conveying the main points within the summary. Comparing system-generated summaries in the fourth sample, it is observed that our system generated a phrase (shown in Gray) by mixing up different regions of the user post. Surprisingly, it appears that "*I don't know how/what to*" is a common phrase used in most human-written summaries that are seeking advice from the community. The experimented summarization systems (i.e., BART and ours) adhere to overgenerating this phrase.

7 Conclusion

Generating short summaries given the users' online posts can save counselors' reading time, and reduce their fatigue. On this basis, they can provide faster responses to community users. While neural Transformers-based summarization models have shown to be promising, they suffer from *inefficient training process* that hinders their potentials for showing a promising performance. To compensate for this issue, in this paper, we incorporated a confidence-aware curriculum learning approach, which uses task-agnostic SUPERLOSS, to the summarization framework in the hope of increasing model's generalization, and ultimately improving model performance. Our automatic evaluations over MENTSUM dataset of mental health posts show the effectiveness of our model, yielding 3.5%, 10.4%, 4.7%, 1.5% relative improvements over BART summarizer on ROUGE-1, ROUGE-2, ROUGE-L, and BERTSCORE, respectively. Our model tailors the new state-of-the-art performance on MENTSUM dataset. We further showed various system-generated summaries to showcase the qualities and limitations of our proposed model.

[7]Faithfulness is defined as generating output text that is supported by the user post.

References

Samy Bengio, Oriol Vinyals, Navdeep Jaitly, and Noam Shazeer. 2015. Scheduled sampling for sequence prediction with recurrent neural networks. In *Advances in Neural Information Processing Systems 28: Annual Conference on Neural Information Processing Systems 2015, December 7-12, 2015, Montreal, Quebec, Canada*, pages 1171–1179.

Yoshua Bengio, Jérôme Louradour, Ronan Collobert, and Jason Weston. 2009. Curriculum learning. In *Proceedings of the 26th Annual International Conference on Machine Learning, ICML 2009, Montreal, Quebec, Canada, June 14-18, 2009*, volume 382 of *ACM International Conference Proceeding Series*, pages 41–48. ACM.

Adrian Benton, Glen A. Coppersmith, and Mark Dredze. 2017. Ethical research protocols for social media health research. In *EthNLP@EACL*.

Lukas Biewald. 2020. Experiment tracking with weights and biases. Software available from wandb.com.

Isabel Cachola, Kyle Lo, Arman Cohan, and Daniel Weld. 2020. TLDR: Extreme summarization of scientific documents. In *Findings of the Association for Computational Linguistics: EMNLP 2020*, pages 4766–4777, Online. Association for Computational Linguistics.

Thibault Castells, Philippe Weinzaepfel, and Jérôme Revaud. 2020. Superloss: A generic loss for robust curriculum learning. In *Advances in Neural Information Processing Systems 33: Annual Conference on Neural Information Processing Systems 2020, NeurIPS 2020, December 6-12, 2020, virtual.*

Ernie Chang, Hui-Syuan Yeh, and Vera Demberg. 2021. Does the order of training samples matter? improving neural data-to-text generation with curriculum learning. In *Proceedings of the 16th Conference of the European Chapter of the Association for Computational Linguistics: Main Volume*, pages 727–733, Online. Association for Computational Linguistics.

Xiaohan Chen, Yu Cheng, Shuohang Wang, Zhe Gan, Zhangyang Wang, and Jingjing Liu. 2021. EarlyBERT: Efficient BERT training via early-bird lottery tickets. In *Proceedings of the 59th Annual Meeting of the Association for Computational Linguistics and the 11th International Joint Conference on Natural Language Processing (Volume 1: Long Papers)*, pages 2195–2207, Online. Association for Computational Linguistics.

Munmun De Choudhury, Michael Gamon, Scott Counts, and Eric Horvitz. 2013. Predicting depression via social media. In *ICWSM*.

Arman Cohan, Bart Desmet, Andrew Yates, Luca Soldaini, Sean MacAvaney, and Nazli Goharian. 2018. Smhd: a large-scale resource for exploring online language usage for multiple mental health conditions. In *COLING*.

Arman Cohan and Nazli Goharian. 2016. Revisiting summarization evaluation for scientific articles. In *Proceedings of the Tenth International Conference on Language Resources and Evaluation (LREC'16)*, pages 806–813, Portorož, Slovenia. European Language Resources Association (ELRA).

Arman Cohan, Sydney Young, Andrew Yates, and Nazli Goharian. 2017. Triaging content severity in online mental health forums. *Journal of the Association for Information Science and Technology*, 68.

Glen A. Coppersmith, Craig Harman, and Mark Dredze. 2014. Measuring post traumatic stress disorder in twitter. In *ICWSM*.

J. Elman. 1993. Learning and development in neural networks: the importance of starting small. *Cognition*, 48:71–99.

Sebastian Gehrmann, Yuntian Deng, and Alexander Rush. 2018. Bottom-up abstractive summarization. In *Proceedings of the 2018 Conference on Empirical Methods in Natural Language Processing*, pages 4098–4109, Brussels, Belgium. Association for Computational Linguistics.

Linyuan Gong, Di He, Zhuohan Li, Tao Qin, Liwei Wang, and Tie-Yan Liu. 2019. Efficient training of BERT by progressively stacking. In *Proceedings of the 36th International Conference on Machine Learning, ICML 2019, 9-15 June 2019, Long Beach, California, USA*, volume 97 of *Proceedings of Machine Learning Research*, pages 2337–2346. PMLR.

Gholamreza Haffari, Maxim Roy, and Anoop Sarkar. 2009. Active learning for statistical phrase-based machine translation. In *Proceedings of Human Language Technologies: The 2009 Annual Conference of the North American Chapter of the Association for Computational Linguistics*, pages 415–423, Boulder, Colorado. Association for Computational Linguistics.

Mike Lewis, Yinhan Liu, Naman Goyal, Marjan Ghazvininejad, Abdelrahman Mohamed, Omer Levy, Veselin Stoyanov, and Luke Zettlemoyer. 2020. BART: Denoising sequence-to-sequence pre-training for natural language generation, translation, and comprehension. In *Proceedings of the 58th Annual Meeting of the Association for Computational Linguistics*, pages 7871–7880, Online. Association for Computational Linguistics.

Chin-Yew Lin. 2004. ROUGE: A package for automatic evaluation of summaries. In *Text Summarization Branches Out*, pages 74–81, Barcelona, Spain. Association for Computational Linguistics.

Ilya Loshchilov and Frank Hutter. 2019. Decoupled weight decay regularization. In *7th International Conference on Learning Representations, ICLR 2019, New Orleans, LA, USA, May 6-9, 2019*. OpenReview.net.

Sean MacAvaney, Bart Desmet, Arman Cohan, Luca Soldaini, Andrew Yates, Ayah Zirikly, and Nazli Goharian. 2018. Rsdd-time: Temporal annotation of self-reported mental health diagnoses. In *CLPsych@NAACL-HTL*.

Sean MacAvaney, Franco Maria Nardini, Raffaele Perego, Nicola Tonellotto, Nazli Goharian, and Ophir Frieder. 2020. Training curricula for open domain answer re-ranking. In *Proceedings of the 43rd International ACM SIGIR conference on research and development in Information Retrieval, SIGIR 2020, Virtual Event, China, July 25-30, 2020*, pages 529–538. ACM.

Sean MacAvaney, Sajad Sotudeh, Arman Cohan, Nazli Goharian, Ish A. Talati, and Ross W. Filice. 2019. Ontology-aware clinical abstractive summarization. In *Proceedings of the 42nd International ACM SIGIR Conference on Research and Development in Information Retrieval, SIGIR 2019, Paris, France, July 21-25, 2019*, pages 1013–1016. ACM.

Jennifer Nicholas, Sandersan Onie, and Mark Erik Larsen. 2020. Ethics and privacy in social media research for mental health. *Current Psychiatry Reports*, 22:1–7.

Emmanouil Antonios Platanios, Otilia Stretcu, Graham Neubig, Barnabas Poczos, and Tom Mitchell. 2019. Competence-based curriculum learning for neural machine translation. In *Proceedings of the 2019 Conference of the North American Chapter of the Association for Computational Linguistics: Human Language Technologies, Volume 1 (Long and Short Papers)*, pages 1162–1172, Minneapolis, Minnesota. Association for Computational Linguistics.

Martin Popel and Ondrej Bojar. 2018. Training tips for the transformer model. *The Prague Bulletin of Mathematical Linguistics*, 110:43 – 70.

Philip Resnik, Anderson Garron, and Rebecca Resnik. 2013. Using topic modeling to improve prediction of neuroticism and depression in college students. In *EMNLP*.

Shreyas Saxena, Oncel Tuzel, and Dennis DeCoste. 2019. Data parameters: A new family of parameters for learning a differentiable curriculum. In *Advances in Neural Information Processing Systems 32: Annual Conference on Neural Information Processing Systems 2019, NeurIPS 2019, December 8-14, 2019, Vancouver, BC, Canada*, pages 11093–11103.

Sajad Sotudeh, Nazli Goharian, and Ross Filice. 2020. Attend to medical ontologies: Content selection for clinical abstractive summarization. In *Proceedings of the 58th Annual Meeting of the Association for Computational Linguistics*, pages 1899–1905, Online. Association for Computational Linguistics.

Sajad Sotudeh, Nazli Goharian, and Zachary Young. 2022. Mentsum: A resource for exploring summarization of mental health online posts. In *Proceedings of the Language Resources and Evaluation Conference*, pages 2682–2692, Marseille, France. European Language Resources Association.

Yi Tay, Shuohang Wang, Anh Tuan Luu, Jie Fu, Minh C. Phan, Xingdi Yuan, Jinfeng Rao, Siu Cheung Hui, and Aston Zhang. 2019. Simple and effective curriculum pointer-generator networks for reading comprehension over long narratives. In *Proceedings of the 57th Annual Meeting of the Association for Computational Linguistics*, pages 4922–4931, Florence, Italy. Association for Computational Linguistics.

Thomas Wolf, Lysandre Debut, Victor Sanh, Julien Chaumond, Clement Delangue, Anthony Moi, Pierric Cistac, Tim Rault, Remi Louf, Morgan Funtowicz, Joe Davison, Sam Shleifer, Patrick von Platen, Clara Ma, Yacine Jernite, Julien Plu, Canwen Xu, Teven Le Scao, Sylvain Gugger, Mariama Drame, Quentin Lhoest, and Alexander Rush. 2020. Transformers: State-of-the-art natural language processing. In *Proceedings of the 2020 Conference on Empirical Methods in Natural Language Processing: System Demonstrations*, pages 38–45, Online. Association for Computational Linguistics.

Benfeng Xu, Licheng Zhang, Zhendong Mao, Quan Wang, Hongtao Xie, and Yongdong Zhang. 2020. Curriculum learning for natural language understanding. In *Proceedings of the 58th Annual Meeting of the Association for Computational Linguistics*, pages 6095–6104, Online. Association for Computational Linguistics.

Andrew Yates, Arman Cohan, and Nazli Goharian. 2017. Depression and self-harm risk assessment in online forums. In *EMNLP*.

Tianyi Zhang, Varsha Kishore, Felix Wu, Kilian Q. Weinberger, and Yoav Artzi. 2020. Bertscore: Evaluating text generation with BERT. In *8th International Conference on Learning Representations, ICLR 2020, Addis Ababa, Ethiopia, April 26-30, 2020*. OpenReview.net.

Improving information fusion on multimodal clinical data in classification settings

Sneha Jha
Imperial College London
sneha.jha@imperial.ac.uk

Erik Mayer
Imperial College London
e.mayer@imperial.ac.uk

Mauricio Barahona
Imperial College London
m.barahona@imperial.ac.uk

Abstract

Clinical data often exists in different forms across the lifetime of a patient's interaction with the healthcare system - structured, unstructured or semi-structured data in the form of laboratory readings, clinical notes, diagnostic codes, imaging and audio data of various kinds, and other observational data. Formulating a representation model that aggregates information from these heterogeneous sources may allow us to jointly model on data with more predictive signal than noise and help inform our model with useful constraints learned from better data. Multimodal fusion approaches help produce representations combined from heterogeneous modalities, which can be used for clinical prediction tasks. Representations produced through different fusion techniques require different training strategies. We investigate the advantage of adding narrative clinical text to structured modalities to classification tasks in the clinical domain. We show that while there is a competitive advantage in combined representations of clinical data, the approach can be helped by training guidance customized to each modality. We show empirical results across binary/multiclass settings, single/multitask settings and unified/multimodal learning rate settings for early and late information fusion of clinical data.

1 Introduction

A variety of clinical use cases emerge where it is not sufficient to use a single data modality as input to a learning or decision making system (Weber et al., 2014). A single data modality is often known to be insufficient for a clinical purpose. For instance, diagnoses that require imaging data as well as lab values or outcomes that depend on values routinely recorded in the narrative text but not elsewhere. An additional modality can be used to characterize additional features. For instance, information in narrative text that conflicts with or adds specificity to diagnosis or procedure codes or imaging data that can indicate severity of a condition not recorded in structured form or qualitatively mentioned in narrative text. Sometimes, a modality with highly predictive or informative features is particularly expensive or invasive and an alternative source is present that may have features unintelligible or hard to parse for humans. Also, most clinical machine learning systems focus on one clinical prediction task at a time (D'Costa et al., 2020; Ji et al., 2020). However, in real-world systems more than one such task are often performed simultaneously and are interrelated (Yang and Wu, 2021). There is a need to investigate task-specific unified representations of multimodal clinical data in both single-task and multi-task settings to improve decisions in the clinical workflow by demonstrating an increase in predictive power, robustness, and confidence over any single mode of data (Tiulpin et al., 2019). Besides creating and combining efficient representations of data from more than one modality, we also need to study the factors that affect the design and evaluation of these multimodal representations.

2 Multimodal Representations

There are various ways modalities of clinical data can be combined. Multimodal deep learning models integrate information at various possible steps. This can occur in the following ways –

- By finding a common representation for input data for a specific task before modeling. e.g., extracting clinical mentions from narrative text and concatenating it with independent diagnostic signals to form a model input.
- By jointly learning intermediate feature representations for one or more additional modalities, besides the basic input e.g., learning text embeddings from narrative text and using that as an additional input besides the structured

Proceedings of the 13th International Workshop on Health Text Mining and Information Analysis (LOUHI), pages 154 - 159
December 7, 2022 ©2022 Association for Computational Linguistics

data to the same neural network. This is designed for the training algorithm to jointly improve the intermediate embeddings along with the task-specific loss.

- By modeling each modality separately and combining predictions from different models under a task-specific scheme. e.g., aggregating diagnostic predictions from a text modeling and an image modeling system through an averaging scheme or a meta classifier.

As detailed in (Baltrušaitis et al., 2018), multimodal models can be categorized by the fusion techniques based on which they learn the join representations of underlying data. The most common approaches are called early fusion (Chen and Jin, 2015) where individual modality features are combined right after feature extraction and late fusion (Atrey et al., 2010) which combines outputs from unimodal predictors jointly. Early fusion is expected to capture some of the feature-level interactions of each modality and often is easier to model and train. On the other hand, late fusion allows for more flexibility and is expected to model individual modalities better and also handles scenarios where one or more modalities are missing. However, it cannot be expected to capture low level interaction between the modalities. While training late fusion models, the simplest choice is to use the same learning rate across all modalities. But it is both intuitive as well as demonstrable through layer analysis (Yao and Mihalcea, 2022) that learning rates for different modalities can differ a lot and must be handled separately to optimize learning from heterogeneous sources.

3 Methodology

3.1 Data source

We use a publicly available clinical data set - Medical Information Mart for Intensive Care (MIMIC) Johnson et al. (2016) - containing data across various modalities for patients admitted and readmitted to the intensive care unit (ICU). MIMIC-III is a large, freely-available database comprising deidentified health-related data associated with over forty thousand patients who stayed in critical care units of the Beth Israel Deaconess Medical Center between 2001 and 2012. The database includes information such as demographics, vital sign measurements made at the bedside (1 data point per hour), laboratory test results, procedures, medications, caregiver notes, imaging reports, and mortality, including post-hospital discharge. It contains highly granular data, including vital signs, laboratory results, and medications.

3.2 Data modalities

The following structured and semi-structured modalities typically found in inpatient settings in clinical data were extracted and compiled at a patient level -

- **Clinical Notes** - Free-form narrative text is entered by clinicians and nurses during the stay of a patient and these usually summarize reasons for admission, details of treatment, nutrition, and the patient's symptoms and diagnoses. These clinical notes are temporally ordered.

- **Tabular Data** – Metadata such as sex, age, height, weight at admission, the type of the ICU, and other tabular inputs were recorded for each patient. Values such as weight, may vary during the patient's stay and are potentially part of the time series data set as well.

- **Time Series data** – Various temporal physiological variables, such as diastolic blood pressure, systolic blood pressure, oxygen saturation, were recorded for each patient. These physiological variables are recorded irregularly, and they are important indicators of the patient's condition during the hospital stay.

We add the two different kinds of structured data from the MIMIC-III dataset to the clinical text. Data is preprocessed as in (Harutyunyan et al., 2019), excluding ICU stays with missing events or missing length-of-stay and excluding patients younger than eighteen years of age since both clinical dynamics and clinical documentation of paediatrics facilities are significantly different from those of adults.

3.3 Experiments

The experiments focus on the following two tasks -

- **In-hospital mortality prediction** : To predict death by the end of the hospital stay based on first 48 hours of observations. To prevent mortality is the primary aim and a number of task formulations as in (Harutyunyan et al., 2019) and (Khadanga et al., 2019) attempt to predict patient survival at the end of the hospital stay. The hypothesis is that observations from the first 48 hours of a patient's stay in the ICU include crucial clues towards the probability of survival.

- **Phenotyping** : To predict a patient's phenotype at the time of discharge in terms of billing codes. This is a multilabel classification task and the target label set is derived from the billing code at a patients discharge, which is then converted to 25 labels following the procedure from (Harutyunyan et al., 2019).

The above two are standard tasks in clinical prediction settings and allowed us to compare directly with prior work such as (Harutyunyan et al., 2019) and (Khadanga et al., 2019). They provide two representative tasks in binary and multiclass, multilabel settings. Multitask learning is a particularly useful direction to explore in clinical settings with potential to capture dependencies between tasks, especially in low-data regimes. It was first proposed in the clinical prediction setting by (Caruana et al., 1995), where they used future lab values as auxiliary targets during training improving prediction of mortality among pneumonia patients.
Multimodal embeddings are used as the input to the two task- specific components. Each layer per task is a fully connected network with h_t hidden units, a dropout layer with dropout probability α_t, a ReLU activation, and an output layer matching the shape of the individual component's respective task. Each task-specific component shares the base multimodal embedding but is independent of the other layers. The multimodal encoder is comprised of one child encoder per input modality. In the early fusion setup, the multimodal embedding is a concatenation of the outputs from each child encoder. In the late fusion setting, for each time step, the model structure is a linear layer with 512 hidden states with ReLU activation projected to a 128-dimensional linear layer to predict the output class. For the phenotyping task each of the 25 output neurons has a sigmoid activation. The results for late fusion multimodal learning have been reported only in single-task settings.

Each task-specific component can employ a task-specific loss. To learn across both tasks simultaneously in the multitask experiments, we take the weighted sum of all the losses resulting to form the multitask loss. The current experiments use cross entropy loss for both tasks. To find multitask weights, we used uncertainty weighting described in (Kendall et al., 2018).

Time series encoder.Given a patient's ICU stay of length of T hours, the time series data is resampled with 1 hour interval to obtain $[TS_t]$ from t = 1 to t = T. The time series encoder is an LSTM (Hochreiter and Schmidhuber, 1997). The input $[TS_t]$ at time step t is directly the input to an LSTM model (Hochreiter and Schmidhuber, 1997) along with the previous states, and the next hidden state is the extracted feature, denoted by f_t^n .

$$f_t^n = LSTM(TS_t, f_{t-1}^n) \quad (1)$$

The experiments use a 1-layer LSTM with 256 hidden units as the time series encoder.

Clinical Text Encoder. For each ICU stay, there are N clinical notes recorded at irregular intervals.The chart time of these notes are $[Time(i)_{i=1}^N]$ where $N <= T$ The convolutional model TextCNN in (Kim, 2014) is used to extract features from textual clinical notes. To create embeddings from N_t notes collected at time $Time_i$, the CNN model gives us the feature matrix z per clinical note. A weighted average of all notes, weighted by recency produces a feature vector for a record.

$$weight(t, i) = exp(-\lambda(t - Time_i)) \quad (2)$$

$$f_t = \gamma \sum_{i=1}^{N} weight(t, i).z_i \quad (3)$$

Here, λ is a scaling factor and γ is a normalization term. The embeddings are generated using word2vec embeddings (Mikolov et al., 2013). The TextCNN model has three 1-D kernels of size 2,3 and 4 with 128 filters each.

Tabular Data Encoder. To process the tabular inputs, we learn an embedding table for each categorical input dimension as in (De Brébisson et al., 2015) and individual features are concatenated to form one tabular embedding. All features are represented as 32-dimensional embeddings.

In the early fusion setup, the default Adam optimizer (Kingma and Ba, 2014) is used with a learning rate of 1e-4 with early stopping. The mortality prediction task uses $h_t = 108$. The phenotyping task uses $h_t = 512$. In the late fusion setup, we ran the following sets of experiments -

- **Unified learning rate across modalities** : We use the default Adam optimizer with a learning rate of $3e$-4 with early stopping. The mortality prediction task uses $h_t = 108$. The phenotyping task uses $h_t = 512$.
- **Adapted learning rate per modality** : We use the best fine-tuned learning rate per modality for each of them while training the late-fusion model.

We observed better results with the AdamW optimizer (Loshchilov and Hutter, 2017) but report results using the default Adam optimizer to be able to at least partially compare with (Harutyunyan et al., 2019) since learning rates can vary a lot based on the optimizer used.

4 Results

Since the mortality prediction is a binary classification problem, we report the AUC-PR numbers, which is a standard evaluation metric. Because diseases can co-occur and a majority of patients often have more than one diagnosis, phenotyping is a multilabel classification problem, which requires the performance to be reported by averaging across labels or examples. These labels have varying base rates. In imbalanced tasks, (Lipton et al., 2014) show that if the predictive features for rare labels are lost, which is possible due to feature selection, macro F1 is an unsuitable metric. We report the macro AUC-ROC, which is the unweighted mean of AUC-ROC for each label. We also add a weighted average AUC-ROC metric accounting for base rate of the diseases in Table 3. The phenotyping task does not categorically benefit from the multitask setting. The model is trained to jointly predict 25 labels which in itself might have a regularizing effect akin to multitask learning and the additional regularization expected from adding the in-hospital mortality prediction task may be unable to provide further significant improvement over the single-task setting.

We follow (Harutyunyan et al., 2019) as the baseline for the MultiTask TimeSeries set-up and (Khadanga et al., 2019) for the SingleTask Notes + TimeSeries set-up. We show results of the early fusion runs in Table 1 and the late fusion runs in Table 2.

We also report error bounds for the experiments by choosing observations at the 2.5th percentile and the 97.5th percentiles and reporting the median. This was computed by drawing 5000 samples with replacement 100 times from the test set.

5 Related Work

We refer to (Harutyunyan et al., 2019) that uses a single modality of time series in a multi-task setting using LSTMs and channel-wise LSTMs. Similarly, (Khadanga et al., 2019) presents a unimodal model with clinical notes only for individual task settings but they also report additional results in a multimodal setting using both time series and text data without using multitask learning. We reuse some of the multitasking configuration for the MIMIC dataset described in (Huang et al., 2020). There are also available comparisons against baseline logistic regression and random forest models in (Zhang et al., 2020). All of these use only a unified learning rate across modalities. A number of works note the need of exploiting modality-specific features such as (Wang et al., 2015; Liu et al., 2018) for combining text with other modalities such as image and audio. In the late-fusion setting, a closely related work is (Yao and Mihalcea, 2022) that investigates modality-specific learning rates. They do not investigate a multitask setting and also study modalities structurally different from ours. Another closely related work is (Fujimori et al., 2019) that take up the issue of potential overfitting to certain modalities. Their approach is via early stopping and is still closer to our unified learning rate set up. It is also worth noting that the typical modalities in clinical data are very domain-specific and even well-studied modalities such as text in general-domain NLP often behave differently in the clinical domain (Rumshisky et al., 2020).

6 Limitations and Future Work

This work investigates one way to adapt learning rates to modalities. There can be more adaptive strategies that take a priori clinical knowledge about a modality into account, which is a possible topic of future work. The late fusion methods discussed are also occasionally unstable during training. It is also conceivable that clinical text with different linguistic structure (e.g. short, more standardised radiology reports vs longer, less structured progress reports) behave differently when combined with other modalities. Further investigation is required to mitigate these issues. The current work aims to show the effect of adding modalities and adapting parameters specific to useful modal-

Task	Modality	IHMortality	Phenotype
SingleTask	Notes	0.517±0.052	0.712±0.004
SingleTask	TimeSeries	0.423±0.052	0.788±0.004
SingleTask	Notes + Tabular	0.519±0.04	0.72±0.007
SingleTask	Notes + TimeSeries	0.580±0.05	0.796±0.005
SingleTask	Notes + TimeSeries + Tabular	0.570±0.051	**0.814**±0.005
MultiTask	TimeSeries	0.423±0.052	0.77±0.005
MultiTask	TimeSeries + Tabular	0.526±0.003	0.781±0.002
MultiTask	Notes + TimeSeries	**0.601**±0.05	0.773±0.005
MultiTask	Notes + TimeSeries + Tabular	0.599±0.051	0.813±0.004

Table 1: Effect of multimodal learning with early fusion

RateAcrossModality	Modality	IHMortality	Phenotype
-	Notes	0.517±0.052	0.712±0.004
-	Time series	0.423±0.052	0.788±0.004
unified	Notes + TimeSeries	0.590±0.049	0.802±0.005
multimodal	Notes + TimeSeries	0.614±0.047	0.803±0.004
unified	Notes + TimeSeries + Tabular	0.601±0.051	0.815±0.005
multimodal	Notes + TimeSeries + Tabular	**0.62**±0.050	**0.817**±0.004

Table 2: Effect of multimodal learning with late fusion with varying learning rate across modalities

Task	Modality	Phenotype
SingleTask	Notes	0.707±0.003
SingleTask	TimeSeries	0.781±0.005
SingleTask	Notes + Tabular	0.73±0.006
SingleTask	Notes + TimeSeries	0.789±0.007
SingleTask	Notes + TimeSeries + Tabular	0.808±0.002
MultiTask	TimeSeries	0.767±0.006
MultiTask	TimeSeries + Tabular	0.772±0.002
MultiTask	Notes + TimeSeries	0.766±0.004
MultiTask	Notes + TimeSeries + Tabular	**0.812**±0.004

Table 3: Effect of multimodal learning with early fusion on phenotype (AUC-ROC weighted by label prevalence)

ities. Future work will also address comparisons not possible with existing baselines. More complex models with advanced architecture can be applied in a modular fashion in both single task and multi-task settings.

References

Pradeep K Atrey, M Anwar Hossain, Abdulmotaleb El Saddik, and Mohan S Kankanhalli. 2010. Multimodal fusion for multimedia analysis: a survey. *Multimedia systems*, 16(6):345–379.

Tadas Baltrušaitis, Chaitanya Ahuja, and Louis-Philippe Morency. 2018. Multimodal machine learning: A survey and taxonomy. *IEEE transactions on pattern analysis and machine intelligence*, 41(2):423–443.

Rich Caruana, Shumeet Baluja, and Tom Mitchell. 1995. Using the future to" sort out" the present: Rankprop and multitask learning for medical risk evaluation. *Advances in neural information processing systems*, 8.

Shizhe Chen and Qin Jin. 2015. Multi-modal dimensional emotion recognition using recurrent neural networks. In *Proceedings of the 5th International Workshop on Audio/Visual Emotion Challenge*, pages 49–56.

Alister D'Costa, Stefan Denkovski, Michal Malyska, Sae Young Moon, Brandon Rufino, Zhen Yang, Taylor Killian, and Marzyeh Ghassemi. 2020. Multiple sclerosis severity classification from clinical text. In *Proceedings of the 3rd Clinical Natural Language Processing Workshop*, pages 7–23, Online. Association for Computational Linguistics.

Alexandre De Brébisson, Étienne Simon, Alex Auvolat, Pascal Vincent, and Yoshua Bengio. 2015. Artificial neural networks applied to taxi destination prediction. ECMLPKDDDC'15, page 40–51, Aachen, DEU. CEUR-WS.org.

Naotsuna Fujimori, Rei Endo, Yoshihiko Kawai, and Takahiro Mochizuki. 2019. Modality-specific learning rate control for multimodal classification. In *Asian Conference on Pattern Recognition*, pages 412–422. Springer.

Hrayr Harutyunyan, Hrant Khachatrian, David C. Kale, Greg Ver Steeg, and Aram Galstyan. 2019. Multitask learning and benchmarking with clinical time series data. *Scientific Data*, 6(1):96. ArXiv: 1703.07771.

Sepp Hochreiter and Jürgen Schmidhuber. 1997. Long Short-Term Memory. *Neural Computation*, 9(8):1735–1780.

Shih-Cheng Huang, Anuj Pareek, Saeed Seyyedi, Imon Banerjee, and Matthew P. Lungren. 2020. Fusion of medical imaging and electronic health records using deep learning: a systematic review and implementation guidelines. *npj Digital Medicine*, 3(1):1–9.

Shaoxiong Ji, Erik Cambria, and Pekka Marttinen. 2020. Dilated convolutional attention network for medical code assignment from clinical text. In *Proceedings of the 3rd Clinical Natural Language Processing Workshop*, pages 73–78, Online. Association for Computational Linguistics.

Alistair EW Johnson, Tom J Pollard, Lu Shen, H Lehman Li-Wei, Mengling Feng, Mohammad Ghassemi, Benjamin Moody, Peter Szolovits, Leo Anthony Celi, and Roger G Mark. 2016. Mimic-iii, a freely accessible critical care database. *Scientific data*, 3(1):1–9.

Alex Kendall, Yarin Gal, and Roberto Cipolla. 2018. Multi-task learning using uncertainty to weigh losses for scene geometry and semantics. In *Proceedings of the IEEE conference on computer vision and pattern recognition*, pages 7482–7491.

Swaraj Khadanga, Karan Aggarwal, Shafiq Joty, and Jaideep Srivastava. 2019. Using clinical notes with time series data for ICU management. In *Proceedings of the 2019 Conference on Empirical Methods in Natural Language Processing and the 9th International Joint Conference on Natural Language Processing (EMNLP-IJCNLP)*, pages 6432–6437, Hong Kong, China. Association for Computational Linguistics.

Yoon Kim. 2014. Convolutional neural networks for sentence classification. In *Proceedings of the 2014 Conference on Empirical Methods in Natural Language Processing (EMNLP)*, pages 1746–1751, Doha, Qatar. Association for Computational Linguistics.

Diederik P Kingma and Jimmy Ba. 2014. Adam: A method for stochastic optimization. *arXiv preprint arXiv:1412.6980*.

Zachary C Lipton, Charles Elkan, and Balakrishnan Naryanaswamy. 2014. Optimal thresholding of classifiers to maximize f1 measure. In *Joint European Conference on Machine Learning and Knowledge Discovery in Databases*, pages 225–239. Springer.

Zhun Liu, Ying Shen, Varun Bharadhwaj Lakshminarasimhan, Paul Pu Liang, Amir Zadeh, and Louis-Philippe Morency. 2018. Efficient low-rank multimodal fusion with modality-specific factors. *arXiv preprint arXiv:1806.00064*.

Ilya Loshchilov and Frank Hutter. 2017. Decoupled weight decay regularization. *arXiv preprint arXiv:1711.05101*.

Tomas Mikolov, Ilya Sutskever, Kai Chen, Greg S Corrado, and Jeff Dean. 2013. Distributed representations of words and phrases and their compositionality. In *Advances in neural information processing systems*, pages 3111–3119.

Anna Rumshisky, Kirk Roberts, Steven Bethard, and Tristan Naumann, editors. 2020. *Proceedings of the 3rd Clinical Natural Language Processing Workshop*. Association for Computational Linguistics, Online.

A. Tiulpin, S. Klein, S. Bierma-Zeinstra, J. Thevenot, Esa Rahtu, J. Meurs, E. Oei, and S. Saarakkala. 2019. Multimodal Machine Learning-based Knee Osteoarthritis Progression Prediction from Plain Radiographs and Clinical Data. *Scientific Reports*.

Jian Wang, Yonghao He, Cuicui Kang, Shiming Xiang, and Chunhong Pan. 2015. Image-text cross-modal retrieval via modality-specific feature learning. In *Proceedings of the 5th ACM on International Conference on Multimedia Retrieval*, pages 347–354.

Griffin M. Weber, Kenneth D. Mandl, and Isaac S. Kohane. 2014. Finding the Missing Link for Big Biomedical Data. *JAMA*, 311(24):2479–2480.

Bo Yang and Lijun Wu. 2021. How to leverage multimodal ehr data for better medical predictions? *arXiv preprint arXiv:2110.15763*.

Yiqun Yao and Rada Mihalcea. 2022. Modality-specific Learning Rates for Effective Multimodal Additive Late-fusion. In *Findings of the Association for Computational Linguistics: ACL 2022*, pages 1824–1834, Dublin, Ireland. Association for Computational Linguistics.

Dongdong Zhang, Changchang Yin, Jucheng Zeng, Xiaohui Yuan, and Ping Zhang. 2020. Combining structured and unstructured data for predictive models: a deep learning approach. *BMC medical informatics and decision making*, 20(1):1–11.

How Long Is Enough? Exploring the Optimal Intervals of Long-Range Clinical Note Language Modeling

Samuel Cahyawijaya[1*], Bryan Wilie[1*], Holy Lovenia[1*], MingQian Zhong[2,3,4], Huan Zhong[2,3], Nancy Y. Ip[2,3,4], Pascale Fung[1]

[1]Center for Artificial Intelligence Research (CAiRE), Department of Electronic and Computer Engineering, The Hong Kong University of Science and Technology, Hong Kong, China
{scahyawijaya, bwilie, hlovenia, pascale}@ust.hk

[2]Division of Life Science, State Key Laboratory of Molecular Neuroscience, Molecular Neuroscience Center, The Hong Kong University of Science and Technology, Clear Water Bay, Kowloon, Hong Kong, China
{mzhongac,dorothyzhong,boip}@ust.hk

[3]Hong Kong Center for Neurodegenerative Diseases, Hong Kong Science Park, Hong Kong, China
{mzhongac,dorothyzhong,boip}@ust.hk

Abstract

Large pre-trained language models (LMs) have been widely adopted in biomedical and clinical domains, introducing many powerful LMs such as bio-lm and BioELECTRA. However, the applicability of these methods to real clinical use cases is hindered, due to the limitation of pre-trained LMs in processing long textual data with thousands of words, which is a common length for a clinical note. In this work, we explore long-range adaptation from such LMs with Longformer, allowing the LMs to capture longer clinical notes context. We conduct experiments on three n2c2 challenges datasets and a longitudinal clinical dataset from Hong Kong Hospital Authority electronic health record (EHR) system to show the effectiveness and generalizability of this concept, achieving 10% F1-score improvement. Based on our experiments, we conclude that capturing a longer clinical note interval is beneficial to the model performance, but there are different cut-off intervals to achieve the optimal performance for different target variables. Our code is available at https://github.com/HLTCHKUST/long-biomedical-model.

1 Introduction

Clinical note is one of the most abundant data available in EHR systems, which records most of the patient interaction with the hospital services, such as consultation with doctors, procedure note, laboratory report, discharge summary, etc.[1] Despite retaining rich clinical information, clinical notes are highly unstructured and composed of non-standardized information, which curbs the potential practicality of such information. Large pre-trained LMs, such as BERT (Devlin et al., 2019), RoBERTa (Liu et al., 2020), GPT-2 (Radford et al., 2019), etc., have been shown to work well in extracting crucial information from clinical notes by utilizing transfer learning and attention mechanism (Ji et al., 2021; Alsentzer et al., 2019; Lewis et al., 2020). The adaptation of these models to biomedical and clinical domain emphasizes this success, establishing many new state-of-the-art performances on multiple biomedical and clinical benchmarks (Peng et al., 2019; Gu et al., 2021; Zhang et al., 2022).

While the attention mechanism embedded in the pre-trained models enables them to achieve great performance, it is to be noted that it also causes a quadratic growth in computation cost with respect to input sequence length (Tay et al., 2022; Wang et al., 2020; Cahyawijaya et al., 2022). This makes efficiently processing long documents with pre-trained LMs difficult, especially in clinical note modeling, in which a single clinical note tends to consist of hundreds or even thousands of words (Uzuner et al., 2008; Uzuner, 2009; Stubbs et al., 2015; Gehrmann et al., 2018; Johnson et al., 2019; Stubbs et al., 2019). Current approaches to this problem commonly involve truncation, chunking, or windowing of the long input sequence, preventing the models from acquiring an entire medical record information provided by a whole clinical note. Considering that clinical note modeling requires capturing and understanding the underlying long-term dependencies in the clinical notes, this certainly puts a limit on their predictive capability.

*These authors contributed equally.

[1]https://www.healthit.gov/isa/uscdi-data-class/clinical-notes

Proceedings of the 13th International Workshop on Health Text Mining and Information Analysis (LOUHI), pages 160 - 172
December 7, 2022 ©2022 Association for Computational Linguistics

For this reason, to maximize the models' capability without sacrificing a part of the input clinical notes, we explore the application of long-range adaptation through linear attention mechanism (Dai et al., 2019; Beltagy et al., 2020; Wang et al., 2020; Choromanski et al., 2021), which reduces the computation cost of attention from quadratic to linear in regards to input sequence length.

In this work, we focus on assessing the benefit of capturing longer clinical notes on large pre-trained LMs to n2c2 (National Clinical NLP Challenges)[2] clinical tasks by adapting a linear attention mechanism, i.e., Longformer (Beltagy et al., 2020). Furthermore, to test the generality of this approach, we evaluate it on a longitudinal clinical note corpus from Hong Kong Hospital Authority EHR system, which covers records from 43 hospitals in Hong Kong. Lastly, we hypothesize that modeling longer interval of clinical notes improves the prediction quality of the models on any clinical task. To prove our hypothesis, we conduct our experiment using different context-length, allowing the model to access various intervals of clinical notes. Our result suggests that a longer interval of clinical notes increases the prediction quality of the models in most cases, but there is a limit of context length required depending on the target variable.

Our contributions in this work can be summarized in three-fold:

- We assess the effectiveness of capturing longer interval of clinical notes on biomedical and clinical large pre-trained LMs on three n2c2 challenges which increase the performance by ~10% F1-score,
- We evaluate the generalization of this approach using longitudinal clinical note data gathered in Hong Kong Hospital Authority EHR system on two clinical tasks, i.e., disease risk and mortality risk predictions, which improve the performance by ~5-10 F1-score,
- We observe that each target variable has a different optimal clinical notes cut-off interval and we conclude that the optimal cut-off interval for mortality risk prediction is ~2-3 months, while for disease risk prediction, it requires 3.5 years or even longer interval to achieve the optimal performance.

[2]https://n2c2.dbmi.hms.harvard.edu/

2 Related Works

Clinical Note Modeling Clinical notes have been utilized for various applications in healthcare. Text mining methods for analyzing pharmacovigilance signals using clinical notes have been explored and yield promising results (Haerian et al., 2012; LePendu et al., 2012, 2013). Clinical notes with other EHR data are also employed for estimating the readmission time and mortality risk of the next patient encounter (Hammoudeh et al., 2018; Rajkomar et al., 2018). Clinical note data is also effective for analyzing disease comorbidity, such as mental illness (Wu et al., 2013), autoimmune diseases (Escudié et al., 2017), and obesity (Pantalone et al., 2017). Predicting disease risk using clinical note data has also been explored (Miotto et al., 2016; Choi et al., 2018; Liu et al., 2019, 2018; Koleck et al., 2019). Despite all the efforts in clinical note modeling, to the best of our knowledge, how clinical note interval impacts the performance of pre-trained LMs has never been studied.

Biomedical and Clinical Pre-trained LMs Self-supervised pre-training LMs employing transformer-based architectures (Vaswani et al., 2017), such as BERT (Devlin et al., 2019), RoBERTa (Liu et al., 2020), and ELECTRA (Clark et al., 2019), have thrived in various general domain NLP benchmarks (Wang et al., 2018; Rajpurkar et al., 2016; Ladhak et al., 2020; Lai et al., 2017; Wilie et al., 2020; Cahyawijaya et al., 2021; Park et al., 2021). To extend the understanding of these LMs to the linguistic properties in biomedical and clinical domain, a generation of LMs exploiting biomedical and clinical corpora emerges.

In 2019, Alsentzer et al. (2019) introduce BioBERT, an extended version of BERT pre-trained on large-scale biomedical data (i.e., PubMed abstracts and PMC full-text articles) which surpasses off-the-shelf BERT in three fundamental downstream tasks in biomedical domain. Due to the linguistic differences exhibited by non-clinical biomedical texts and clinical texts, Alsentzer et al. (2019) introduce ClinicalBERT by fine-tuning BERT and BioBERT on the MIMIC-III corpus, and improve the performance over five clinical NLP tasks.

Unlike prior works, PubMedBERT (Gu et al., 2020) performs biomedical pre-training from scratch, which offers larger performance gains over various biomedical downstream tasks in the

BLURB benchmark. Similarly, bio-lm (Lewis et al., 2020) employs recent pre-training advances, utilizes various biomedical and clinical corpora for pre-training, and achieves the highest performance on 9 biomedical and clinical NLP tasks. In 2021, BioELECTRA (Kanakarajan et al., 2021), a general domain ELECTRA (Clark et al., 2019) pre-trained on biomedical corpora, sets the new state-of-the-art performance for all datasets in the BLURB benchmark and 4 datasets in the BLUE benchmark (Peng et al., 2019).

Long Sequence Language Modeling Recent progress in language modeling is dominated by transformer-based models which shows a remarkable results on numerous tasks. Nevertheless, these models have limited capability to process long-range clinical notes data due to its quadratic attention complexity. Various approaches have been introduced to reduce this complexity problem, such as recurrence approach (Dai et al., 2019; Rae et al., 2020), sparse and local attention patterns (Kitaev et al., 2020; Qiu et al., 2020; Child et al., 2019; Zaheer et al., 2020; Beltagy et al., 2020), low-rank approximation (Wang et al., 2020; Winata et al., 2020), and kernel methods (Katharopoulos et al., 2020; Choromanski et al., 2021). Adaptation from existing pre-trained models to some of these methods have also been explored and show the potential for knowledge transfer (Beltagy et al., 2020; Choromanski et al., 2021). In this work, we utilize Longformer (Beltagy et al., 2020) to enable the model to capture long-range clinical note information.

3 Methodology

3.1 Problem Definition

Clinical notes are narrative patient data relevant to the context identified by note types[3]. There are multiple types of clinical notes, e.g., discharge summary, consultation note, progress note, lab report, etc. In general, a single clinical note consists of a text narrative and additional metadata defining the clinical note, e.g., note identifier, recording timestamp of the note, etc. In n2c2 challenges, a single clinical note is presented in a textual format with the metadata written on top of the text narrative, while a longitudinal clinical note is presented as a concatenation of several clinical notes with a separator text placed between two clinical notes. This clinical note is usually long, ranging from several hundreds to thousands words, while most existing biomedical and clinical pre-trained LMs can only capture up to 512 subwords, which is insufficient to capture the whole content of most clinical notes.

3.2 Long-Range Clinical Note LMs

We increase the capacity of LMs to process longer clinical notes by adapting Longformer (Beltagy et al., 2020) to the existing biomedical and clinical pre-trained LMs. Longformer enables linear attention mechanism by dividing single quadratic all-to-all attention into two attention steps, i.e., sliding-window and global attentions. Sliding-window attention allows each token to attend to neighboring tokens, while global attention allows some, usually a few, tokens to attend to all tokens, hence has a better computation complexity compared to the quadratic attention mechanism. It is to be noted that when extending an original transformer-based model into a Longformer, some new parameters are introduced, i.e., the new positional embeddings, the sliding-window projection parameters, and global attention projection parameters. For the positional embeddings, following (Beltagy et al., 2020), we copy the weights of the pre-trained positional embeddings to initialize the new positional embeddings. For the sliding-window and global attention parameters, we initialize both projection parameters with the pre-trained projection parameters.

4 Long-Range Clinical Note LMs on n2c2 Challenges

We assess the effectiveness of long-range clinical note LMs on US-based clinical note datasets from three n2c2 challenges. Additionally, we also evaluate six different pre-trained LMs without long-range adaptation to benchmark the performance of the biomedical and clinical LMs.

4.1 Dataset

We use three clinical datasets concentrating on classifying diverse clinical problems from n2c2. These datasets are: 1) n2c2 2006 smoking challenge, focusing on predicting smoking status of patients based on their discharge summary; 2) n2c2 2008 obesity challenge, focusing on recognizing obesity and its comorbidities of patients through their discharge summary; and 3) n2c2 2018 cohort selection challenge, focusing on determining if a patient meets selection criteria of certain clinical trials co-

[3] `https://www.healthit.gov/isa/uscdi-data-class/clinical-notes`

Dataset	|Train|	|Test|	Word count				Longitudinal?	#Label	#Class
			Median	Q3	95%	Max			
2006 Smoking	398	104	677	1096	1775	3023	No	5	1
2008 Obesity (Textual)	730	507	1084	1425	2094	4280	No	16	4
2008 Obesity (Intuitive)	730	507	1084	1425	2094	4280	No	16	4
2018 Cohort Selection	202	86	2550	3235	4578	7070	Yes	13	1

Table 1: The overall statistics of the n2c2 datasets used in our experiment.

horts through longitudinal clinical notes. We utilize BigBIO framework (Fries et al., 2022)[4] to load the n2c2 datasets. We provide overview of these datasets in Table 1.

n2c2 2006 Smoking Challenge We utilize the smoking prediction subtask from n2c2 2006 challenge (Uzuner et al., 2008), where each data instance consists of a de-identified discharge summary annotated by two pulmonologists with smoking status. This smoking status can be either past smoker" (when it is explicitly stated that the patient is a past smoker or that the patient used to smoke but has stopped for at least a year), "current smoker" (when it is explicitly stated that the patient is a current smoker or that the patient has smoked within the past year), "smoker" (when there is not enough temporal information to classify whether a patient is a "past smoker" or "current smoker"), "non-smoker" (when a patient's discharge summary indicates an absence of smoking habit), or "unknown" (when there is no mention of smoking).

n2c2 2008 Obesity Challenge The n2c2 2008 obesity challenge (Uzuner, 2009) consists of 1027 pairs of de-identified discharge summaries and 16 disease labels. The disease labels include obesity and its 15 comorbidities, e.g., asthma, atherosclerotic cardiovascular disease (CAD), congestive heart failure (CHF), depression, diabetes mellitus (DM), gallstones/cholecystectomy, gastroesophageal reflux disease (GERD), gout, hypercholesterolemia, hypertension (HTN), hypertriglyceridemia, obstructive sleep apnea (OSA), osteoarthritis (OA), peripheral vascular disease (PVD), and venous insufficiency.

The annotation for each discharge summary is done by providing each disease label with either "present", "absent", "questionable", or "unmentioned". The dataset has two types of annotations, i.e., textual judgement (only based on related explicit statements) and intuitive judgement (based on everything written in the discharge summaries). We use both annotations in our experiments and report the evaluation scores for each annotation.

n2c2 2018 Cohort Selection Challenge The 2018 Shared Task 1: Clinical Trial Cohort Selection (Stubbs et al., 2019) reuses 288 patient records from the 2014 n2c2 shared task dataset (Stubbs et al., 2015) and reframes it as a cohort selection task, which requires an automatic evaluation of whether a patient fits or does not fit in certain cohorts according to their longitudinal de-identified clinical notes, ranging between 2-5 clinical notes.

The cohorts or selection criteria used in the dataset as labels are: DRUG-ABUSE (current or past usage of drugs), ALCOHOL-ABUSE (current alcohol intake over weekly recommended limit), ENGLISH (English-speaking patient), MAKES-DECISIONS (patients required to make their own medical decisions), ABDOMINAL (history of related surgery), MAJOR-DIABETES (major diabetes-related complication), ADVANCED-CAD (advanced cardiovascular disease), MI-6MOS (myocardial infarction in the past 6 months), KETO-1YR (diagnosis of ketoacidosis in the past year), DIETSUPP-2MOS (dietary supplement intake in the past 2 months, excluding vitamin D), ASP-FOR-MI (usage of aspirin to prevent MI), HBA1C (any hemoglobin A1c value between 6.5% and 9.5%), and CREATININE (serum creatinine above the upper limit of normal). Two annotators with medical expertise classify each label of a patient's set of clinical notes as either "met" or "not met".

4.2 Models

In this experiment, we compare several pre-trained LMs, covering two variants of BERT model representing general domain LMs, i.e., uncased BERT[5] and cased BERT[6], two variants of biomedical do-

[4]https://github.com/bigscience-workshop/biomedical

[5]https://huggingface.co/bert-base-uncased

[6]https://huggingface.co/bert-base-cased

	2006 Smoking		2008 Obesity (Text.)		2008 Obesity (Intui.)		2018 Cohort Selection	
	micro-f1	macro-f1	micro-f1	macro-f1	micro-f1	macro-f1	micro-f1	macro-f1
				Baseline				
Top-5 scorer	88.00%	69.00%	97.04%	77.18%	95.58%	63.44%	90.30%	
Top-10 scorer	86.00%	58.00%	96.39%	61.40%	95.08%	62.87%	87.70%	
				Pre-trained Language Model				
BERT-cased	61.63%	31.79%	82.47%	38.73%	81.69%	51.71%	72.80%	48.45%
BERT-uncased	65.63%	41.12%	85.73%	40.83%	83.46%	53.28%	74.86%	51.32%
clinicalBERT	56.59%	39.34%	85.64%	40.64%	85.20%	54.88%	72.83%	49.99%
PubMedBERT	69.38%	41.65%	**88.98%**	46.27%	**87.11%**	**56.47%**	74.78%	49.94%
bio-lm	**71.44%**	**49.43%**	86.57%	43.15%	84.92%	54.73%	**75.03%**	**52.18%**
BioELECTRA	70.72%	48.26%	86.71%	**48.26%**	85.31%	55.00%	74.32%	49.10%
				Long-range Pre-trained Language Model				
bio-lm (1024)	82.12%	55.72%	92.52%	50.36%	90.36%	59.13%	77.03%	53.94%
bio-lm (2048)	**86.01%**	62.30%	96.44%	55.99%	94.76%	62.61%	76.76%	52.93%
bio-lm (4096)	84.52%	57.76%	**97.11%**	55.68%	**95.48%**	63.19%	79.42%	57.85%
bio-lm (8192)	84.66%	59.49%	97.07%	55.08%	**95.48%**	**63.20%**	81.43%	**61.95%**
BioELECTRA (1024)	82.98%	63.35%	93.54%	54.47%	90.40%	59.12%	74.95%	51.69%
BioELECTRA (2048)	82.84%	61.09%	96.03%	56.08%	91.69%	60.21%	77.59%	54.39%
BioELECTRA (4096)	80.40%	57.22%	95.81%	56.06%	92.88%	61.12%	79.10%	56.38%
BioELECTRA (8192)	85.21%	**64.32%**	96.20%	**59.59%**	92.78%	61.09%	**81.63%**	58.44%

Table 2: Evaluation results of our experiments on the n2c2 datasets. Top-5 and Top-10 scorers are retrieved from the submission benchmark of corresponding challenge. The number inside the bracket denotes the length of context that can be captured by the model. **Bold** and underline denotes the first and second best scores within a group.

main LMs, i.e. PubMedBERT (Gu et al., 2021)[7] and BioELECTRA (Kanakarajan et al., 2021)[8], one variant of clinical domain LM, i.e., ClinicalBERT (Alsentzer et al., 2019)[9], and one variant of mixed biomedical and clinical domains LM, i.e., bio-lm (Lewis et al., 2020)[10].

To enable longer context clinical note modeling, we adapt Longformer (Beltagy et al., 2020) with the initialization strategy specified in §3.2. We conduct experiments with four different context lengths, i.e., $\{1024, 2048, 4096, 8192\}$ on two pre-trained LMs variants, i.e., BioELECTRA and bio-lm.

4.3 Training and Evaluation

Following BERT, RoBERTa, and bio-lm experiments, we tune the learning rate for all BERT and RoBERTa models from [1e-5, 2e-5, 3e-5]. While for the BioELECTRA model, following ELECTRA (Clark et al., 2019) and BioELECTRA (Kanakarajan et al., 2021), we tune the learning rate from [5e-5, 1e-4, 2e-4]. In all experiments, we use a batch size of 8, and a linear learning rate decay. For the n2c2 2006 and n2c2 2008 tasks, we train the models for 50 epochs, while for the n2c2 2018 task, we train the models for 80 epochs. For the evaluation, we incorporate the official evaluation metrics defined for each challenge. All of them report micro-F1 and macro-F1 scores.

4.4 Result and Analysis

As shown in Table 2, in general, domain-specific LMs yield higher performance compared to general domain LMs, except for ClinicalBERT which performs on a par with the general domain BERT models. PubMedBERT, bio-lm, and BioELECTRA produce comparable evaluation performances across all tasks, with ~2-5% higher F1-score compared to the general domain BERT and ClinicalBERT. Nevertheless, the scores are much lower compared to the Top-10 scorer on the challenge benchmark since the models can only capture partial information of the clinical note data.

By increasing the context length of the model, the performance rises significantly. Comparing with the original pre-trained versions of the models, the best performing long-range pre-trained LM improves the evaluation performance by ~10% F1-score in all datasets. As shown in Figure 1, models with longer context length tend to perform better, but the performance gain is limited to the length of

[7]https://huggingface.co/microsoft/BiomedNLP-PubMedBERT-base-uncased-abstract

[8]https://huggingface.co/kamalkraj/bioelectra-base-discriminator-pubmed

[9]https://huggingface.co/emilyalsentzer/Bio_ClinicalBERT

[10]https://huggingface.co/EMBO/bio-lm

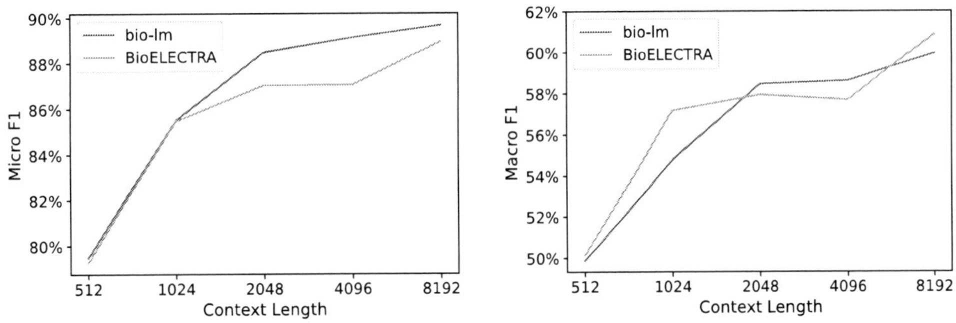

Figure 1: Effect of capturing longer clinical notes context to the evaluation performance, i.e., on micro-F1 **(Left)** and macro-F1 **(Right)**, averaged over the context length across the evaluated n2c2 tasks.

the clinical notes in the dataset. For instance, on the n2c2 2006 dataset, the performance improvement of both bio-lm and BioELECTRA models are steeper from context length 512 to 1024 rather than from context length 1024 to 2048, 2048 to 4096, and 4096 to 8192. This is because a huge portion of the notes in the datasets can be sufficiently captured within 1024 subwords. In contrast, the performance improvement on the n2c2 2018 dataset is more linear per context length step since most of the length of the clinical notes is much longer than the other two datasets. Every step of extending the context length provides more information to the model, which is likely to improve the model performance considerably.

On the n2c2 2006 and 2008 challenges, our best performing models mostly achieve a comparable score to the Top-10 or Top-5 scorer of the corresponding challenge benchmark. This is a remarkable feat since our models neither utilize any ensemble method, incorporate any clinical expert, nor exploit external data–common practices used by the top scorers in the challenge benchmarks.

5 Long-Range Clinical Note LMs on Hong Kong Longitudinal Dataset

We assess the generalization and effectiveness of long-range clinical notes LMs on Hong Kong longitudinal clinical note data. We construct a longitudinal dataset with two target variables, i.e., disease risk and mortality risk, and evaluate long-range LMs on the dataset. In addition, we add a baseline model, which takes high-level features extracted from the corresponding tabular data provided by the EHR system as the input, to assess the effectiveness of clinical note modeling.

Split	# Patients	# Seen patient records	# Unseen patient records
Train	278,253	2,027,561	-
Valid	3,621	3,177	-
Test	17,903	15,541	2,362
Total	299,777	2,046,279	2,362

Table 3: The overall statistics of our Hong Kong longitudinal dataset. **# Seen patient records** and **# Unseen patient records** indicate the number of records on the *seen* and *unseen* test set respectively.

5.1 Dataset Construction

We construct a longitudinal clinical note dataset for disease risk and mortality risk predictions from anonymized cancer cohort patient records gathered in the Hong Kong Hospital Authority EHR system covering 43 hospitals in Hong Kong. The patient records span across the year 2000 and 2018. We exclude all patients having less than two clinical notes and gather a total of $\sim$300,000 patients. To construct labelled data for the supervised learning, from patient P_i with T clinical records, we build $T-1$ labelled autoregressive data $\mathcal{D}^{Pi} = \{\{C_k^{P_i}\}_{k=1}^t, Y_{t+1}^{P_i}\}_{t=1}^{T-1}$, where $\{C_k^{P_i}\}_{k=1}^t$ denotes t prior clinical notes of the patient P_i, and $Y_{t+1}^{P_i}$ denotes the target criterion retrieved from the $t+1^{th}$ clinical record of the patient P_i. We collect over $\sim$2M labelled clinical notes from all patients with two targets: disease risk and mortality risk.

We take the last two health records from all patient records in the year 2018 as the validation and test sets. To assess the generalization to new patient data, we omit some patient data from the training set and only used the last labelled record of those patients as the *unseen* test set. The remaining test data becomes the *seen* test set. The dataset statistics

Test set	Models	Diagnosis				Mortality	
		Top-1	Top-3	Top-5	F1	F1	AUC
Seen	**EHR-FFN**	64.3%	75.7%	80.3%	40.6%	49.5%	78.1%
	BioELECTRA (512)	76.2%	88.6%	91.8%	51.6%	61.5%	**92.0%**
	BioELECTRA (2048)	<u>79.8%</u>	<u>91.5%</u>	<u>94.3%</u>	<u>54.2%</u>	**65.3%**	<u>91.9%</u>
	BioELECTRA (8192)	**81.3%**	**92.9%**	**95.5%**	**55.7%**	<u>64.9%</u>	91.8%
Unseen	**EHR-FFN**	17.8%	32.9%	43.1%	9.5%	49.6%	73.9%
	BioELECTRA (512)	63.4%	78.6%	83.7%	43.1%	<u>52.2%</u>	84.8%
	BioELECTRA (2048)	<u>66.3%</u>	<u>81.2%</u>	<u>85.9%</u>	<u>45.1%</u>	52.0%	<u>85.8%</u>
	BioELECTRA (8192)	**69.1%**	**84.0%**	**88.2%**	**46.8%**	**52.3%**	**88.1%**

Table 4: Evaluation results of our experiments on the *seen* patient test set and the *unseen* patient test set. **Bold** and <u>underline</u> denotes the first and the second best score on each test set, respectively.

is shown in Table 3. For the disease risk estimation, we take the final disease diagnosis on the next clinical record as the label. For cancer diseases, we group the diagnosis based on the cancer site categorization from the Hong Kong Cancer Registry[11], while for other diseases, we take the first three digits of the ICD-10 codes. In total, there are 79 classes for disease risk estimation. For the mortality label, we retrieve the mortality status from the discharge code from the next clinical record of the corresponding patient. The label distribution of the dataset is shown in Appendix A.

5.2 Models

We experiment using Longformer with three variants of sequence length, i.e., $\{512, 2048, 8192\}$. We initialized all models with the same pre-trained BioELECTRA (Kanakarajan et al., 2021) checkpoint as in §4.3. To assess the effectiveness of clinical note modeling, we employ another baseline using a 4-layer feedforward model ($\sim$5M parameters), which takes an input of 3,942 dimension high-level features from the EHR database (EHR-FFN). Similar to DeepPatient (Miotto et al., 2016), we extract high-level features from the diagnoses, medications, procedures, and laboratory test records by counting the occurrence of each feature type. In addition, we also add other features such as length of stay, the indicator for emergency unit admission, age group, etc. The details of EHR-FFN and the extracted features are shown in Appendix B.

5.3 Training and Evaluation

We train all of the models with an initial learning rate of 5e-5, batch size of 48, and a linear learning rate decay. We train the model for 3 epochs and test the model with the best validation score. For evaluating the diagnosis label, we incorporate the F1-score along with the Top-1, Top-3,and Top-5 accuracy scores. For the mortality label, we incorporate F1-score and AUC. The evaluation is conducted on two different test sets: (i) the *seen* patient test set and (ii) the *unseen* patient test set.

5.4 Results and Analysis

Effect of Clinical Note Modeling We show our experiment results for the *seen* and the *unseen* test sets in Table 4. All BioELECTRA models yield higher results than the EHR-FFN for both test sets, showing the effectiveness of clinical note modeling for disease risk and mortality risk predictions using EHR data. From the comparison between different clinical notes interval of the BioELECTRA model, we found that modeling longer clinical note interval will likely increase the performance on both tasks. This behavior aligns with the results reported in §2. Nevertheless, this behavior does not apply to the mortality risk prediction on the *seen* test set. We describe this phenomenon further in §5.4.

Generalization to New Patient Data We observe that there is a huge gap of performance for the baseline EHR-FFN model, especially in the diagnosis predictions of *seen* and *unseen* test set ($\sim$40 p.p.). In this case, utilizing clinical note modeling closes the performance gap on the *seen* and *unseen* test sets to be much narrower ($\sim$10 p.p.) on either label, especially for the BioELECTRA model with longer context length. This suggests that longer clinical notes interval not only improves the performance of the model on the similar patient record distribution, but also improves the performance on

[11]`https://www3.ha.org.hk/cancereg/allages.asp`

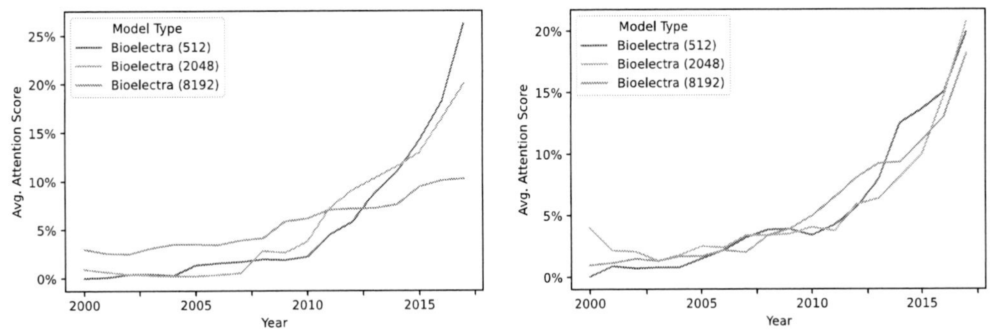

Figure 2: **(Left)** and **(Right)** show the clinical notes **time importance** of the disease risk prediction and mortality risk prediction, respectively.

the out-of-distribution patient records.

Optimal Cut-off Interval for Disease Risk and Mortality Risk Prediction We measure the number of clinical notes that can be processed by the models to analyze the optimal cut-off interval. Using the length statistics on our dataset, we find that our BioELECTRA (512), BioELECTRA (2048), and BioELECTRA (8192) models can encode 4, 17, 66 clinical notes on average, which correspond to the average clinical note intervals of 2-3 months, ~1 year, and 3.5 years, respectively. As shown in Table 4, for the disease risk prediction label, the utilization of longer clinical notes intervals always yields better performance, while the same trend is not observed for the mortality risk label. This evidence suggests that there are different optimal interval of clinical notes required to infer the correct prediction for different target labels.

To verify this phenomenon, we analyze the input fractions considered to be important by the models. Specifically, we retrieve 1,000 correctly-predicted samples with the highest confidence values from each of the models and collect the clinical note timestamps corresponding to the high-magnitude (>5% of the total input gradient magnitude) input gradient with respect to the output prediction by using saliency map (Simonyan et al., 2014; Yosinski et al., 2015; Wallace et al., 2019). The timestamps from all samples are then aggregated with yearly granularity. We denote the number of year occurrences divided by the total number of timestamps collected as **time importance** to show how likely the model attends to the clinical note from the corresponding year given the label prediction in 2018.

As shown in Figure 2, for the disease risk label, the slope of the **time importance** curves over the years become more flattened as the utilized clinical note interval widens, indicating that the **time importance** spreads more uniformly on longer clinical note intervals. Whereas for the mortality risk label, the **time importance** curve has a similar slope over different clinical notes intervals. This evidence supports that for modeling an accurate disease risk prediction, a long clinical note interval ($\geq$ 3.5 years) is required. While for mortality risk prediction, a shorter clinical note interval (~2-3 months) is sufficient to reach optimal performance.

6 Conclusion

In this paper, we show the importance of capturing longer clinical notes for biomedical and clinical large pre-trained LMs on 6 clinical NLP tasks on the United States and Hong Kong clinical note data. Our result suggests that utilizing longer clinical notes can significantly increase the performance of LMs by ~5-10% F1-score without the loss of generalization to the unseen data. We also observe that incorporating a longer interval of clinical notes does not always entail performance improvement and there is an optimal cut-off interval depending on the target variable. Based on our analysis, we conclude that an interval of ~2-3 months is the optimal cut-off for mortality risk prediction, while 3.5 years or an even longer interval of clinical notes is required to achieve the optimal performance for disease risk prediction. Future work in long-range clinical note modeling would open up opportunities towards a general solution in clinical NLP.

7 Limitation

Although there are many linear attention mechanisms that have been proposed (Dai et al., 2019; Kitaev et al., 2020; Beltagy et al., 2020; Zaheer et al., 2020), the exploration of linear attention in our experiments is currently limited to Longformer (Beltagy et al., 2020). Furthermore, the constructed longitudinal clinical note dataset from the Hong Kong Hospital Authority EHR system cannot be made public due to the data-sharing policy. Lastly, due to the limited computational power, we only conduct the long-range clinical notes experiment for bio-lm and BioELECTRA for the n2c2 experiment and BioELECTRA for the Hong Kong longitudinal dataset. We conjecture that the performance of the long-range versions of other pre-trained models will follow similar trends to the result on existing biomedical and clinical benchmarks.

Acknowledgements

This work has been partially funded by School of Engineering PhD Fellowship Award, the Hong Kong University of Science and Technology and PF20-43679 Hong Kong PhD Fellowship Scheme, Research Grant Council, Hong Kong.

References

Emily Alsentzer, John Murphy, William Boag, Wei-Hung Weng, Di Jindi, Tristan Naumann, and Matthew McDermott. 2019. Publicly available clinical BERT embeddings. In *Proceedings of the 2nd Clinical Natural Language Processing Workshop*, pages 72–78, Minneapolis, Minnesota, USA. Association for Computational Linguistics.

Iz Beltagy, Matthew E. Peters, and Arman Cohan. 2020. Longformer: The long-document transformer. *CoRR*, abs/2004.05150.

Samuel Cahyawijaya, Genta Indra Winata, Bryan Wilie, Karissa Vincentio, Xiaohong Li, Adhiguna Kuncoro, Sebastian Ruder, Zhi Yuan Lim, Syafri Bahar, Masayu Khodra, Ayu Purwarianti, and Pascale Fung. 2021. IndoNLG: Benchmark and resources for evaluating Indonesian natural language generation. In *Proceedings of the 2021 Conference on Empirical Methods in Natural Language Processing*, pages 8875–8898, Online and Punta Cana, Dominican Republic. Association for Computational Linguistics.

Samuel Cahyawijaya, Tiezheng Yu, Zihan Liu, Xiaopu Zhou, Tze Wing Tiffany Mak, Yuk Yu Nancy Ip, and Pascale Fung. 2022. SNP2Vec: Scalable self-supervised pre-training for genome-wide association study. In *Proceedings of the 21st Workshop on Biomedical Language Processing*, pages 140–154, Dublin, Ireland. Association for Computational Linguistics.

Rewon Child, Scott Gray, Alec Radford, and Ilya Sutskever. 2019. Generating long sequences with sparse transformers. *arXiv preprint arXiv:1904.10509*.

Edward Choi, Cao Xiao, Walter F. Stewart, and Jimeng Sun. 2018. Mime: Multilevel medical embedding of electronic health records for predictive healthcare. *arXiv:1810.09593 [cs, stat]*. ArXiv: 1810.09593.

Krzysztof Marcin Choromanski, Valerii Likhosherstov, David Dohan, Xingyou Song, Andreea Gane, Tamas Sarlos, Peter Hawkins, Jared Quincy Davis, Afroz Mohiuddin, Lukasz Kaiser, David Benjamin Belanger, Lucy J Colwell, and Adrian Weller. 2021. Rethinking attention with performers. In *International Conference on Learning Representations*.

Kevin Clark, Minh-Thang Luong, Quoc V Le, and Christopher D Manning. 2019. Electra: Pre-training text encoders as discriminators rather than generators. In *International Conference on Learning Representations*.

Zihang Dai, Zhilin Yang, Yiming Yang, Jaime Carbonell, Quoc Le, and Ruslan Salakhutdinov. 2019. Transformer-XL: Attentive language models beyond a fixed-length context. In *Proceedings of the 57th Annual Meeting of the Association for Computational Linguistics*, pages 2978–2988, Florence, Italy. Association for Computational Linguistics.

Jacob Devlin, Ming-Wei Chang, Kenton Lee, and Kristina Toutanova. 2019. BERT: Pre-training of deep bidirectional transformers for language understanding. In *Proceedings of the 2019 Conference of the North American Chapter of the Association for Computational Linguistics: Human Language Technologies, Volume 1 (Long and Short Papers)*, pages 4171–4186, Minneapolis, Minnesota. Association for Computational Linguistics.

Jean-Baptiste Escudié, Bastien Rance, Georgia Malamut, Sherine Khater, Anita Burgun, Christophe Cellier, and Anne-Sophie Jannot. 2017. A novel data-driven workflow combining literature and electronic health records to estimate comorbidities burden for a specific disease: a case study on autoimmune comorbidities in patients with celiac disease. *BMC Med. Inform. Decis. Mak.*, 17(1).

Jason Alan Fries, Leon Weber, Natasha Seelam, Gabriel Altay, Debajyoti Datta, Samuele Garda, Myungsun Kang, Ruisi Su, Wojciech Kusa, Samuel Cahyawijaya, et al. 2022. Bigbio: A framework for data-centric biomedical natural language processing. *arXiv preprint arXiv:2206.15076*.

Sebastian Gehrmann, Franck Dernoncourt, Yeran Li, Eric T. Carlson, Joy T. Wu, Jonathan Welt, John Foote, Edward T. Moseley, David W. Grant, Patrick D. Tyler, and Leo A. Celi. 2018. Comparing deep learning and concept extraction based methods for patient phenotyping from clinical narratives. *PLOS ONE*, 13(2):e0192360.

Yu Gu, Robert Tinn, Hao Cheng, Michael Lucas, Naoto Usuyama, Xiaodong Liu, Tristan Naumann, Jianfeng Gao, and Hoifung Poon. 2020. Domain-specific language model pretraining for biomedical natural language processing.

Yu Gu, Robert Tinn, Hao Cheng, Michael Lucas, Naoto Usuyama, Xiaodong Liu, Tristan Naumann, Jianfeng Gao, and Hoifung Poon. 2021. Domain-specific language model pretraining for biomedical natural language processing. *ACM Trans. Comput. Healthcare*, 3(1).

K Haerian, D Varn, S Vaidya, L Ena, H S Chase, and C Friedman. 2012. Detection of pharmacovigilance-related adverse events using electronic health records and automated methods. *Clin. Pharmacol. Ther.*, 92(2):228–234.

Ahmad Hammoudeh, Ghazi Al-Naymat, Ibrahim Ghannam, and Nadim Obeid. 2018. Predicting hospital readmission among diabetics using deep learning. *Procedia Computer Science*, 141:484–489.

Shaoxiong Ji, Matti Hölttä, and Pekka Marttinen. 2021. Does the magic of bert apply to medical code assignment? a quantitative study. *Computers in Biology and Medicine*, 139:104998.

Alistair Johnson, Tom Pollard, and Roger Mark. 2019. Mimic-iii clinical database demo.

Kamal raj Kanakarajan, Bhuvana Kundumani, and Malaikannan Sankarasubbu. 2021. BioELECTRA:pretrained biomedical text encoder using discriminators. In *Proceedings of the 20th Workshop on Biomedical Language Processing*, pages 143–154, Online. Association for Computational Linguistics.

Angelos Katharopoulos, Apoorv Vyas, Nikolaos Pappas, and Franccois Fleuret. 2020. Transformers are rnns: Fast autoregressive transformers with linear attention. In *ICML*.

Nikita Kitaev, Łukasz Kaiser, and Anselm Levskaya. 2020. Reformer: The efficient transformer. *arXiv:2001.04451 [cs, stat]*. ArXiv: 2001.04451.

Theresa A Koleck, Caitlin Dreisbach, Philip E Bourne, and Suzanne Bakken. 2019. Natural language processing of symptoms documented in free-text narratives of electronic health records: a systematic review. *Journal of the American Medical Informatics Association*, 26(4):364–379.

Faisal Ladhak, Esin Durmus, Claire Cardie, and Kathleen Mckeown. 2020. Wikilingua: A new benchmark dataset for cross-lingual abstractive summarization. In *Findings of the Association for Computational Linguistics: EMNLP 2020*, pages 4034–4048.

Guokun Lai, Qizhe Xie, Hanxiao Liu, Yiming Yang, and Eduard Hovy. 2017. Race: Large-scale reading comprehension dataset from examinations. In *Proceedings of the 2017 Conference on Empirical Methods in Natural Language Processing*, pages 785–794.

P LePendu, S V Iyer, A Bauer-Mehren, R Harpaz, J M Mortensen, T Podchiyska, T A Ferris, and N H Shah. 2013. Pharmacovigilance using clinical notes. *Clinical Pharmacology & Therapeutics*, 93(6):547–555.

Paea LePendu, Srini Iyer, Cedric Fairon, and Nigam Haresh Shah. 2012. Annotation analysis for testing drug safety signals using unstructured clinical notes. *Journal of Biomedical Semantics*, 3:S5 – S5.

Patrick Lewis, Myle Ott, Jingfei Du, and Veselin Stoyanov. 2020. Pretrained language models for biomedical and clinical tasks: Understanding and extending the state-of-the-art. In *Proceedings of the 3rd Clinical Natural Language Processing Workshop*, pages 146–157, Online. Association for Computational Linguistics.

Feifan Liu, Abhyuday Jagannatha, and Hong Yu. 2019. Towards drug safety surveillance and pharmacovigilance: Current progress in detecting medication and adverse drug events from electronic health records. *Drug Saf.*, 42(1):95–97.

Jingshu Liu, Zachariah Zhang, and Narges Razavian. 2018. Deep ehr: Chronic disease prediction using medical notes. *Journal of Machine Learning Research (JMLR)*.

Yinhan Liu, Myle Ott, Naman Goyal, Jingfei Du, Mandar Joshi, Danqi Chen, Omer Levy, Mike Lewis, Luke Zettlemoyer, and Veselin Stoyanov. 2020. Ro{bert}a: A robustly optimized {bert} pretraining approach.

Riccardo Miotto, Li Li, Brian A Kidd, and Joel T Dudley. 2016. Deep patient: An unsupervised representation to predict the future of patients from the electronic health records. *Sci. Rep.*, 6:26094.

Kevin M Pantalone, Todd M Hobbs, Kevin M Chagin, Sheldon X Kong, Brian J Wells, Michael W Kattan, Jonathan Bouchard, Brian Sakurada, Alex Milinovich, Wayne Weng, Janine Bauman, Anita D Misra-Hebert, Robert S Zimmerman, and Bartolome Burguera. 2017. Prevalence and recognition of obesity and its associated comorbidities: cross-sectional analysis of electronic health record data from a large us integrated health system. *BMJ Open*, 7(11).

Sungjoon Park, Jihyung Moon, Sungdong Kim, Won Ik Cho, Ji Yoon Han, Jangwon Park, Chisung Song, Junseong Kim, Youngsook Song, Taehwan Oh, Joohong Lee, Juhyun Oh, Sungwon Lyu, Younghoon Jeong, Inkwon Lee, Sangwoo Seo, Dongjun Lee, Hyunwoo Kim, Myeonghwa Lee, Seongbo Jang, Seungwon Do, Sunkyoung Kim, Kyungtae Lim, Jongwon Lee, Kyumin Park, Jamin Shin, Seonghyun Kim, Lucy Park, Alice Oh, Jung-Woo Ha, and Kyunghyun Cho. 2021. KLUE: Korean language understanding evaluation. In *Thirty-fifth Conference on Neural Information Processing Systems Datasets and Benchmarks Track (Round 2)*.

Yifan Peng, Shankai Yan, and Zhiyong Lu. 2019. Transfer learning in biomedical natural language processing: An evaluation of bert and elmo on ten benchmarking datasets. In *Proceedings of the 2019 Workshop on Biomedical Natural Language Processing (BioNLP 2019)*.

Jiezhong Qiu, Hao Ma, Omer Levy, Wen-tau Yih, Sinong Wang, and Jie Tang. 2020. Blockwise self-attention for long document understanding. In *Findings of the Association for Computational Linguistics: EMNLP 2020*, pages 2555–2565, Online. Association for Computational Linguistics.

Alec Radford, Jeffrey Wu, Rewon Child, David Luan, Dario Amodei, Ilya Sutskever, et al. 2019. Language models are unsupervised multitask learners. *OpenAI blog*, 1(8):9.

Jack W. Rae, Anna Potapenko, Siddhant M. Jayakumar, Chloe Hillier, and Timothy P. Lillicrap. 2020. Compressive transformers for long-range sequence modelling. In *International Conference on Learning Representations*.

Alvin Rajkomar, Eyal Oren, Kai Chen, Andrew M. Dai, Nissan Hajaj, Michaela Hardt, Peter J. Liu, Xiaobing Liu, Jake Marcus, Mimi Sun, Patrik Sundberg, Hector Yee, Kun Zhang, Yi Zhang, Gerardo Flores, Gavin E. Duggan, Jamie Irvine, Quoc Le, Kurt Litsch, Alexander Mossin, Justin Tansuwan, De Wang, James Wexler, Jimbo Wilson, Dana Ludwig, Samuel L. Volchenboum, Katherine Chou, Michael Pearson, Srinivasan Madabushi, Nigam H. Shah, Atul J. Butte, Michael D. Howell, Claire Cui, Greg S. Corrado, and Jeffrey Dean. 2018. Scalable and accurate deep learning with electronic health records. *npj Digital Medicine*, 1(1):18.

Pranav Rajpurkar, Jian Zhang, Konstantin Lopyrev, and Percy Liang. 2016. Squad: 100,000+ questions for machine comprehension of text. In *Proceedings of the 2016 Conference on Empirical Methods in Natural Language Processing*, pages 2383–2392.

Karen Simonyan, Andrea Vedaldi, and Andrew Zisserman. 2014. Deep inside convolutional networks: Visualising image classification models and saliency maps. In *2nd International Conference on Learning Representations, ICLR 2014, Banff, AB, Canada, April 14-16, 2014, Workshop Track Proceedings*.

Amber Stubbs, Michele Filannino, Ergin Soysal, Samuel Henry, and Özlem Uzuner. 2019. Cohort selection for clinical trials: n2c2 2018 shared task track 1. *J. Am. Med. Inform. Assoc.*, 26(11):1163–1171.

Amber Stubbs, Christopher Kotfila, Hua Xu, and Özlem Uzuner. 2015. Identifying risk factors for heart disease over time: Overview of 2014 i2b2/UTHealth shared task track 2. *J. Biomed. Inform.*, 58 Suppl:S67–S77.

Yi Tay, Mostafa Dehghani, Dara Bahri, and Donald Metzler. 2022. Efficient transformers: A survey. *ACM Comput. Surv.* Just Accepted.

Özlem Uzuner. 2009. Recognizing Obesity and Comorbidities in Sparse Data. *Journal of the American Medical Informatics Association*, 16(4):561–570.

Özlem Uzuner, Ira Goldstein, Yuan Luo, and Isaac Kohane. 2008. Identifying Patient Smoking Status from Medical Discharge Records. *Journal of the American Medical Informatics Association*, 15(1):14–24.

Ashish Vaswani, Noam Shazeer, Niki Parmar, Jakob Uszkoreit, Llion Jones, Aidan N Gomez, Ł ukasz Kaiser, and Illia Polosukhin. 2017. Attention is all you need. In *Advances in Neural Information Processing Systems*, volume 30. Curran Associates, Inc.

Eric Wallace, Jens Tuyls, Junlin Wang, Sanjay Subramanian, Matt Gardner, and Sameer Singh. 2019. Allennlp interpret: A framework for explaining predictions of NLP models. *CoRR*, abs/1909.09251.

Alex Wang, Amanpreet Singh, Julian Michael, Felix Hill, Omer Levy, and Samuel Bowman. 2018. GLUE: A multi-task benchmark and analysis platform for natural language understanding. In *Proceedings of the 2018 EMNLP Workshop BlackboxNLP: Analyzing and Interpreting Neural Networks for NLP*, pages 353–355, Brussels, Belgium. Association for Computational Linguistics.

Sinong Wang, Belinda Li, Madian Khabsa, Han Fang, and Hao Ma. 2020. Linformer: Self-attention with linear complexity. Cite arxiv:2006.04768.

Bryan Wilie, Karissa Vincentio, Genta Indra Winata, Samuel Cahyawijaya, X. Li, Zhi Yuan Lim, S. Soleman, R. Mahendra, Pascale Fung, Syafri Bahar, and A. Purwarianti. 2020. Indonlu: Benchmark and resources for evaluating indonesian natural language understanding. In *Proceedings of the 1st Conference of the Asia-Pacific Chapter of the Association for Computational Linguistics and the 10th International Joint Conference on Natural Language Processing*.

Genta Indra Winata, Samuel Cahyawijaya, Zhaojiang Lin, Zihan Liu, and Pascale Fung. 2020. Lightweight and efficient end-to-end speech recognition using low-rank transformer. *ICASSP 2020 - 2020 IEEE International Conference on Acoustics, Speech and Signal Processing (ICASSP)*, pages 6144–6148.

Li-Tzy Wu, Kenneth R Gersing, Marvin S Swartz, Bruce Burchett, Ting-Kai Li, and Dan G Blazer. 2013. Using electronic health records data to assess comorbidities of substance use and psychiatric diagnoses and treatment settings among adults. *J. Psychiatr. Res.*, 47(4):555–563.

Jason Yosinski, Jeff Clune, Anh Nguyen, Thomas Fuchs, and Hod Lipson. 2015. Understanding neural networks through deep visualization. In *Deep Learning Workshop, International Conference on Machine Learning (ICML)*.

Manzil Zaheer, Guru Guruganesh, Kumar Avinava Dubey, Joshua Ainslie, Chris Alberti, Santiago Ontanon, Philip Pham, Anirudh Ravula, Qifan Wang, Li Yang, et al. 2020. Big bird: Transformers for longer sequences. In *NeurIPS*.

Ningyu Zhang, Mosha Chen, Zhen Bi, Xiaozhuan Liang, Lei Li, Xin Shang, Kangping Yin, Chuanqi Tan, Jian Xu, Fei Huang, Luo Si, Yuan Ni, Guotong Xie, Zhifang Sui, Baobao Chang, Hui Zong, Zheng Yuan, Linfeng Li, Jun Yan, Hongying Zan, Kunli Zhang, Buzhou Tang, and Qingcai Chen. 2022. CBLUE: A Chinese biomedical language understanding evaluation benchmark. In *Proceedings of the 60th Annual Meeting of the Association for Computational Linguistics (Volume 1: Long Papers)*, pages 7888–7915, Dublin, Ireland. Association for Computational Linguistics.

A Label Distribution

Our Hong Kong longitudinal clinical notes dataset is extracted from Hong Kong Hospital Authority EHR system which covers records from 43 hospitals in Hong Kong. For the diagnosis, to reduce the dimensionality, we group the diagnosis labels into 79 classes. For cancer diseases, we group the diagnosis based on the cancer site categorization from the Hong Kong Cancer Registry[12]. While for other diseases, we take the first three digits of the ICD-10 codes as the grouping. We show the label distribution of our Hong Kong longitudinal clinical notes dataset in Figure 3.

B Detail of EHR-FFN Model

We derive 3,942 features from the tabular data for each encounter. We derive these features from 4 data tables: diagnosis, procedure, prescription, and inpatient data. Specifically, we generate one-hot representations for each derived feature and concatenate all the one-hot representation into a single vector . The detail of each one-hot feature is shown in Table 5. We extract the feature vectors per patient encounter. To aggregate all the historical tabular feature vectors, we aggregate the vectors into a single feature vector by summing up all the vectors producing a single high-level feature vector per patient. To learn the high-level feature vector, we employ a feed forward network with 3 hidden layers with a total size of ~5M parameters. The hyperparameters of the feed forward model is shown in Table 6.

Feature Name	Length	Description
Diagnosis Type	1699	Diagnosis type based on ICD-10 code
Procedure Type	127	Procedure type based on ICD-9 code
Prescription Type	1271	Type of presribed drug based on regional standard
Prescription BNF	73	Type of presribed drug based on BNF Therapeutic Classification
Emergency Indicator	1	Indicator for emergency unit admission
Length of Stay	5	Length of stay in the hospital
Age Group	5	Age of the patient during admission to the hospital
Ward Type	4	Type of hospital ward
Ward Sub-Care Type	6	Sub-type of hospital ward

Table 5: Details of the tabular features

[12]https://www3.ha.org.hk/cancereg/allages.asp

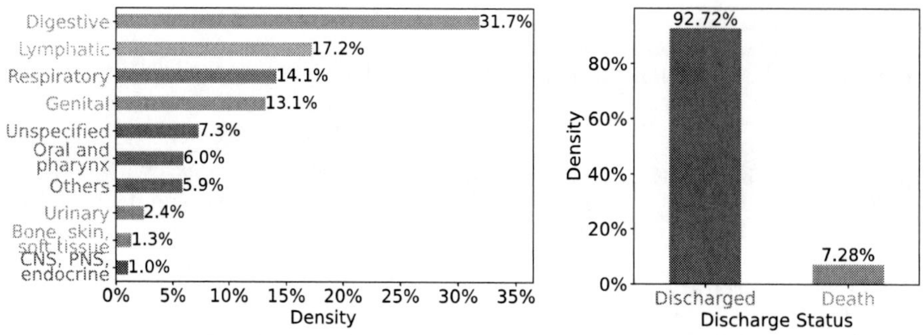

Figure 3: Label statistics of our dataset. **(Left)** shows the aggregated distribution of diagnosis based on the cancer ICD-10's site grouping[13]. `Unspecified` denotes all cancer diagnoses with unspecified site. `Others` denotes diseases other than cancer. **(Right)** shows the distribution of the discharge status (discharged/death) gathered from all inpatient records, which is used to define the mortality label.

Hyperparameter settings	Value
Tabular Encoder	
#hidden layers	3
hidden size	[1024, 512, 256]
input size	3942
layer activation	ReLU
drop out	0.1

Table 6: Details of the model hyperparameters

A Quantitative and Qualitative Analysis of Schizophrenia Language

Amal Alqahtani [1,2], Efsun Kayi [1,3], Sardar Hamidian [1], Michael Compton[4], and Mona Diab[1,5]

[1]The George Washington University, DC, USA
[2]King Saud University, Riyadh, KSA
[3]Johns Hopkins University Applied Physics Laboratory, Laurel, USA
[4]Medical Center, Columbia University, New York, USA
[5]Meta AI, USA
{amalqahtani,sardar}@gwu.edu, efsun.kayi@jhuapl.edu
mtc2176@cumc.columbia.edu, mdiab@meta.com

Abstract

Schizophrenia is one of the most disabling mental health conditions to live with. Approximately one percent of the population has schizophrenia which makes it fairly common, and it affects many people and their families. Patients with schizophrenia suffer different symptoms: formal thought disorder (FTD), delusions, and emotional flatness. In this paper, we quantitatively and qualitatively analyze the language of patients with schizophrenia measuring various linguistic features in two modalities: speech and written text. We examine the following features: coherence and cohesion of thoughts, emotions, specificity, level of committed belief (LCB), and personality traits. Our results show that patients with schizophrenia score high in fear and neuroticism compared to healthy controls. In addition, they are more committed to their beliefs, and their writing lacks details. They score lower in most of the linguistic features of cohesion with significant p-values.

1 Introduction

Schizophrenia is a mental illness that can disrupt thought processes and perception (Kerns and Berenbaum, 2002). It can impair people's ability to manage their emotions, and can cause motor and behavioral disorders (Elvevag and Goldberg, 2000).

Understanding and identifying the underlying signs of schizophrenia is critical in early detection and intervention before the malady becomes severely disabling if left untreated (Seeber and Cadenhead, 2005). Moreover, it is vital to support mental health practitioners as well as policymakers to eliminate barriers to treating mental illnesses such as schizophrenia.

Gradual decline in functioning and cognition are some common characteristics of schizophrenia patients. Symptoms may include delusions, which are fixed false beliefs, as well as hallucinations but also importantly, they tend to have strong convictions regardless of the veridicality of the beliefs themselves. Another symptom that some individuals with schizophrenia exhibit is formal thought disorder (FTD), where a patient becomes unable to form coherent or logical thoughts (Kuperberg, 2010). Moreover, they suffer in some cases from lack of motivation and/or emotional response.

One way to capture mental disorders and related symptomatology is by analyzing patients' linguistic cues. Hence, we map the aforementioned symptoms to linguistic features that we can measure. To date, most of the employed measures used by clinicians measure superficial linguistic cues and they tend to be more qualitative. We hypothesize that advances in pragmatic NLP tools allow us to measure many of these symptoms via analyzing language cues used by patients. We surmise that given such tools, we help create objective quantitative measures for clinicians beyond what they are using today for diagnostics. Moreover having such tools could help them discover and codify further studies allowing for even more signals in detecting such mental health disorders.[1] Accordingly, we present the first comprehensive study of deep pragmatically oriented linguistic modeling tools for diagnostic purposes. We leverage an emotion detection model to assess the lack of emotional response. We also employ a personality detection model to measure lack of motivation, which is one of the negative symptoms they may exhibit. We use a level of committed belief detection model to identify the level of committed belief corresponding to strength of conviction. Formal Thought Disorder (FTD) is measured by using language model-based sentence scoring as well as other coherence features such as LSA, connectives, lexical diversity, syntactic complexity, word information, and level

[1]Despite our focus in this work on schizophrenia, we believe that many of the tools we use here could be applicable to other mental disorders.

Proceedings of the 13th International Workshop on Health Text Mining and Information Analysis (LOUHI), pages 173 - 183
December 7, 2022 ©2022 Association for Computational Linguistics

of linguistic specificity. Finally, we employ the *Coh-metrix* computational tool for analyzing texts for a variety of cohesion measures.(Graesser et al., 2004).

Accordingly, we investigate the following metrics: cohesion, level of committed belief, emotion, and personality and their corresponding correlation with symptomatic patients' language use. We examine both speech and text modalities, comparing patients vs. a matched set of controls.

Our results show that when patients express their emotions in writing or speech, they tend to show fear more often than other emotions. The findings also detect a neuroticism personality as they may suffer from feelings such as anger, and anxiety more frequently and severely. Furthermore, the results indicate that their writings lack specificity (details), and they are more committed to their beliefs in contrast with healthy controls. In addition, our results show that writings of healthy controls are more coherent demonstrated via the high scores of language model probabilities of their writing. To the best of our knowledge, our findings present the first set of measurable pragmatic linguistic cues that significantly correlate with contrastive mental health patients' language use that goes beyond the typical superficial metrics used in the literature to date. Our study provides a set of objective linguistic measures that can serve as metrics that further assist clinicians and policy makers in the mental health domain. The contributions of this paper are as follows:

1. It provides a comprehensive set of cognitive and linguistics *quantitative* metrics for schizophrenia patients language use;

2. We provide a translation of clinical observations of patient language use onto specific measurable linguistic cues that are mapped into advance NLP technology;

3. For the first time, our work leverages advances in the pragmatic NLP to measure patients' cognitive state (namely their levels of committed beliefs), personality traits, emotions, specificity and coherence;

4. We use LM with perplexity scores to measure both coherence and cohesion.

2 Related Work

Language provides significant insight into the content of thought. It also reflects the presence of impairments resulting from mental disorders such as schizophrenia. The predominent reflection of mental impairment for schizophrena is the lack of coherent text or speech. Accordingly, cohesion scores were first proposed as an indicator of predicting schizophrenia (Elvevåg et al., 2007) where they used Latent Semantic Analysis (LSA) as a feature extractor. This was further amplified by (Bedi et al., 2015) where they measured the semantic coherence in disorganized speech captured by LSA, specifically where large amounts of language overlap was interpreted as coherent language. The study found that these features, together with syntactic markers of complexity, could predict later development of psychosis with 100% accuracy using a convex hull algorithm. Later, Corcoran et al. (2018) used a logistic regression model to predict the onset of psychosis using coherence as measured by LSA combined with the usage of possessive pronouns. This approach showed an accuracy of 83% in predicting the onset of psychosis with a cross-validation accuracy of 79%.

Metrics for Schizophrenia detection were investigated by (AlQahtani et al., 2019) where they used linguistic features such as referential cohesion, text ease, situation model, and readability in patients' and controls' writing or speech to classify presence or absence of the disorder. The researchers trained Support Vector Machine (SVM) and Random Forests (RF) models. The study results showed that the situation model and readability performed the best among all cohesion features for the SVM model yielding a 72% F-score in the binary classification task of detecting whether a person (through their writing or speech) is a schizophrenia patient.

Different from previous studies of schizophrenia, we propose measuring cohesion using language model perplexity. Moreover, we provide a comprehensive exploration of the language of patients relative to that of controls along the following linguistics cues: coherence, emotion, personality, level of specificity, and level of committed belief.

3 Data

Our study comprises two datasets speech, *Lab-Speech*, and written text, *LabWriting*. The data is obtained from schizophrenia patients and healthy controls. Both datasets are described in detail in (Kayi et al., 2018).[2] *LabWriting* has 188 participants who are native English-speakers between the ages of 18 – 50 years, corresponding to 93 patients and 95 healthy controls. All participants are asked to write two paragraph-long essays: the first one is about their average Sunday and the second essay is about what makes them the angriest. The total number of writing samples collected from both patients and controls is 373 pieces of text.

The second dataset, *LabSpeech*, includes three questions that prompt participants to describe some emotional and social events. Patients and controls are asked to describe (1) a picture, (2) their ideal day, and (3) their scariest experience. The total number of speech script samples collected from both patients and controls is 431. Speech data is transcribed to text and a punctuation tool (Tilk and Alumäe, 2016) is used to add the missing punctuation.

3.1 Superficial Descriptive statistics

Table 1 illustrates various descriptive statistics comparing and contrasting the *LabWriting* and *LabSpeech* datasets. The results indicate that healthy controls in both datasets are more verbose (produce more words and sentences) when answering questions in both modalities, i.e. writing or speech. The mean values of the number of words and the number of sentences generated by Controls in *LabWriting* are 141 and 7, respectively. However, the mean values are lower for Patients, with 110 and 6 for the same analysis. The Patients in *LabSpeech* also score lower averages in all the descriptive features. These results are in line with a previous study (De Boer et al., 2020) that individuals with schizophrenia speak less and use less complex sentences. * in Table 1 indicates the higher results and statistically significant.

Descriptive	LabWriting		LabSpeech	
	P	C	P	C
Avg. # words	110	**141***	220	**277***
Avg. #sent.	6	**7***	11	**14***
sent./paragraph	5.6	**6.6***	11	**14***

Table 1: Descriptive statistics for *LabWriting* and *LabSpeech* datasets. We present overall average number of words, overall average number of sentences and a finer grained average number of sentences per paragraph. *P* denotes patient, and *C* denotes control.

4 Pragmatic Cues

4.1 Emotion

Emotion refers to a person's internal or external reaction to an event. This reaction can be expressed verbally, outwardly/visibly (e.g., frowning), or physiologically (e.g., crying) (Kring and Caponigro, 2010). Schizophrenia patients are often characterized as having disorganized thinking; however, according to (Kring and Elis, 2013), they still report their emotional experiences using the same general definitions of emotions (happy, sad, etc.) as persons who do not have schizophrenia. We use the EmoNet [3] (Abdul-Mageed and Ungar, 2017) to obtain the eight core emotions (PL8), which are trust, anger, anticipation, disgust, joy, fear, sadness, and surprise.

4.2 Specificity

Specificity in computational linguistic measures how much detail exists in a text (Louis and Nenkova, 2011). This is an important pragmatic concept and a characteristic of any text (Li and Nenkova, 2015). We quantify this feature because schizophrenia may impacts one's language specificity. Hence, our hypothesis is that patients tend to write less specific paragraphs which lack references to any specific person, object, or event. We use (Ko et al., 2019) to measure a sentence specificity by indicating how many details exist in each sentence. This tool generates a rate for each sentence between 0 (general sentence) and 1 (detailed sentence). We also use Coh-Metrix (Graesser et al., 2004) to measure word hyponyms (i.e., word specificity) in a text. A higher value reflects an overall use of more specific words, which increases the ease and speed of text processing.

4.3 Level of Committed belief (LCB)

In natural language, the level of committed belief is a linguistic modality that indicates the author's belief in a given proposition (Diab et al., 2009). We measure this feature as it can detect an individual's cognitive state. We want to explore this feature to test our hypothesis that patients with schizophrenia may hold strong beliefs towards their own propositions. We rely on a belief tagger (Rambow et al., 2016) to label each sentence with the

[2]The authors of (Kayi et al., 2018) kindly shared the data after we obtained IRB permission.

[3]`https://github.com/UBC-NLP/EmoNet`

committed belief tags as *(CB)* where someone *(SW)* strongly believes in a proposition, Non-committed belief *(NCB)* where SW reflects a weak belief in the proposition, and Non-Attributable Belief *(NA)* where SW is not (or could not be) expressing a belief in the proposition (e.g., desires, questions, etc.). There is also the *ROB* tag where SW's intention is to report on someone else's stated belief, regardless of whether or not they themselves believe it. The feature values are set to a binary 0 or 1 for each *CB, NCB, NA, and ROB* corresponding to unseen or observed. The following text is an example from *LabWriting*.

> Every Sunday I usually <cb−I> get </cb−I> up and <cb−I> watch </cb−I> gospel shows on TV. I <cb−I> do </cb−I> my house chores and then <cb−I> watch </cb−I> other things on TV. Then later on I <cb−I> go </cb−I> down the street to the food resturants to <na−I> eat </na−I> something to eat.[4]

We calculate the LCB as:

$$\textbf{\textit{LCB}} < tag > = \textit{total} < tag > \textit{in a text} \; / \; \textit{all LCB tags in the same text}$$

where $< tag >$ is one of the 4 LCB features: CB, NCB, NA, or ROB.

4.4 Personality

In psychology, personality is the distinctive sets of behaviors, cognitions, and emotional patterns that derive from biological and environmental influence (Major et al., 2000). We study the personality of patient and healthy controls in our datasets based on the famous Big-Five (Digman, 1990) personality measure, which are the following five traits: Extraversion (EXT), Neuroticism (NEU), Agreeableness (AGR), Conscientiousness (CON), and Openness (OPN). Neuroticism is characterized by a proclivity for negative emotions (Bono and Vey, 2007). Individuals with high scores for neuroticism experience feelings such as anxiety, worry, fear, anger, frustration, depressed mood, and loneliness (Widiger, 2009). Extraversion indicates how outgoing and social a person is (Smelser et al., 2001). A low score in extraversion means an individual prefers to stay alone. We explore personality to test our hypothesis that patients with schizophrenia are high in neuroticism (emotionally unstable), especially if delusional, and low in extraversion (Horan et al., 2008). We use (Kazameini et al., 2020) to predict personality traits for each text in our datasets. The model makes binary predictions of the author's personality.

[4]Typos are in the original text.

5 Cohesion Linguistic Features

5.1 Information Structure (Givenness)

Latent Semantic Analysis (LSA) measures the semantic similarity/overlap between sentences or between paragraphs (Dennis et al., 2003). We use LSA to evaluate givenness, which is an information structure defined as a phenomenon where a speaker presumes that the listener is already familiar with the context of a discussion topic (Féry and Ishihara, 2016). The sentence is considered to be coherent when the average givenness score is high (Graesser et al., 2004).

5.2 Lexical Diversity

Lexical diversity of a text is a measure of unique words (types), and consequently a measurement of different words that appear in the text compared to the total number of words (tokens) in that text (Durán et al., 2004) (Johansson, 2008). Type-token ratio (TTR), i.e., the ratio of types to tokens, is the most basic metric of lexical diversity (Durán et al., 2004). When the number of types equals that of tokens in a text, all words are different, with TTR being equal to 1, and the lexical diversity of the text reaches its maximum possible value. Such a text, i.e., one with very high lexical diversity, is likely to be either low in cohesion because cohesion requires repetition of words or very short in length. After all, a naturally occurring longer text implies a greater frequency of the same word (Graesser et al., 2004).

5.3 Connectives

The use of connecting words creates cohesive links between ideas and clauses and provides clues about text organization (Graesser et al., 2004). We evaluate two types of connectives which are logic and temporal. The logic connectives are used to connect two or more ideas (such as *and, or*). In contrast, temporal connectives are words or phrases that are used to indicate when something is taking place (such as *first, until*).

5.4 Syntactic Complexity

Syntax refers to the arrangements of words and morphemes in forming larger units, such as phrases and clauses, ultimately resulting in well-formed sentences in a language (Crowhurst, 1983). A tree-like structure, a syntactic tree, can visualize the arrangement of words in a sentence. A tree can be simple: containing basic structure like actor-action-object; or complex, larger in size, with significant number of branches, and a complicated relationship among its different parts (Graesser et al., 2004).

5.5 Word Information

All words in a sentence can be categorized as one of two types: a) Content words, such as nouns, verbs, adjectives, and adverbs, which primarily carry the semantic substance of the sentence and contribute to its meaning; and, b) Function words, such as prepositions, determiners, and pronouns, which primarily express the grammatical relationships among content words without significant semantic content (Wilks, 1998).[5] Word Information refers to the notion that each word can be assigned a syntactic part-of-speech category and, with this assignment, be further rendered as a content or a function word, thus carrying either substantive or "inconsequential" meaning (Graesser et al., 2004).

5.6 Language Model (LM)

A Language model (LM) is the probability distribution over text (Bengio et al., 2003). To analyze coherence in free text, we propose an approach based on LMs. We use a python library *LM-scorer* (Simone, 2020) to calculate probabilities of each word in a text and score sentences. The library uses the GPT2 model (Radford et al., 2019) internally to provide a probability score for each next word.The sentence score (probability) is computed as the mean of tokens' probabilities. For a given sentence, the LM predicts a higher score for a sentence that is more grammatically correct. Performance of LMs is commensurate with word information, content words tend to have lower probabilities compared to function words.

We calculate multiple LM scores: the perplexity scores at sentence and paragraph level. Moreovere, we analyze the LM probabilities (scores) across two segmentation/levels: paragraph level and sentence level. We compare the performance of both levels using the means of statistical hypothesis testing.

[5]We contend that this view is controversial since function words are critical to the meaning of utterances, however we would like to emphasize the qualitative difference between content words and function words.

5.6.1 Analysis at Paragraph level

1. **Mean Sentence Probability:** For a given sentence, the LM predicts a higher score/probability for a sentence that is more grammatically and logically sound. We calculate the mean sentences probability in a text for each observation in each group (control/patient).

2. **Median Sentence Probability:** This statistic is calculated by taking the median of the probabilities of sentences. The justification for using this score is that the median, compared to the mean, is more robust to outliers.

5.6.2 Analysis at Sentence level

1. **Sentence probabilities:** This statistic is extracted by aggregating LM individual sentence scores. Sentences scores for all patients and all controls are compared. The number of sentence probability scores analyzed is equivalent to the number of all sentences in the sample.

2. **Mean of the deltas in sentence probabilities:** By using the sentences scores, the changes between the consecutive probability scores of the sentences in the paragraphs are extracted (deltas), and their average is calculated. The total number of this statistic is equivalent to the number of instances in the dataset. Our aim here is to check if the patient group has more fluctuations in their sentence probabilities.

3. **Minimum deltas in sentence probabilities:** The minimum of changes in the sentence probabilities of consecutive sentences in each paragraph is calculated and compared. The total number of this statistic equals the number of instances in the dataset.

4. **Maximum of deltas in sentence probabilities:** Similar to the last statistic, the maximum of changes in the sentence probabilities of consecutive sentences in each paragraph are calculated and compared. The total number of this statistic equals the number of instances in the dataset.

6 Discussion of the Results

Table 2 and Table 3 illustrate the results of emotion analysis and specificity, respectively. Table 4 reports LCB averages and Table 5 summarizes

Emotion	LabWriting		LabSpeech	
	P	C	P	C
Anger	0.159	0.162	**0.104***	0.095
Anticip.	0.043	0.039	0.035	0.039
Disgust	0.086	0.093	0.117	0.112
Fear	**0.117***	0.103	**0.181***	0.167
Joy	0.372	0.381	0.297	0.311
Sadness	0.127	0.127	0.139	0.138
Surprise	0.077	0.079	0.117	0.127
Trust	**0.019***	0.015	0.010	0.010

Table 2: Emotion results. The bold values above indicate the high means, and * indicates only the statistically significant values.

personality percentages. Table 6 in appendix A summarize the values of the cohesion linguistic features: Information Structure (Givenness), Connectives, Lexical Diversity, Syntactic Complexity, Syntactic Pattern Density, and Word Information. Table 7 and Table 8 in appendix A show the values of the language model and perplexity scores. For each comparison criteria we compare the p-value to a significance level α = **0.05** to make conclusions about our hypotheses. * is used to indicate the results with a statistically significant p-value.

1. **Descriptive features** The p-values of the total number of sentences in both datasets are significant. There is a noticeable difference between the distribution of this statistic between the two groups and it shows that Controls, on average, generate more sentences.

2. **Emotion** We hypothesise that Patients score high in fear. Our results show that Patients in both *LabWriting* and *LabSpeech* score high in fear (p–value = **0.002**) and (p-value=**0.004**), respectively. This result is consistent with a previous study (Suslow et al., 2003) which states that Patients tend to feel fear more often. Patients in *LabWriting* score high in trust, and this may be due to interviewing them in a trustful environment.

3. **Specificity** We hypothesise that Patients write less specific paragraphs. In the score of word hyponyms (Noun) as a measure of specificity, our results show that the Controls score significantly **higher** in *LabWriting* (p-value = **0.03**). Furthermore, Controls score higher in *LabSpeech*, though not significantly. Specificity

Specificity	LabWriting		LabSpeech	
	P	C	P	C
Sent. level	0.47	**0.48***	0.39	0.39
Hyponym	5.85	**6.06***	6.36	**6.45**

Table 3: Specificity results. The above table shows the average specificity at sentence level as well as word hyponyms (Noun).

LCB	LabWriting		LabSpeech	
	P	C	P	C
CB	**0.52**	0.51	**0.60**	0.58
NCB	0.013	**0.020***	0.04	**0.05**
NA	0.45	0.46	0.34	0.35
ROB	0.008	0.010	0.010	0.012

Table 4: LCB results.

at sentence level is also significantly **higher** in *LabWriting* for Controls (p-value = **0.009**). However, there is no difference between Controls and Patients in *LabSpeech*. It should be noted that the speech data are faithfully transcribed where pauses and filler words such as *um, er, uh* can lower the quality of the speech relative the specificity model which is trained on native textual input hence making it challenging to capture specificity.

4. **LCB** The hypothesis of this study states that Patients show more commitment to their beliefs. Table 4 shows the results of LCB. It can be noticed that Patients in both datasets score **higher** in committed belief (CB) and Controls score **higher** in Non-committed belief (NCB). It confirms our hypothesis, and these findings coincide with a previous study (Kayi et al., 2018) that patients with schizophrenia may show more commitment of their belief to propositions expressed in either modality, writing or speech.

5. **Personality** The hypothesis of this study states that Patients score high levels of neuroticism and low levels of extraversion. Table 5 reports the results of personality analysis. The results show that Patients in both datasets score **lower** in extroversion (EXT) (p-value = **0.03**) in *LabWriting* and score **higher** in neuroticism (NEU) (p-value = **0.04**) in *LabWriting*. These results are in line with previ-

Personality	LabWriting		LabSpeech	
	P	C	P	C
EXT	34%	**50%***	27%	**30%**
NEU	**52%***	40%	**35%**	27%
AGR	60%	**63%**	81%	**85%**
CON	46%	**54%**	6.6%	**7.3%**
OPN	45%	40%	84%	75%

Table 5: Frequency Distribution of Personality Traits.

ous studies (Camisa et al., 2005),(Horan et al., 2008), (Smeland et al., 2017) which show that schizophrenia is associated with high levels of neuroticism and low levels of extraversion. We report all other personality traits in table 5; However, our analysis mainly focuses on neuroticism and extraversion.

6. **Information Structure (Givenness)** The average givenness per sentence of the schizophrenia patients is statistically significantly **lower** than that of the Controls in both *LabWriting* (p-value =**0.001**) and *LabSpeech* (p-value = **0.01**). Patients demonstrate challenges in recognizing things that others would find obvious and consequently question or repeat those. In addition, they present something that they have already mentioned earlier as completely new, compromising givenness.

7. **Lexical Diversity** In the metric Type-token ratio (TTR) for all words, Patients scored **higher** than Controls, with the difference being statistically significant in both *LabWriting* (p-value = **0.004**) and *LabSpeech* (p-value = **0.0001**). The higher proportion of types by Patients stems from the fact that they produce more incomplete, indistinct, inaudible, or incomprehensible words or sounds and shorter sentences and utterances, struggling to reorganize their thoughts (Hinzen et al., 2019) (Merrill et al., 2017). These non-words, particularly shorter sentences, contribute to the higher TTRs for Patients.

   Schizophrenic patients, however, are known to repeat words and phrases (Manschreck et al., 1985), and hence a basic TTR in itself is not a reliable indicator for distinguishing between Controls and Patients. TTR is only possible to apply when text or speech are of equal length. We thus compute two more metrics of lexical diversity, namely measure of textual lexical diversity (MTLD) and measure D vocabulary diversity (VocD), which allow comparison of lexical diversity of texts of unequal lengths. By these measures, we find text and speech of Controls to be lexically much more diverse, with p-values in the order of 10^{-4}.

8. **Connectives** In the uses of logical, temporal, and extended temporal connectives in text and speech, Controls consistently score **higher**. The difference in scores is statistically significant in all three cases of speech which are logic, temporal, and extended temporal connectives with p-values respectively **0.03**, **0.04**, and **0.03**. In *LabWriting*, the difference is, however, found to be statistically significant (p-value = **0.03**) only in the case of logical connectives. Our findings validate one of the decisive signs of schizophrenia, deficits of logical reasoning among patients (Willits et al., 2018) (Mackinley et al., 2021).

9. **Syntactic Complexity** In addition to phonetic anomalies in terms of more pauses, loss of prosody, and mumbled sounds, syntactic and semantic conventions that govern the formation of sentences and ultimately the language are routinely violated by schizophrenia patients (Stein, 1993). One of the manifestations of these violations is the decrease in the syntactic complexity of their writing and speech, resulting in disorganized language with poor content. According to all our three measures of syntactic complexity – SYNMEDpos, SYNMEDwrd, and SYNMEDlem – Controls demonstrate much **higher** syntactically complex text, with statistically significant differences from Patients in all cases, except in *LabSpeech*, in which the difference is nevertheless nearly significant. These results concur with previous studies (Kayi et al., 2018) (Hinzen et al., 2019) which showed that a patient with schizophrenia alters the patterns of linguistic organization, which leads to increased syntactic errors.

10. **Word Information** In the usage of pronouns, our results show that Patients use the first-person pronouns, e.g., **I**, my, me, comparatively more, while Controls prefer first person plural, second-person, and third-person more. The differences are statistically significant in

text but not in speech. This result is in line with the previous study (Kayi et al., 2018) (Tang et al., 2021). One metric in which Controls score significantly **higher** in both *LabWriting* and *LabSpeech* is the average minimum word frequency in sentences. With Controls producing significantly longer writings or speeches, a greater frequency of words is necessary to maintain coherence and a logical flow in the text.

11. **Language Model Analysis at Paragraph-Level** We measure the mean of the probabilities of the sentences and the corresponding medians to account for outlier effects. Since both Patients and Controls produced an appreciable number of tokens per sentence, we find these probabilities lower for both groups. We are primarily interested in the comparison of the probabilities and find that the mean and median probabilities are significantly **lower** for Patients than for Controls in *LabWriting*, with mean p-values of **0.01** and median p-values of **0.02**. The findings are in line with previous studies (Kuperberg, 2010), (Hinzen and Rosselló, 2015),(De Boer et al., 2020) that schizophrenia patients often produce idiosyncratic expressions and hence less probable naturally occurring sentences.

    While the probabilities in *LabSpeech* are **lower** for Patients, the differences in corresponding probabilities are not statistically significant at mean p-values of **0.524** and median p-values of **0.237**. This can be explained by the fact that Controls can exploit the time during writing better to their advantage to produce more organized and coherent text. Speech, on the other hand, is swift and spontaneous.

12. **Language Model Analysis at Sentence Level**

    In line with the mean and median of the probabilities of the sentences at the paragraph level, we compute the average of the probabilities of all sentences. This metric, average sentence probabilities, is also significantly **lower** for Patients (**0.109**) than for Controls (**0.117**) with (p-value=**0.0007**). The difference in *LabSpeech* dataset, like that in the paragraph level, is again not statistically significant at (p-value=**0.175**).

    The mean of changes in the sentence probabilities, computed to evaluate how strongly the sentence probabilities change from one sentence to another in a paragraph and consequently how much the sentences deviate from a coherent and logical flow, is **higher** for Controls (p-value= **0.05**) in *LabWriting*. Two other metrics related to this, the minimum and the maximum of changes in sentence probabilities, provide mixed, hence inconclusive, results. These probabilities, therefore may not be consistent indicators for the fluctuations we expected.

13. **Perplexity** Table 8 in appendix A shows the results of perplexity. We compute it at two levels: the sentence level and the paragraph level, to determine how predictable the language of Patients is compared to that of Controls. In *LabWriting*, the model is more perplexed for Patients in both levels, and the difference between the two groups is highly significant (p-value =**0.01**) at the paragraph level while (p-value =**0.00005**) at the sentence level. However, the results are not significant for *LabSpeech* for any of the two levels.

7 Conclusion

Patients with schizophrenia experience different symptoms, some of which involve problems with concentration and memory, which in return may lead to disorganization in speech or behavior. Therefore, diagnosing this disorder early and correctly is extremely important as it may help alleviate the adverse effects on patients.

Among the linguistic features of cohesion investigated in this study, we found that Patients' scores are lower, with significant p-values in information structure (givenness), lexical diversity except for Type-token ratio (TTR), connectives, and syntactic complexity in both datasets. Among the pragmatic cues, we found that Patients' score high in fear, and their personality is associated with elevated neuroticism. They also show more commitment to their beliefs, and their average specificity at sentence and word levels is lower than Controls.

In the future, we plan to expand our analysis to other related mental health disorders. We also plan to explore the pragmatically motivated linguistics features of schizophrenia in other languages.

Limitations

One of the main limitations of this study is the size of the sample, and this is due to the data privacy and the cost associated with collecting scripts written by patients with schizophrenia.

References

Muhammad Abdul-Mageed and Lyle Ungar. 2017. Emonet: Fine-grained emotion detection with gated recurrent neural networks. In *Proceedings of the 55th annual meeting of the association for computational linguistics (volume 1: Long papers)*, pages 718–728.

Amal AlQahtani, Efsun Kayi, and Mona Diab. 2019. Understanding cohesion in writings and speech of schizophrenia patients. In *2019 18th IEEE International Conference On Machine Learning And Applications (ICMLA)*, pages 364–369. IEEE.

Gillinder Bedi, Facundo Carrillo, Guillermo A Cecchi, Diego Fernández Slezak, Mariano Sigman, Natália B Mota, Sidarta Ribeiro, Daniel C Javitt, Mauro Copelli, and Cheryl M Corcoran. 2015. Automated analysis of free speech predicts psychosis onset in high-risk youths. *npj Schizophrenia*, 1(1):1–7.

Yoshua Bengio, Réjean Ducharme, Pascal Vincent, and Christian Jauvin. 2003. A neural probabilistic language model. *Journal of machine learning research*, 3(Feb):1137–1155.

Joyce E Bono and Meredith A Vey. 2007. Personality and emotional performance: Extraversion, neuroticism, and self-monitoring. *Journal of occupational health psychology*, 12(2):177.

Kathryn M Camisa, Marcia A Bockbrader, Paul Lysaker, Lauren L Rae, Colleen A Brenner, and Brian F O'Donnell. 2005. Personality traits in schizophrenia and related personality disorders. *Psychiatry research*, 133(1):23–33.

Cheryl M Corcoran, Facundo Carrillo, Diego Fernández-Slezak, Gillinder Bedi, Casimir Klim, Daniel C Javitt, Carrie E Bearden, and Guillermo A Cecchi. 2018. Prediction of psychosis across protocols and risk cohorts using automated language analysis. *World Psychiatry*, 17(1):67–75.

Marion Crowhurst. 1983. Syntactic complexity and writing quality: A review. *Canadian Journal of Education/Revue canadienne de l'education*, pages 1–16.

JN De Boer, M van Hoogdalem, RCW Mandl, J Brummelman, AE Voppel, MJH Begemann, E van Dellen, FNK Wijnen, and IEC Sommer. 2020. Language in schizophrenia: relation with diagnosis, symptomatology and white matter tracts. *npj Schizophrenia*, 6(1):1–10.

Simon Dennis, Tom Landauer, Walter Kintsch, and Jose Quesada. 2003. Introduction to latent semantic analysis. In *25th Annual Meeting of the Cognitive Science Society. Boston, Mass*, page 25.

Mona Diab, Lori Levin, Teruko Mitamura, Owen Rambow, Vinodkumar Prabhakaran, and Weiwei Guo. 2009. Committed belief annotation and tagging. In *Proceedings of the Third Linguistic Annotation Workshop (LAW III)*, pages 68–73.

John M Digman. 1990. Personality structure: Emergence of the five-factor model. *Annual review of psychology*, 41(1):417–440.

Pilar Durán, David Malvern, Brian Richards, and Ngoni Chipere. 2004. Developmental trends in lexical diversity. *Applied Linguistics*, 25(2):220–242.

Brita Elvevåg, Peter W Foltz, Daniel R Weinberger, and Terry E Goldberg. 2007. Quantifying incoherence in speech: an automated methodology and novel application to schizophrenia. *Schizophrenia research*, 93(1-3):304–316.

Brita Elvevag and Terry E Goldberg. 2000. Cognitive impairment in schizophrenia is the core of the disorder. *Critical Reviews™ in Neurobiology*, 14(1).

Caroline Féry and Shinichiro Ishihara. 2016. *The Oxford handbook of information structure*. Oxford University Press.

Arthur C Graesser, Danielle S McNamara, Max M Louwerse, and Zhiqiang Cai. 2004. Coh-metrix: Analysis of text on cohesion and language. *Behavior research methods, instruments, & computers*, 36(2):193–202.

Wolfram Hinzen, Derya Çokal, Vitor C Zimmerer, Douglas Turkington, I Nicol Ferrier, Rosemary Varley, and Stuart Watson. 2019. Disturbing the rhythm of thought: speech pausing patterns in schizophrenia, with and without formal thought disorder. *Plos One. 2019; 14 (5): e0217404. DOI: 10.1371/journal. pone. 0217404.*

Wolfram Hinzen and Joana Rosselló. 2015. The linguistics of schizophrenia: thought disturbance as language pathology across positive symptoms. *Frontiers in psych ology*, 6:971.

William P Horan, Jack J Blanchard, Lee Anna Clark, and Michael F Green. 2008. Affective traits in schizophrenia and schizotypy. *Schizophrenia bulletin*, 34(5):856–874.

Victoria Johansson. 2008. Lexical diversity and lexical density in speech and writing: A developmental perspective. *Working papers/Lund University, Department of Linguistics and Phonetics*, 53:61–79.

Efsun Sarioglu Kayi, Mona Diab, Luca Pauselli, Michael Compton, and Glen Coppersmith. 2018. Predictive linguistic features of schizophrenia. *arXiv preprint arXiv:1810.09377.*

Amirmohammad Kazameini, Samin Fatehi, Yash Mehta, Sauleh Eetemadi, and Erik Cambria. 2020. Personality trait detection using bagged svm over bert word embedding ensembles. *arXiv preprint arXiv:2010.01309*.

John G Kerns and Howard Berenbaum. 2002. Cognitive impairments associated with formal thought disorder in people with schizophrenia. *Journal of abnormal psychology*, 111(2):211.

Wei-Jen Ko, Greg Durrett, and Junyi Jessy Li. 2019. Domain agnostic real-valued specificity prediction. In *AAAI*.

Ann M Kring and Janelle M Caponigro. 2010. Emotion in schizophrenia: where feeling meets thinking. *Current directions in psychological science*, 19(4):255–259.

Ann M Kring and Ori Elis. 2013. Emotion deficits in people with schizophrenia. *Annual review of clinical psychology*, 9:409–433.

Gina R Kuperberg. 2010. Language in schizophrenia part 1: an introduction. *Language and linguistics compass*, 4(8):576–589.

Junyi Li and Ani Nenkova. 2015. Fast and accurate prediction of sentence specificity. In *AAAI*.

Annie Louis and Ani Nenkova. 2011. Automatic identification of general and specific sentences by leveraging discourse annotations. In *Proceedings of 5th international joint conference on natural language processing*, pages 605–613.

Michael Mackinley, Jenny Chan, Hanna Ke, Kara Dempster, and Lena Palaniyappan. 2021. Linguistic determinants of formal thought disorder in first episode psychosis. *Early intervention in psychiatry*, 15(2):344–351.

Brenda Major, Catherine Cozzarelli, Mardi J Horowitz, Peter J Colyer, Lynn S Fuchs, Edward S Shapiro, Karen Callan Stoiber, Ulrik Fredrik Malt, Thomas Teo, David G Winter, et al. 2000. Encyclopedia of psychology: 8 volume set. *New York and Washington: Oxford University Press and the American Psychological Association.*

Theo C Manschreck, Brendan A Maher, Toni M Hoover, and Donna Ames. 1985. Repetition in schizophrenic speech. *Language and Speech*, 28(3):255–268.

Anne M Merrill, Nicole R Karcher, David C Cicero, Theresa M Becker, Anna R Docherty, and John G Kerns. 2017. Evidence that communication impairment in schizophrenia is associated with generalized poor task performance. *Psychiatry research*, 249:172–179.

Alec Radford, Jeffrey Wu, Rewon Child, David Luan, Dario Amodei, Ilya Sutskever, et al. 2019. Language models are unsupervised multitask learners. *OpenAI blog*, 1(8):9.

Owen Rambow, Tao Yu, Axinia Radeva, Alexander R Fabbri, Christopher Hidey, Tianrui Peng, Kathleen R McKeown, Smaranda Muresan, Sardar Hamidian, Mona T Diab, et al. 2016. The columbia-gwu system at the 2016 tac kbp best evaluation. In *TAC*.

Katherine Seeber and Kristin S Cadenhead. 2005. How does studying schizotypal personality disorder inform us about the prodrome of schizophrenia? *Current Psychiatry Reports*, 7(1):41–50.

Primarosa Simone. 2020. lm-scorer. `https://github.com/simonepri/lm-scorer`.

Olav B Smeland, Yunpeng Wang, Min-Tzu Lo, Wen Li, Oleksandr Frei, Aree Witoelar, Martin Tesli, David A Hinds, Joyce Y Tung, Srdjan Djurovic, et al. 2017. Identification of genetic loci shared between schizophrenia and the big five personality traits. *Scientific reports*, 7(1):1–9.

Neil J Smelser, Paul B Baltes, et al. 2001. *International encyclopedia of the social & behavioral sciences*, volume 11. Elsevier Amsterdam.

Johanna Stein. 1993. Vocal alterations in schizophrenic speech. *Journal of Nervous and Mental Disease*.

Thomas Suslow, Cornelia Roestel, Patricia Ohrmann, and Volker Arolt. 2003. The experience of basic emotions in schizophrenia with and without affective negative symptoms. *Comprehensive psychiatry*, 44(4):303–310.

Sunny X Tang, Reno Kriz, Sunghye Cho, Suh Jung Park, Jenna Harowitz, Raquel E Gur, Mahendra T Bhati, Daniel H Wolf, João Sedoc, and Mark Y Liberman. 2021. Natural language processing methods are sensitive to sub-clinical linguistic differences in schizophrenia spectrum disorders. *npj Schizophrenia*, 7(1):1–8.

Ottokar Tilk and Tanel Alumäe. 2016. Bidirectional recurrent neural network with attention mechanism for punctuation restoration. In *Interspeech 2016*.

Thomas A Widiger. 2009. *Neuroticism*, volume 11. The Guilford Press.

Yorick Wilks. 1998. D. arnold, l. balkan, r. lee humphries, s. meijer and l. sadler. machine translation: an introductory guide. ncc blackwell, oxford, 1994.(hardback isbn 1-85554-246-3 $49.95/£40.00; paperback isbn 1-85554-217-x$19.95/£ 18.99) viii+ 240 pages. *Natural Language Engineering*, 4(4):363–382.

Jon A Willits, Timothy Rubin, Michael N Jones, Kyle S Minor, and Paul H Lysaker. 2018. Evidence of disturbances of deep levels of semantic cohesion within personal narratives in schizophrenia. *Schizophrenia Research*, 197:365–369.

A Appendix

Cohesion Linguistic Featuress	LabWriting		LabSpeech	
	P	C	P	C
1. LSA				
Avg. givenness of each sentence	0.20	**0.23***	0.31	**0.32***
2. Lexical Diversity				
Type token ratio (TTR) for all words	**0.63***	0.60	**0.46***	0.43
MTLD lexcical diversity measure for all words	59.9	**68.82***	40.10	**44.95***
VOC lexical diversity measure for all words	40.6	**67.7***	42.73	**52.20***
3. Connectives				
Score of logic connectives	47.1	**53.1***	33.66	**38.31***
Score of temporal connectives	24.5	**26.7**	12.24	**14.74***
Score of extended temporal connectives	24.3	**27.2**	14.44	**18.04***
4. Syntactic Complexity				
SYNMEDpos*	0.56	**0.61***	0.66	**0.68**
SYNMEDwrd*	0.73	**0.81***	0.84	**0.87***
SYNMEDlem*	0.71	**0.79***	0.82	**0.84***
5. Word Information				
Score of pronouns, first person, single form	**96.6***	86.11	**56.68**	55.03
Score of pronouns, first person, plural form	6.3	**10.2***	5.24	**7.26**
Score of pronouns, second person	3.37	**6.22***	**7.99**	7.33
Score of pronouns, third person, plural form	7.90	**12.25***	7.20	**8.65**
Avg. minimum word frequency in sentences	0.83	**1.01***	1.30	**1.45***

SYNMEDpos*: mean minimum editorial distance score between adjacent sentences computed from POS.
SYNMEDwrd*: minimum editorial distance score between adjacent sentences computed from words.
SYNMEDlem*: This is the minimum editorial distance score between adjacent sentences from lemmas.

Table 6: Coh-Metrix Linguistic Features Results

Cohesion Linguistic Features	LabWriting		LabSpeech	
	P	C	P	C
1. Analysis at Paragraph level				
- Mean of probabilities of sentences	0.110	**0.119***	0.106	**0.107**
- Median of probabilities of sentences	0.107	**0.117***	0.104	**0.107**
2. Analysis at Sentence level				
-Sentence Probabilities	0.109	**0.117***	0.103	**0.106**
-Mean of changes in sentence probabilities	-0.106	**-0.045***	**-0.060**	-0.062
-Minimum of changes in sentence probabilities	-1.283	**-1.036***	**-1.835***	-2.107
-Maximum of changes in sentence probabilities	**1.036**	0.986	1.572	**1.902***

Table 7: The language model scores (probabilities) across different segmentation (levels)

Levels	LabWriting		LabSpeech	
	P	C	P	C
Sentence	1.12	**1.10***	**1.11**	1.12
Paragraph	203.9	**150.4***	245.5	**230.1**

Table 8: Perplexity across different segmentation (levels)

Exploring Hybrid and Ensemble Models for Multiclass Prediction of Mental Health Status on Social Media

Sourabh Zanwar
RWTH Aachen University
sourabh.zanwar@rwth-aachen.de

Daniel Wiechmann
University of Amsterdam
d.wiechmann@uva.nl

Yu Qiao
RWTH Aachen University
yu.qiao@rwth-aachen.de

Elma Kerz
RWTH Aachen University
elma.kerz@ifaar.rwth-aachen.de

Abstract

In recent years, there has been a surge of interest in research on automatic mental health detection (MHD) from social media data leveraging advances in natural language processing and machine learning techniques. While significant progress has been achieved in this interdisciplinary research area, the vast majority of work has treated MHD as a binary classification task. The multiclass classification setup is, however, essential if we are to uncover the subtle differences among the statistical patterns of language use associated with particular mental health conditions. Here, we report on experiments aimed at predicting six conditions (anxiety, attention deficit hyperactivity disorder, bipolar disorder, post-traumatic stress disorder, depression, and psychological stress) from Reddit social media posts. We explore and compare the performance of hybrid and ensemble models leveraging transformer-based architectures (BERT and RoBERTa) and BiLSTM neural networks trained on within-text distributions of a diverse set of linguistic features. This set encompasses measures of syntactic complexity, lexical sophistication and diversity, readability, and register-specific ngram frequencies, as well as sentiment and emotion lexicons. In addition, we conduct feature ablation experiments to investigate which types of features are most indicative of particular mental health conditions.

1 Introduction

Mental health is a major challenge in healthcare and in our modern societies at large, as evidenced by the topic's inclusion in the United Nations' 17 Sustainable Development Goals. The World Health Organization estimates that 970 million people worldwide suffer from mental health issues, the most common being anxiety and depressive disorders[1]. The problem is compounded by the fact that the rate of undiagnosed mental disorders has been estimated to be as high as 45% (La Vonne et al., 2012). The societal impact of mental health disorders requires prevention and intervention strategies focused primarily on screening and early diagnosis. In keeping with the WHO Mental Health Action Plan (Saxena et al., 2013), natural language processing and machine learning can make an important contribution to gathering more comprehensive information and knowledge about mental illness. In particular, an increasing use of social media platforms by individuals is generating large amounts of high-quality behavioral and textual data that can support the development of computational solutions for the study of mental disorders. An emerging, interdisciplinary field of research at the intersections of computational linguistics, health informatics and artificial intelligence now leverages natural language processing techniques to analyze such data to develop models for early detection of various mental health conditions.

Systematic reviews of this research show that the vast majority of the existing work has focused primarily on automatic identification of specific disorders, with depression and anxiety being the most commonly studied target conditions (Calvo et al., 2017; Chancellor and De Choudhury, 2020; Zhang et al., 2022). As a result, existing work has focused on developing binary classifiers that aim to distinguish between individuals with a particular mental illness and control users.

The current work addresses the more complex problem of distinguishing between multiple mental states, which is essential if we are to uncover the subtle differences among the statistical patterns of language use associated with particular disorders. Specifically, in this paper we make the following contributions to the existing literature on health text mining based on social media data: (1) We frame the MHC detection tasks as a multiclass prediction task aimed to determine to what

[1] https://www.who.int/news-room/fact-sheets/detail/mental-disorders

Proceedings of the 13th International Workshop on Health Text Mining and Information Analysis (LOUHI), pages 184 - 196
December 7, 2022

extent six mental health conditions (anxiety, attention deficit hyperactivity disorder, bipolar disorder, post-traumatic stress disorder, depression, and psychological stress) can be predicted on the basis of social media posts from Reddit. (2) We explore and compare the performance of hybrid and ensemble models leveraging transformer-based architectures (BERT and RoBERTa) and BiLSTM neural networks trained on within-text distributions of a diverse set of linguistic features. (3) We conduct feature ablation experiments to investigate which types of features are most indicative of particular mental health conditions.

This paper is organized into five sections. Section 2 provides a concise overview of the current state of research on mental health detection from Reddit social media posts. Section 3 presents the experimental setup including descriptions of the data, the type of linguistic features used and their computation, and the modeling approach. The main results are presented and discussed in Section 4. In Section 5 general conclusions are drawn and an outlook is given.

2 Related work

A growing body of research has demonstrated that NLP techniques in combination with text data from social media provide a valuable approach to understanding and modeling people's mental health and have the potential to enable more individualized and scalable methods for timely mental health care (see Calvo et al. (2017); Chancellor and De Choudhury (2020); Zhang et al. (2022), for systematic reviews). A surge in the number of research initiatives by way of workshops and shared tasks, such as Computational Linguistics and Clinical Psychology (CLPsych) Workshop, Social Media Mining for Health Applications (SMMH) and International Workshop on Health Text Mining and Information Analysis (LOUHI), are advancing this research area: It fosters an interdisciplinary approach to automatic methods for the collection, extraction, representation, and analysis of social media data for health informatics and text mining that tightly integrates insights from clinical and cognitive psychology with natural language processing and machine learning. It actively contributes to making publicly available large labeled and high quality datasets, the availability of which has a significant impact on modeling and understanding mental health.

While earlier research on social media mining for health applications has been conducted primarily with Twitter texts (Braithwaite et al., 2016; Coppersmith et al., 2014), a more recent stream of research has turned towards leveraging Reddit as a richer source for constructing mental health benchmark datasets (Cohan et al., 2018; Turcan and McKeown, 2019). Reddit is an interactive, discussion-oriented platform without any length constraints like Twitter, where posts are limited to 280 characters. Its users, the Redditors, are anonymous and the site is clearly organized into more than two million different topics, subreddits. Another crucial fact that makes Reddit more suitable for health text mining is that, unlike Twitter (with its limited text length), extended text production provides a richer linguistic signal that allows analysis at all levels of organization (morpho-syntactic complexity, lexical and phrasal variety, and sophistication and readability). Yates et al. (2017), for instance, proposed an approach for automatically labeling the mental health status of Reddit users. Reflecting the topic organization of Reddits with its subreddits, the authors created high precision patterns to identify users who claimed to have been diagnosed with a mental health condition (diagnosed users) and used exclusion criteria to match them with control users. To prevent easy identification of diagnosed users, the resulting dataset excluded all obvious expressions used to construct it. This approach was also adapted to other mental health conditions (Cohan et al., 2018).

Previous research on health text mining from social media posts has primarily focused on the automatic identification of specific mental disorders and has treated it as a binary classification task aimed at distinguishing between users with a target mental condition and control ones (see the systematic reviews mentioned above). To the best of our knowledge, the only two exceptions are Gkotsis et al. (2017) and Murarka et al. (2021). Gkotsis et al. (2017) proposed an approach to classify mental health-related posts according to theme-based subreddit groupings using deep learning techniques. The authors constructed a dataset of 458,240 posts from mental health related subreddits paired with a control set approximately matched in size (476,388 posts). The mental health-related posts were grouped into 11 MHC themes (addiction, autism, anxiety, bipolar, BPD, depression, schizophrenia, selfharm, SuicideWatch, cripplingalcoholism, opiates) based on a combination of manual assessment steps and automated topic detection. Their best performing model, a convolutional neural network classifier trained on word embeddings, was able to identify the correct theme with a weighted average accuracy of 71.37%. The approach taken in this work was primarily aimed at identifying posts that are relevant to a mental health subreddit, as well as the actual mental health topic to which they relate. Another more recent exception similar to our work is Murarka et al. (2021). The authors used RoBERTa (Robustly Optimized BERT Pretraining Approach, Liu et al. (2019)) to build multiclass models to identify five mental health conditions from Reddit posts (ADHD, anxiety, bipolar disorder, depression, and PTSD). The model was trained on a dataset consisting of Reddit subreddits with 17,159 posts. The RoBERTa-based model achieved a macro-averaged F1 value of 89%, with F1 values for individual conditions ranging from 84% for depression to 91% for ADHD. Although these results appear impressive, they should be interpreted with caution: To obtain data for each of the mental health conditions, the authors extracted posts from five subreddits (r/adhd, r/anxiety, r/bipolar r/disorder, r/depression, r/ptsd) and assigned them a class label corresponding to the name of the condition with which they were associated. Posts for the control group were selected from subreddits with a wide range of general topics (music, travel, India, politics, English, datasets, mathematics and science). The way the datasets in Gkotsis et al. (2017) and Murarka et al. (2021) are constructed rendered the classification tasks relatively easy, as it allows the classifier to use explicit mentions of mental health terms associated with a particular mental health condition. However, there is growing recognition that careful dataset construction is critical to developing robust and generalizable models for detecting mental health status on social media. This requires the removal of expressions indicating mental health status for both diagnosed and control users (see (Yates et al., 2017) or SMHD (Cohan et al., 2018); see also Chancellor and De Choudhury (2020) and Harrigian et al. (2021) for discussions on obtaining ground truth labels for the positive classes and data preprocessing/selection).

The existing research on the detection of mental health conditions in social media mainly follows one of two approaches: One focuses on linguistic features, mainly in the form of unigrams with TF-IDF (term frequency-inverse document frequency) weighting, or on specialized dictionaries, especially the categories from the Linguistic Inquiry and Word Count (LIWC) dictionaries (De Choudhury et al., 2013; Nguyen et al., 2014; Sekulic and Strube, 2019; Zomick et al., 2019). The second centers on leveraging contextualized embedding techniques and pre-trained language models such as BERT (Devlin et al., 2019), ELMo (Peters et al., 2018), and RoBERTa (**?**), minimizing the need for tasks such as feature engineering or feature selection (Gkotsis et al. (2017); Murarka et al. (2021), see also Su et al. (2020) for a review). However, less work has been undertaken to date to explore hybrid and ensemble models for mental illness recognition that integrate engineered features with transformer-based language models. Such hybrid models have recently been successfully applied in the neighboring research area of personality recognition (Mehta et al., 2020; Kerz et al., 2022).

3 Experimental setup

3.1 Dataset

The dataset used in this work was constructed from two recent corpora used for the detection of MHC: (1) the Self-Reported Mental Health Diagnoses (SMHD) dataset (Cohan et al., 2018) and (2) the Dreaddit dataset (Turcan and McKeown, 2019). Both SMHD and Dreaddit were compiled from Reddit, a social media platform consisting of individual topic communities called subreddits, including those relevant to MHC detection. The length of Reddit posts makes them a particularly valuable resource, as it allows modeling of the distribution of linguistic features in the text.

SMHD is a large dataset of social media posts from users with nine mental health conditions (MHC) corresponding to branches in the DSM-5 (APA, 2013), an authoritative taxonomy for psychiatric diagnoses. User-level MHC labels were obtained through carefully designed distantly supervised labeling processes based on diagnosis pattern matching. The pattern matching leveraged a seed list of diagnosis keywords collected from the corresponding DSM-5 headings and extended by synonym mappings. To prevent that target labels can be easily inferred from the presence of MHC indicating words/phrases in the posts, all posts made to mental health-related subreddits or containing keywords related to a mental health condition were removed from the diagnosed users' data. Dread-

Table 1: Datasets statistics (number of posts, means and standard deviations of post length (in words) across mental health conditions and control groups.

MHC	Dataset	N posts	M length	SD
Stress	Dreaddit	1857	91	35
ADHD	SMHD	1849	91.4	57
Anxiety	SMHD	1846	91.7	56.3
Bipolar	SMHD	1848	93	57.7
Depression	SMHD	1846	92.4	58.7
PTSD	SMHD	1600	95.7	59.9
Control	Dreaddit	1696	83.6	29.7
	SMHD	1805	78.8	48.6

dit is a dataset of lengthy social media posts from subreddits in five domains that include stressful and non-stressful text. For a subset of 3.5k users employed in this paper, binary labels (+/- stressful) were obtained from aggregated ratings of five crowdsourced human annotators.

Based on these two corpora, we constructed a dataset with the goal of obtaining sub-corpora of equal size for the six MHCs targeted in this paper. To this end, we downsampled SMHD to match the size of Dreaddit and to be balanced in terms of class distributions. The sampling procedure from the SMHD dataset was such that each post was produced by a distinct user. In doing so, we addressed a concerning trend described in recent review articles that points to the presence of a relatively small number of unique individuals, which may hinder the generalization of models to platforms that are already demographically skewed (Chancellor and De Choudhury, 2020; Harrigian et al., 2021). These constraints were met for five of the nine MHC in the SMHD dataset (attention deficit hyperactivity disorder (ADHD), anxiety, bipolar, depression, post-traumatic stress disorder (PTSD)). The data for the control groups contained the full Dreaddit control subset, which comtains just under 1700 posts, plus an additional 1805 control posts from the SMHD dataset that were matched in terms of post length. The control subset was intentionally designed as a majority class to reduce false positive (overdiagnosis) rates (see Merten et al. (2017) for discussion). Statistics for these datasets are presented in Table 1.

3.2 Measurement of within-text distributions of engineered features

A diverse set of features used in this work fall into the following eight broad categories: (1) features of morpho-syntactic complexity (N=19), (2) features of lexical richness (N=52), (3) register-based n-gram frequency features (N=25), (4) readability features (N=14), and lexicon features designed to detect sentiment, emotion and/or affect (N=325). These features were subdivided into four categories: (5) Emotion/Sentiment, (6) LIWC, (7) Affect, and (8) General Inquirer. An overview of these features can be found in Table 4 in the appendix. All measurements of these features were calculated using an automated text analysis (ATA) system that employs a sliding window technique to compute sentence-level measurements (for recent applications of the ATA system in the context of text classification, see Qiao et al. (2021) and Kerz et al. (2022)). These measurements capture the within-text distributions of scores for a given feature. Tokenization, sentence splitting, part-of-speech tagging, lemmatization and syntactic PCFG parsing were performed using Stanford CoreNLP (Manning et al., 2014).

Figure 1 provides **some examples** of within-text distributions for four selected features for twelve randomly selected Reddit posts from two datasets used in our work. Each of panels in Figure 1 shows the distributions of four of the 436 textual features for one 24 randomly selected texts. The panels on top show the within-text distributions for 12 randomly selected Reddit posts categorized as exhibiting stress from the Dreaddit dataset. The panels panels on the bottom show the within-text distributions for 12 randomly selected posts from the SMHD daatset from users diagnosed with depression. We note that the distribution of feature values is generally not uniform, but shows large fluctuations over the course of the text. Furthermore, high values in one feature are often counterbalanced by low values in another feature. The classification models described in Section 3.3 are designed to detect local peaks of particular features and exploit the fluctuations for the detection of specific MHCs.

3.3 Modeling approach

We built five multiclass classification models to predict six mental health conditions (depression, anxiety, bipolar, ADHD, stress and PTSD): Two of these models leverage transformer-based architectures: BERT (Devlin et al., 2019) and RoBERTa (?). These serve as the baseline models and components of our hybrid model. We used the pretrained 'bert-base-uncased' and 'roberta-base' models from the Huggingface Transformers library (Wolf et al.,

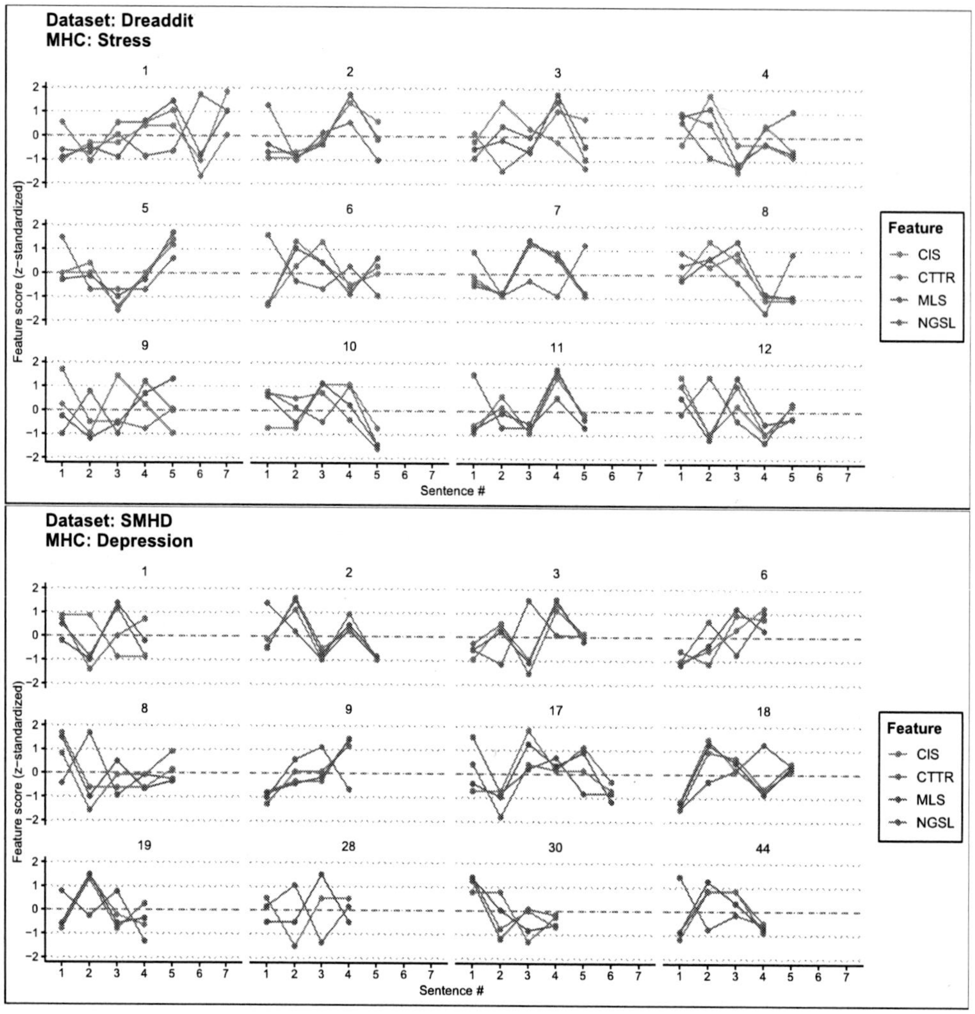

Figure 1: Within-text distributions of ClS (Clauses per Sentence), CTTR (corrected Type/Token Ratio), MLS (Mean Length of Sentence in Words), NGSL (Number of Sophisticated Words). Panels on top show the within-text distributions for 12 randomly selected Reddit posts categorized as exhibiting stress from the Dreaddit dataset. Bottom panels show the within-text distributions for 12 randomly selected posts from the SMHD daatset from users diagnosed with depression.

2020), each with an intermediate bidirectional long short-term memory (BiLSTM) layer with 256 hidden units (Al-Omari et al., 2020). The third model is a BiLSTM classifier (Psyling-BiLSTM) trained solely on the eight feature groups described in Section 3.2. Specifically, we constructed a 4-layer BiLSTM with a hidden state dimension of 1024. The input to that model was a sequence $CM_1^N = (CM_1, CM_2 \ldots, CM_N)$, where CM_i, the output of ATA for the ith sentence of a post, is a 436 dimensional vector and N is the sequence length. To predict the labels of a sequence, we concatenate the last hidden states of the last layer in forward ($\overrightarrow{h_n}$) and backward directions ($\overleftarrow{h_n}$). The result vector of concatenation $h_n = [\overrightarrow{h_n}|\overleftarrow{h_n}]$ is then transformed through a 2-layer feedforward neural network, whose activation function is Rectifier Linear Unit (Agarap, 2018). The output of this is then passed to a Fully Connected (FC) layer with ReLu activation function and dropout of 0.2 and it is fed to a final FC layer. The output is passed through sigmoid function and finally a threshold is used to determine the labels. We trained these models for 500 epochs, and saved the model that performs best on validation set, with a batch size of 256 and a sequence length of 10. The fourth model (Hybrid) is a hybrid classification model that integrates (i) a pretrained RoBERTa model whose output is

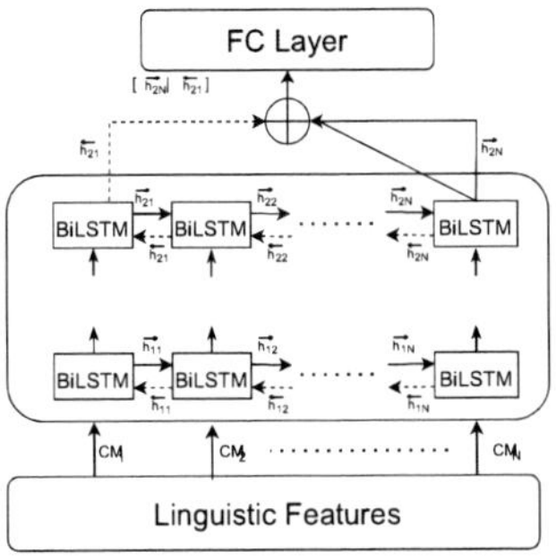

Figure 2: Structure diagram of BiLSTM mental health classification model trained on linguistic features

passed through a BiLSTM layer and a subsequent FC layer with (ii) a BiLSTM network of linguistic features of the text with a subsequent FC layer. The FC layers of both components take as input the concatenation of last hidden states of the last BiLSTM layer in forward and backward direction. We concatenated the outputs of these components before finally feeding them into a final FC layer with a sigmoid activation function. Specifically, the component with the pretrained RoBERTa model comprised a 2-layer BiLSTM with 256 hidden units and a dropout of 0.2. The component with the with the linguistic features consists of a 3-layer BiLSTM with a hidden size of 512 and a dropout of 0.2. We trained this model for 12 epochs, saving the model with the best performance (F1-Score) on the development set. The optimizer used is AdamW with a learning rate of 2e-5 and a weight decay of 1e-4. Structure diagrams of the model based solely on linguistic features and the hybrid architectures are presented in Figures 2 and 3. In order to reduce the variance of the estimates, we trained all models in a 5-fold CV setup. Reported values represent averages over five runs. The fifth model (Stacking) applied a stacking approach to ensemble all models (Wolpert, 1992).

The training procedure consisted of two stages (see Figure 4). In Stage 1, each of the four models was trained independently using 5-fold cross-validation. For each text sample in the test fold, we obtained a prediction vector from each of the four component models. These predictions vectors were then concatenated and constituted the input data in a subsequent training stage (Stage 2). The final predictions of the ensemble model were derived from another logistic regression model trained on the concatenated prediction vectors from Stage 1. To perform inference on the test set, the predic-

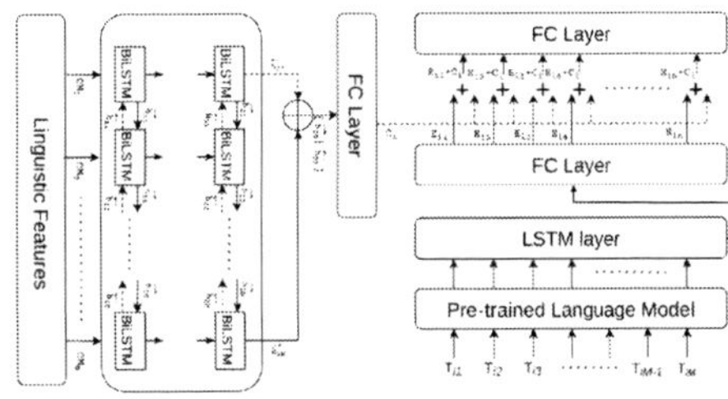

Figure 3: Structure diagram of the hybrid mental health classification models

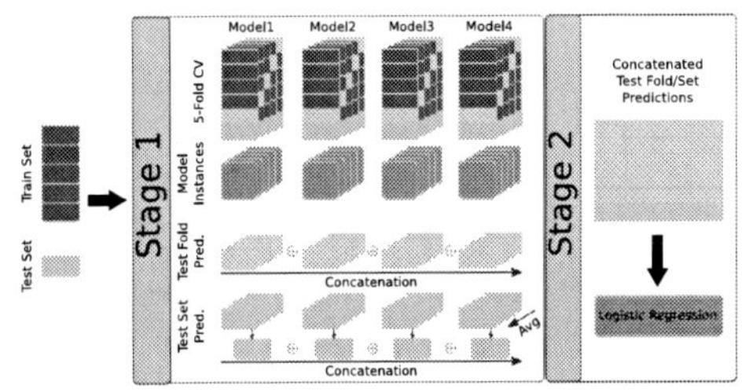

Figure 4: Schematic representation of ensembling by stacking.

tions of all model instances trained in Phase 1 were taken and averaged by model to serve as input to Phase 2 after concatenation. All hyperparameters for the training of each of the ensembled models were selected as specified above.

3.4 Feature ablation

To assess the relative importance of the feature groups in predicting six mental health conditions, we used Submodular Pick Lime (SP-LIME; (Ribeiro et al., 2016)). SP-LIME is a method to construct a global explanation of a model by aggregating the weights of linear models, that locally approximate the original model. To this end, we first constructed local explanations using LIME. Analogous to super-pixels for images, we categorized our features into eight groups (see section 3.2). We used binary vectors $z \in \{0, 1\}^d$ to denote the absence and presence of feature groups in the perturbed data samples, where d is the number of feature groups. Here, 'absent' means that all values of the features in the feature group are set to 0, and 'present' means that their values are retained. For simplicity, a linear regression model was chosen as the local explanatory model. An exponential kernel function with Hamming distance and kernel width $\sigma = 0.75\sqrt{d}$ was used to assign different weights to each perturbed data sample. After constructing their local explanation for each data sample in the original dataset, the matrix $W \in \mathbb{R}^{n \times d}$ was obtained, where n is the number of data samples in the original dataset and W_{ij} is the jth coefficient of the fitted linear regression model to explain data sample x_i. The global importance score of the SP-LIME for feature j can then be derived by: $I_j = \sqrt{\sum_{i=1}^{n} |W_{ij}|}$

4 Results and Discussion

Table 2 gives an overview of the results of the five multiclass classification models described in Section 3.2 Our overall best-performing model (Stacking) achieved a macro F1 score of 31.4%, corresponding to an increase in performance of +3.4% F1 over the BERT baseline and +3.95% F1 over the RoBERTa baseline. In terms of class-wise performance, the highest prediction accuracy was achieved in the detection of stress with a maximum average F1 score of 77%. The second highest prediction accuracy was achieved for the control class with a maximum average F1 score of 53.58%. The next highest classification accuracies were observed for depression (27.48% F1) and ADHD (24.84% F1). Anxiety and bipolar exhibited maximum prediction accuracies greater than 18% F1. Lowest accuracy (14%) was obtained for PTSD. Our Psyling-BiLSTM-model trained exclusively on within-text distributions of eight feature groups achieved a macro F1 score of 22.20%, a decrease of -5.8% F1 from the BERT baseline and -5.25% F1 from the RoBERTa baseline. Another key finding of our experiments is that mental health state prediction benefits immensely from a hybrid approach: The results show that a hybrid model integrating a RoBERTa-based model with text-internal distributions of eight feature groups outperforms the transformer-based models by +1.8% (vs. BERT) and +2.35% (vs. RoBERTa) macro-F1. Moreover, the hybrid model efficiently combined the strengths of the two transformer models (BERT and RoBERTa) and Psyling-BiLSTM, which significantly increased the robustness of the model predictions: Both the transformer-based baseline models and the Psyling-BiLSTM showed below chance performance ($<$ 12.5 % F1) for two of the seven classes. The hybrid model compensated for such drawbacks in an effective manner.

As for the error analysis, Figure 5 shows the confusion matrix of our best model (Stacking) normalized over the actual classes (in rows). We found that for five of the seven mental health conditions,

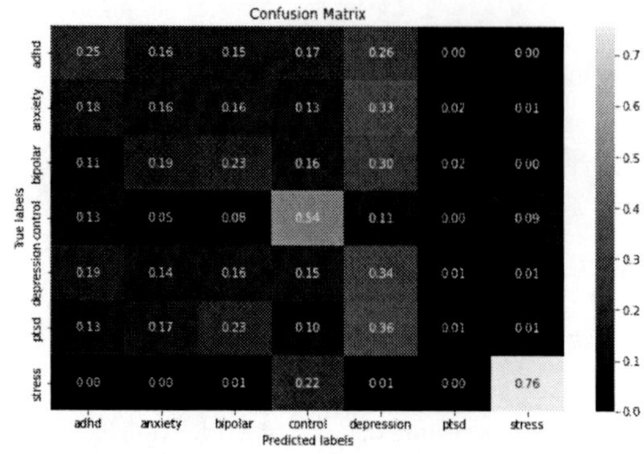

Figure 5: Confusion matrix of the stacking model on multi-class mental health status prediction.

the majority of model predictions applied to the correct class (ADHD 25%, bipolar 23%, depression 34%, stress 76%, control 54%). Bipolar disorder was frequently misclassified as PTSD (23%). Anxiety was most often classified as ADHD (18%), followed by bipolar disorder and correct classification (both 16%). Depression posts were most frequently confused with ADHD (19%), bipolar disorder (16%) and anxiety (14%). At the same time, depression was by far the most frequently predicted class overall, with an average prediction rate of 24.4%.

These findings reflect evidence in the psychiatric literature indicating that there is considerable overlap in clinical symptoms and pathophysiological processes and that depressive symptoms may also occur in the context of another psychiatric disorder (e.g., bipolar disorder) (Baldwin et al., 2002). Furthermore, psychiatric data suggest that depressive disorders (i.e., major depressive disorder and dysthymia) are highly comorbid with other common mental disorders (Rohde et al., 1991; Gold et al., 2020). In contrast, misclassifications in the stress category were almost exclusively controls (22% of all predictions), indicating that statistical patterns of language use reflecting stress differ from those for diagnosed mental health disorders. Controls were in turn most frequently confused with ADHD (13% of all predictions). This finding is consistent with the prevalence of overdiagnosis of ADHD in children and adolescents (Kazda et al., 2019). Finally, PTSD was correctly classified in only 1% of the cases, and typically misclassified as depression (36%) or bipolar (23%). That said, user posts were predicted by the stacking model to be PTSD only 6.5% (21/320) of the time, suggesting that the classifier is sensitive to the slightly lower frequency of

Table 2: Results of the multiclass classification. All numbers represent F1 scores averaged across 5 folds.

Models	Mental Health Condition							Average
	Depression	Anxiety	Bipolar	ADHD	Stress	PTSD	Control	
BERT	17.40	15.20	19.80	5.80	71.20	7.60	48.00	28.00
RoBERTa	27.48	12.83	3.46	17.88	76.22	1.46	52.85	27.45
Psyling-BiLSTM	19.40	15.80	9.60	14.60	51.80	4.00	36.60	22.20
Hybrid	18.40	17.00	11.80	19.40	77.00	14.00	50.60	29.80
Stacking	27.23	18.55	18.21	24.84	76.61	0.96	53.58	31.40

Table 3: Results of the feature ablation. Values represents I scores of a feature group in percent. Values in parentheses indicate the rank of a feature groups per MHC.

Feature Group	Importance					
	Depression	Anxiety	Adhd	Bipolar	Stress	Ptsd
Readability (N=14)	37.06 (1)	38.83 (1)	34.68 (1)	40.1 (1)	25.14 (2)	41.86 (1)
Reg.-spec. Ngram (N=25)	21.85 (2)	21.11 (2)	24.02 (2)	20.56 (2)	21.43 (3)	20 (2)
Lexical richness (N=52)	15.92 (3)	15.17 (3)	15.48 (3)	14.73 (3)	26.15 (1)	14.27 (3)
EmoSent (N=39)	12.09 (4)	11.98 (4)	11.79 (4)	11.87 (4)	12.18 (4)	11.46 (4)
MorphSyn complexity (N=19)	8.01 (5)	7.94 (5)	8.69 (5)	7.81 (5)	9.47 (5)	7.7 (5)
LIWC (N=61)	2.48 (6)	2.42 (6)	2.61 (6)	2.41 (6)	2.71 (6)	2.29 (6)
General Inquirer (N=188)	1.98 (7)	1.94 (7)	2.08 (7)	1.91 (7)	2.22 (7)	1.84 (7)
GALC (N=38)	0.62 (8)	0.61 (8)	0.66 (8)	0.6 (8)	0.69 (8)	0.58 (8)

this mental disorder. In view of the model's tendency to avoid predictions for the less populated class, we conducted additional multiclass experiments without the PTSD class to determine how this would affect the overall pattern of findings. The results of these experiments revealed that the exclusion of PTSD yielded a slight improvement in overall classification accuracy, with the improvement over chance increasing from 18.9% F1 to 23.65% F1. In regards to rank order, the performances of the models mirror those of the models with PTSD: the hybrid model still outperformed both transformer-based models (+3.6% F1 over BERT and +3.37% F1 over RoBERTa) and the stacked generalization still yielded highest classification accuracy (+2.05% F1 over the hybrid model). The general patterns of misclassification remained the same (for further details, see Table 5 in the appendix).

The results of the feature ablation experiments are presented in Table 3. We found that the three most important feature groups across all six mental health conditions are rather general in nature: Readability, lexical richness, and register-specific n-gram frequencies. In comparison, the feature groups representing closed vocabulary approaches (EmoSent, LIWC, General Inquirer, GALC), which have been prominently used in previous work on health text mining, play a minor role. This is particularly striking given that these groups comprise a much greater number of features that have repeatedly been identified as mental health signals (see, e.g., Resnik et al., 2013; Alvarez-Conrad et al., 2001; Tausczik and Pennebaker, 2010, Coppersmith et al., 2014). It is noteworthy that the ranking of the three most important feature groups is consistent across all five mental disorders assessed, with readability features being the most important group. In contrast, stress is strongly associated with features of lexical richness, which includes measures of lexical sophistication, variety, and density. Taken together, these results suggest that research in health text mining and automatic prediction of mental health conditions should move beyond lexicon-based feature groups and place a greater emphasis on more general text features.

5 Conclusion and Outlook

In this paper, we reported on multiclass classification experiments aimed at predicting six mental health conditions from Reddit social media posts. We explored and compared the performance of hybrid and ensemble models leveraging transformer-based architectures (BERT and RoBERTa) and BiLSTM networks trained on within-text distributions of a diverse set of linguistic features. Our results show that the proposed hybrid models significantly improve both model robustness and model accuracy compared to transformer-based baseline models. The use of model stacking proved to be an effective technique to further improve model accuracy. Ablation experiments revealed that the importance of textual features concerning readability, register-specific n-gram frequency and lexical richness far outweighs the importance of closed vocabulary features. In future work, we intend to perform comprehensive feature analysis based on within-text distribution to identify most distinctive indicators of diverse depressive disorders. We also intend to extend the approach presented here to incorporate features of textual cohesion. In addition, we intend to integrate the proposed approach with data on the behavioral activity of the individual, such as the frequency of posting and the temporal distribution of posting histories.

References

Abien Fred Agarap. 2018. Deep learning using rectified linear units (RELU). *arXiv preprint arXiv:1803.08375*.

Hani Al-Omari, Malak A. Abdullah, and Samira Shaikh. 2020. Emodet2: Emotion detection in English textual dialogue using bert and bilstm models. In *2020 11th International Conference on Information and Communication Systems (ICICS)*, pages 226–232.

APA. 2013. Diagnostic and statistical manual of mental disorders. *American Psychiatric Association*, 21(21):591–643.

Oscar Araque, Lorenzo Gatti, Jacopo Staiano, and Marco Guerini. 2019. Depechemood++: a bilingual emotion lexicon built through simple yet powerful techniques. *IEEE transactions on affective computing*.

David S Baldwin, Dwight L Evans, RM Hirschfeld, and Siegfried Kasper. 2002. Can we distinguish anxiety from depression? *Psychopharmacology Bulletin*, 36:158–165.

Margaret M Bradley and Peter J Lang. 1999. Affective norms for English words (ANEW): Instruction manual and affective ratings. Technical report, Technical report C-1, the center for research in psychophysiology ....

Scott R Braithwaite, Christophe Giraud-Carrier, Josh West, Michael D Barnes, and Carl Lee Hanson. 2016. Validating machine learning algorithms for Twitter data against established measures of suicidality. *JMIR mental health*, 3(2):e4822.

Marc Brysbaert, Paweł Mandera, Samantha F McCormick, and Emmanuel Keuleers. 2019. Word prevalence norms for 62,000 English lemmas. *Behavior research methods*, 51(2):467–479.

Rafael A Calvo, David N Milne, M Sazzad Hussain, and Helen Christensen. 2017. Natural language processing in mental health applications using non-clinical texts. *Natural Language Engineering*, 23(5):649–685.

Erik Cambria, Robyn Speer, Catherine Havasi, and Amir Hussain. 2010. Senticnet: A publicly available semantic resource for opinion mining. In *2010 AAAI fall symposium series*.

Stevie Chancellor and Munmun De Choudhury. 2020. Methods in predictive techniques for mental health status on social media: a critical review. *NPJ digital medicine*, 3(1):1–11.

Arman Cohan, Bart Desmet, Andrew Yates, Luca Soldaini, Sean MacAvaney, and Nazli Goharian. 2018. SMHD: a large-scale resource for exploring online language usage for multiple mental health conditions. In *Proceedings of the 27th International Conference on Computational Linguistics*, Santa Fe, New Mexico, USA. Association for Computational Linguistics.

Glen Coppersmith, Mark Dredze, and Craig Harman. 2014. Quantifying mental health signals in Twitter. In *Proceedings of the workshop on computational linguistics and clinical psychology: From linguistic signal to clinical reality*, pages 51–60.

Mark Davies. 2008. The Corpus of Contemporary American English (COCA): 560 million words, 1990-present.

Munmun De Choudhury, Scott Counts, and Eric Horvitz. 2013. Social media as a measurement tool of depression in populations. In *Proceedings of the 5th annual ACM web science conference*, pages 47–56.

Jacob Devlin, Ming-Wei Chang, Kenton Lee, and Kristina Toutanova. 2019. BERT: Pre-training of deep bidirectional transformers for language understanding. In *Proceedings of the 2019 Conference of the North American Chapter of the Association for Computational Linguistics: Human Language Technologies, Volume 1 (Long and Short Papers)*, Minneapolis, Minnesota. Association for Computational Linguistics.

George Gkotsis, Anika Oellrich, Sumithra Velupillai, Maria Liakata, Tim JP Hubbard, Richard JB Dobson, and Rina Dutta. 2017. Characterisation of mental health conditions in social media using informed deep learning. *Scientific reports*, 7(1):1–11.

Stefan M Gold, Ole Köhler-Forsberg, Rona Moss-Morris, Anja Mehnert, J Jaime Miranda, Monika Bullinger, Andrew Steptoe, Mary A Whooley, and Christian Otte. 2020. Comorbid depression in medical diseases. *Nature Reviews Disease Primers*, 6(1):1–22.

Keith Harrigian, Carlos Aguirre, and Mark Dredze. 2021. On the state of social media data for mental health research. In *Proceedings of the Seventh Workshop on Computational Linguistics and Clinical Psychology: Improving Access*, Online. Association for Computational Linguistics.

Brendan T Johns, Melody Dye, and Michael N Jones. 2020. Estimating the prevalence and diversity of words in written language. *Quarterly Journal of Experimental Psychology*, 73(6):841–855.

Luise Kazda, Katy Bell, Rae Thomas, Kevin McGeechan, and Alexandra Barratt. 2019. Evidence of potential overdiagnosis and overtreatment of attention deficit hyperactivity disorder (ADHD) in children and adolescents: protocol for a scoping review. *BMJ open*, 9(11):e032327.

Elma Kerz, Yu Qiao, Sourabh Zanwar, and Daniel Wiechmann. 2022. Pushing on personality detection from verbal behavior: A transformer meets text contours of psycholinguistic features. In *Proceedings of the 12th Workshop on Computational Approaches to Subjectivity, Sentiment & Social Media Analysis*, Dublin, Ireland. Association for Computational Linguistics.

Victor Kuperman, Hans Stadthagen-Gonzalez, and Marc Brysbaert. 2012. Age-of-acquisition ratings for 30,000 English words. *Behavior research methods*, 44(4):978–990.

A Downey La Vonne, Leslie S Zun, and Trena Burke. 2012. Undiagnosed mental illness in the emergency department. *The Journal of emergency medicine*, 43(5):876–882.

Yinhan Liu, Myle Ott, Naman Goyal, Jingfei Du, Mandar Joshi, Danqi Chen, Omer Levy, Mike Lewis, Luke Zettlemoyer, and Veselin Stoyanov. 2019. RoBERTa: A robustly optimized BERT pretraining approach. *arXiv preprint arXiv:1907.11692*.

Christopher Manning, Mihai Surdeanu, John Bauer, Jenny Finkel, Steven Bethard, and David McClosky. 2014. The Stanford CoreNLP natural language processing toolkit. In *Proceedings of 52nd annual meeting of the association for computational linguistics: system demonstrations*, pages 55–60.

Yash Mehta, Samin Fatehi, Amirmohammad Kazameini, Clemens Stachl, Erik Cambria, and Sauleh Eetemadi. 2020. Bottom-up and top-down: Predicting personality with psycholinguistic and language model features. In *2020 IEEE International Conference on Data Mining (ICDM)*, pages 1184–1189. IEEE.

Eva Charlotte Merten, Jan Christopher Cwik, Jürgen Margraf, and Silvia Schneider. 2017. Overdiagnosis of mental disorders in children and adolescents (in developed countries). *Child and adolescent psychiatry and mental health*, 11(1):1–11.

Saif Mohammad. 2018. Obtaining reliable human ratings of valence, arousal, and dominance for 20,000 English words. In *Proceedings of the 56th Annual Meeting of the Association for Computational Linguistics (Volume 1: Long Papers)*, pages 174–184.

Saif Mohammad, Svetlana Kiritchenko, and Xiaodan Zhu. 2013. NRC-Canada: Building the state-of-the-art in sentiment analysis of tweets. In *Proceedings of the seventh international workshop on Semantic Evaluation Exercises (SemEval-2013)*, Atlanta, Georgia, USA.

Saif M Mohammad and Peter D Turney. 2013. Crowdsourcing a word–emotion association lexicon. *Computational intelligence*, 29(3):436–465.

Ankit Murarka, Balaji Radhakrishnan, and Sushma Ravichandran. 2021. Classification of mental illnesses on social media using RoBERTa. In *Proceedings of the 12th International Workshop on Health Text Mining and Information Analysis*, pages 59–68.

Thin Nguyen, Dinh Phung, Bo Dao, Svetha Venkatesh, and Michael Berk. 2014. Affective and content analysis of online depression communities. *IEEE Transactions on Affective Computing*, 5(3):217–226.

James W Pennebaker, Martha E Francis, and Roger J Booth. 2001. Linguistic inquiry and word count: Liwc 2001. *Mahway: Lawrence Erlbaum Associates*, 71(2001):2001.

Matthew E. Peters, Mark Neumann, Mohit Iyyer, Matt Gardner, Christopher Clark, Kenton Lee, and Luke Zettlemoyer. 2018. Deep contextualized word representations. In *Proceedings of the 2018 Conference of the North American Chapter of the Association for Computational Linguistics: Human Language Technologies, Volume 1 (Long Papers)*, New Orleans, Louisiana. Association for Computational Linguistics.

Yu Qiao, Xuefeng Yin, Daniel Wiechmann, and Elma Kerz. 2021. Alzheimer's Disease Detection from Spontaneous Speech Through Combining Linguistic Complexity and (Dis)Fluency Features with Pretrained Language Models. In *Proceedings of Interspeech 2021*, pages 3805–3809.

Marco Tulio Ribeiro, Sameer Singh, and Carlos Guestrin. 2016. " why should i trust you?" explaining the predictions of any classifier. In *Proceedings of the 22nd ACM SIGKDD international conference on knowledge discovery and data mining*, pages 1135–1144.

Paul Rohde, Peter M Lewinsohn, and John R Seeley. 1991. Comorbidity of unipolar depression: Ii. comorbidity with other mental disorders in adolescents and adults. *Journal of abnormal psychology*, 100(2):214.

Shekhar Saxena, Michelle Funk, and Dan Chisholm. 2013. World health assembly adopts comprehensive mental health action plan 2013–2020. *The Lancet*, 381(9882):1970–1971.

Klaus R Scherer. 2005. What are emotions? and how can they be measured? *Social science information*, 44(4):695–729.

Ivan Sekulic and Michael Strube. 2019. Adapting deep learning methods for mental health prediction on social media. In *Proceedings of the 5th Workshop on Noisy User-generated Text (W-NUT 2019)*. Association for Computational Linguistics.

Ryan A Stevenson, Joseph A Mikels, and Thomas W James. 2007. Characterization of the affective norms for English words by discrete emotional categories. *Behavior research methods*, 39(4):1020–1024.

Philip J Stone, Dexter C Dunphy, and Marshall S Smith. 1966. The general inquirer: A computer approach to content analysis.

Chang Su, Zhenxing Xu, Jyotishman Pathak, and Fei Wang. 2020. Deep learning in mental health outcome research: a scoping review. *Translational Psychiatry*, 10(1):1–26.

Elsbeth Turcan and Kathy McKeown. 2019. Dreaddit: A Reddit dataset for stress analysis in social media. In *Proceedings of the Tenth International Workshop on Health Text Mining and Information Analysis (LOUHI 2019)*, pages 97–107. Association for Computational Linguistics.

Thomas Wolf, Julien Chaumond, Lysandre Debut, Victor Sanh, Clement Delangue, Anthony Moi, Pierric Cistac, Morgan Funtowicz, Joe Davison, Sam Shleifer, et al. 2020. Transformers: State-of-the-art natural language processing. In *Proceedings of the 2020 Conference on Empirical Methods in Natural Language Processing: System Demonstrations*, pages 38–45.

David H Wolpert. 1992. Stacked generalization. *Neural networks*, 5(2):241–259.

Andrew Yates, Arman Cohan, and Nazli Goharian. 2017. Depression and self-harm risk assessment in online forums. In *Proceedings of the 2017 Conference on Empirical Methods in Natural Language Processing*, Copenhagen, Denmark. Association for Computational Linguistics.

Tianlin Zhang, Annika M Schoene, Shaoxiong Ji, and Sophia Ananiadou. 2022. Natural language processing applied to mental illness detection: a narrative review. *NPJ digital medicine*, 5(1):1–13.

Jonathan Zomick, Sarah Ita Levitan, and Mark Serper. 2019. Linguistic analysis of schizophrenia in reddit posts. In *Proceedings of the Sixth Workshop on Computational Linguistics and Clinical Psychology*, pages 74–83.

A Appendix

Table 4: Overview of the 436 features investigated in the work.

Feature group	Number of features	Features	Example/Description
Morpho-syntactic	19	MLC	Mean length of clause (words)
		MLS	Mean length of sentence (words)
		MLT	Mean length of T-unit (words)
		C/S	Clauses per sentence
		C/T	Clauses per T-unit
		DepC/C	Dependent clauses per clause
		T/S	T-units per sentence
		CompT/T	Complex T-unit per T-unit
		DepC/T	Dependent Clause per T-unit
		CoordP/C	Coordinate phrases per clause
		CoordP/T	Coordinate phrases per T-unit
		NP.PostMod	NP post-mod (word)
		NP.PreMod	NP pre-mod (word)
		CompN/C	Complex nominals per clause
		CompN/T	Complex nominals per T-unit
		VP/T	Verb phrases per T-unit
		BaseKolDef	Kolmogorov Complexity
		MorKolDef	Morphological Kolmogorov Complexity
		SynKolDef	Syntactic Kolmogorov Complexity
Lexical richness	52	MLWc	Mean length per word (characters)
		MLWs	Mean length per word (sylables)
		LD	Lexical density
		NDW	Number of different words
		CNDW	NDW corrected by Number of words
		TTR	Type-Token Ration (TTR)
		cTTR	Corrected TTR
		rTTR	Root TTR
		AFL	Sequences Academic Formula List
		ANC	LS (ANC) (top 2000)
		BNC	LS (BNC) (top 2000)
		NAWL	LS New Academic Word List
		NGSL	LS (General Service List)
		NonStopWordsRate	Ratio of words in NLTK non-stopword list
		WordPrevalence	See Brysbaert et al. (2019)
		Prevalence	Word prevalence list incl. 35 categories (Johns et al. (2020))
		AoA-mean	avg. age of acquisition (Kuperman et al. (2012))
		AoA-max	max. age of acquisition

(continued)			
Register-based N-gram	25	Spoken ($n \in [1,5]$) Fiction ($n \in [1,5]$) Magazine ($n \in [1,5]$) News ($n \in [1,5]$) Academic ($n \in [1,5]$)	Frequencies of uni-, bi- tri-, four-, five-grams from the five sub-components (genres) of the COCA, see Davies (2008)
Readability	14	ARI	Automated Readability Index
		ColemanLiau	Coleman-Liau Index
		DaleChall	Dale-Chall readability score
		FleshKincaidGradeLevel	Flesch-Kincaid Grade Level
		FleshKincaidReadingEase	Flesch Reading Ease score
		Fry-x	x coord. on Fry Readability Graph
		Fry-y	y coord. on Fry Readability Graph
		Lix	Lix readability score
		SMOG	Simple Measure of Gobbledygook
		GunningFog	Gunning Fog Index readability score
		DaleChallPSK	Powers-Sumner-Kearl Variation of the Dale and Chall Readability score
		FORCAST	FORCAST readability score
		Rix	Rix readability score
		Spache	Spache readability score
Lexicons:	325		
EmoSent	39	ANEW-Emo lexicons	(Stevenson et al., 2007)
		Affective Norms for English Words	(Bradley and Lang, 1999)
		DepecheMood++	(Araque et al., 2019)
		NRC Word-Emotion Association	(Mohammad and Turney, 2013)
		NRC Valence, Arousal, and Dominance	(Mohammad, 2018)
		SenticNet	(Cambria et al., 2010)
		Sentiment140	(Mohammad et al., 2013)
GALC	38	Geneva Affect Label Coder	(Scherer, 2005)
LIWC	61	LIWC	(Pennebaker et al., 2001)
Inquirer	188	General Inquirer	(Stone et al., 1966)

Table 5: Results of the multiclass classification of MHCs (without PTSD).

	Mental Health Condition						
Models	Depression	Anxiety	Bipolar	ADHD	Stress	Control	**Average**
BERT	4.36	29.12	3.47	28.88	77.37	52.22	32.2
RoBERTa	8.07	6.40	26.00	18.84	82.8	52.26	32.43
Psyling-BiLSTM	11.48	6.88	11.43	21.25	59.00	38.32	24.84
Hybrid	20.80	16.00	14.2	26.8	81.60	52.6	35.80
Model Stacking	21.93	18.96	21.93	19.10	83.14	55.22	37.85

A Knowledge-Graph-Based Intrinsic Test for Benchmarking Medical Concept Embeddings and Pretrained Language Models

Claudio Aracena[1], **Fabián Villena**[1,2], **Matías Rojas**[1,2], **and Jocelyn Dunstan**[1,2]
[1]Faculty of Physical and Mathematical Sciences, University of Chile
[2]Center for Mathematical Modeling, University of Chile
{claudio.aracena,fabian.villena,jdunstan}@uchile.cl
matias.rojas.g@ug.uchile.cl

Abstract

Using language models created from large data sources has improved the performance of several deep learning-based architectures, obtaining state-of-the-art results in several NLP extrinsic tasks. However, little research is related to creating intrinsic tests that allow us to compare the quality of different language models when obtaining contextualized embeddings. This gap increases even more when working on specific domains in languages other than English. This paper proposes a novel graph-based intrinsic test that allows us to measure the quality of different language models in clinical and biomedical domains in Spanish. Our results show that our intrinsic test performs better for clinical and biomedical language models than a general one. Also, it correlates with better outcomes for a NER task using a probing model over contextualized embeddings. We hope our work will help the clinical NLP research community to evaluate and compare new language models in other languages and find the most suitable models for solving downstream tasks.

1 Introduction

In healthcare, text plays a role of enormous importance. One of the media that a medical practitioner can persist is the text in clinical records (Dalianis, 2018). Text is one of the richest forms of information inside the electronic health record, so it is fundamental to develop tools to extract information from these text sources. To create these tools in this field, we must pay special attention to ensuring quality and reproducibility.

Analyzing unstructured texts written by humans is challenging since it is complex to formally understand and describe the rules governing human language, as it is ambiguous and constantly evolving. Natural Language Processing (NLP) is an interdisciplinary field of artificial intelligence that seeks to develop algorithms capable of understanding, interpreting, and manipulating these unstructured texts (Jurafsky and Martin, 2000).

In the medical context, using NLP helps to address tasks such as extracting medical entities, disease coding, text classification, and relation extraction, among others. However, one of the steps before solving any of these tasks is to create robust numerical representations of the text so that the computer can handle this data.

Word embeddings are dense, semantically meaningful vector representations of a word. These models have proven to be a fundamental building block of neural network-based architectures (Lample et al., 2016). Although these models have obtained excellent results for several NLP tasks, their main drawback is that they provide a single-word representation in a given document. This is not optimal since a word meaning may depend on the sentence in which it appears. This type of word embedding is known as static word embeddings.

Contextual representation models handle this issue by creating word representations based on sentence-level context. These representations are commonly retrieved from pretrained language models (PLM). Classic examples of these models are ELMO, BERT, RoBERTa, Flair, ALBERT, among others. However, contextualized word embeddings may not represent words as well as static ones, as results obtained in Reimers and Gurevych (2019) suggest.

Although contextualized word embeddings have these drawbacks, we can use these numeric representations of words to understand PLM representations. Specifically, we are interested in studying how domain-specific and general-domain PLM represent clinical and biomedical concepts. In this study, we aim to create a simple and efficient test for measuring concept embeddings' quality and comparing clinical and biomedical PLM performance using a relevant knowledge base and graph, the Unified Medical Language System (UMLS).

A knowledge graph is an extensive network of

Proceedings of the 13th International Workshop on Health Text Mining and Information Analysis (LOUHI), pages 197 - 206
December 7, 2022

entities relevant to a specific domain. The network describes each entity's semantic types, properties, and relationships. Knowledge graphs represent real-world entities and their relations in a graph, define possible classes, and allow to relate arbitrary entities with each other (Ehrlinger and Wöß, 2016).

The UMLS is a knowledge graph that combines many clinical and biomedical vocabularies and standards to enable interoperability between computer systems (Bodenreider, 2004). The UMLS consists of multiple knowledge sources. One is the metathesaurus, a large, multi-purpose, and multilingual vocabulary database that contains information about biomedical and clinical-related concepts, their various names, and their relationships. Another source is the semantic network, a consistent categorization of all concepts represented in the metathesaurus, providing a set of valuable relationships between these concepts. In this work, we used both knowledge sources.

Two testing frameworks have been developed to measure the quality of language representations. First, an extrinsic test framework that uses the language representations to construct a more complex architecture to solve a specific downstream task. Second, an intrinsic test framework that measures the capacity of the language representation to resolve semantic questions regarding the language domain it represents (Zhai et al., 2016; Wang et al., 2019; Bakarov, 2018).

To construct intrinsic tests, we must compose questions based on a source of truth. This source can be expert knowledge, where we ask human experts to write each one of these questions manually, or we can use a knowledge base to compose these questions automatically. We used the UMLS knowledge graph to automatically derive a concept similarity intrinsic test using the length of the shortest path in the graph to compute a true similarity measure between concepts.

This intrinsic test will be used as a metric to check how good language representations are, but also as a comparison measure of whether clinical and biomedical PLM are better compared to general ones in downstream tasks such as Named Entity Recognition (NER).

2 Related work

PLM such as BERT (Devlin et al., 2019), ELMo (Peters et al., 2018), and GPT-2 (Radford et al., 2019) are able to produce contextualized word embeddings. It has been shown that contextualized word embeddings can achieve near state-of-the-art performance in tasks such as POS tagging or NER using probing models (Liu et al., 2019). Additionally, contextualized word embeddings from top layers of PLM produce more context-specific and anisotropic representations (Ethayarajh, 2019).

Regarding the clinical and biomedical domain in English, there are several models to obtain contextualized embeddings, such as BioELMo (Jin et al., 2019), Clinical BERT (Alsentzer et al., 2019), SciBERT (Beltagy et al., 2019), BioBERT (Lee et al., 2020), among others. However, there remains a significant lack of language models in Spanish. The only models available are SciELO Flair (Akhtyamova et al., 2020), Clinical Flair (Rojas et al., 2022b), and clinical and biomedical versions of RoBERTa (Carrino et al., 2022). Although these studies have shown that incorporating domain-specific contextualized embeddings significantly improves the models' performance in several extrinsic tasks, comparing their performances with intrinsic tests is still necessary.

Since PLM creates word-level contextual representations, it is necessary to define a method for combining these vectors to create sentence-level embeddings. For this purpose, a popular technique is the mean pooling of contextual word embeddings (Reimers and Gurevych, 2019). However, this method may lead to poor results if the PLM is not explicitly trained for similarity. Another study has proposed transforming the distribution of sentence-level embeddings to generate isotropic and smooth representations (Li et al., 2020). Creating these sentence-level representations is fundamental for testing the intrinsic tests proposed in this research.

Common approaches to evaluate biomedical PLM performance are benchmarks such as BLUE (Peng et al., 2019) and BLURB (Gu et al., 2021), which are built for the English language. There is no relevant benchmark in Spanish, and every author selects some annotated datasets to evaluate PLM performance on specific downstream tasks. Although the amount of annotated datasets in Spanish is growing, there is a lack of intrinsic tasks that can help to understand if a PLM is improving, and this research tries to fill that gap.

3 Methods

Our proposed method creates a semantic similarity intrinsic test with medical concept pairs and their semantic distances. We extracted these concept pairs from the UMLS[1] term graph and computed their distances as the length of the shortest directed path of parent relationships between the concepts. We measured the correlation of the knowledge-graph-derived distance to the cosine similarity of the terms string descriptions on an embedding space projected using different language representations. Finally, we compare these correlations with the performance on downstream tasks of each language representation.

3.1 Concept pair selection and its graph distances

In this vocabulary database, a concept is simply the meaning of a medical entity. Each concept in the metathesaurus has a unique and permanent concept identifier (CUI).

A UMLS concept can have multiple names because the same meaning can be described with numerous strings, for example, in different languages or source vocabularies. Each concept named description is called an atom and is identified by an atom identifier (AUI). To select a single concept description, we filtered out the atoms marked as non-preferred in the metathresaurus. With this filter and by only selecting atoms in Spanish, we assigned a single string describing each medical concept. In the UMLS Semantic Network, concepts are related using multiple relation types. The only relation type we used to connect the concepts was the parent relationship (PAR). We tried other relationship types but continued with PAR relationships because they are the most frequent. Child relationships (CHD) have the same frequency as PAR relationships, given they are the inverse relation type of PAR. Thus we can choose any of them.

After the previous step, we imported concepts and their PAR relations into a graph database[2]. Next, we queried the graph to select several random concepts and recursively extracted direct or related concepts at multiple distances. This means there is a path of one or more PAR relations of distance between pairs of concepts, as shown in Figure 1. Given that sometimes it is possible to find multiple paths between two concepts, we only used the shortest path between them. This process allowed us to extract the path length between two concepts. We select 20,000 concepts for this study to conduct the intrinsic tests rapidly. However, we can choose more concepts if necessary.

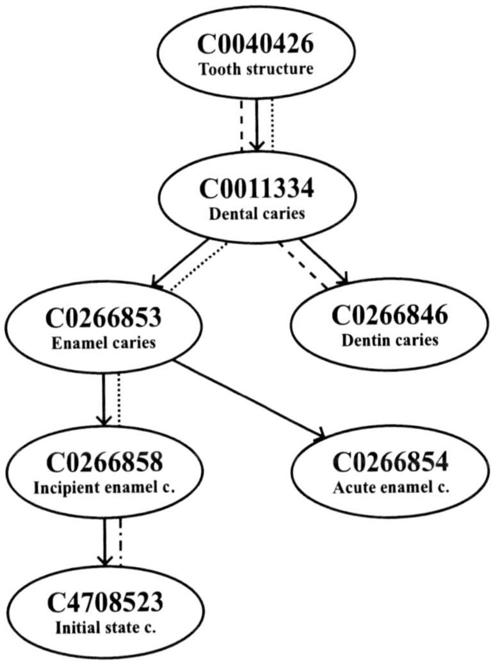

Figure 1: PAR-related concepts from C0040426 (Tooth structure). We highlight multiple paths,

- A dash-dot line represents the path between C0266858 (Incipient enamel caries) and C4708523 (Initial state caries) with a distance of 1 PAR edge.
- A dash-dash line represents the path between C0040426 (Tooth structure) and C0266846 (Dentin caries) with a distance of 2 PAR edges.
- A dot-dot line represents the path between C0040426 (Tooth structure) and C0266858 (Incipient enamel caries) with a distance of 3 PAR edges.

3.2 Generation of UMLS concepts' embeddings

After selecting the pairs of concepts and their descriptions, we generate concepts' embeddings using PLM. As UMLS concepts may contain more than one token, extracting embeddings that can represent the whole concept and not just one

[1] version 2022AA

[2] Neo4j (https://neo4j.com/)

word is essential. To do this, we used mean pooling of embeddings obtained for concept tokens from a PLM. For models hosted in the Huggingface Model Repository[3], we used the Python library `sentence-transformers`[4] (Reimers and Gurevych, 2019), and for models hosted in the Flair repository, we used the Python library `flair`[5] (Akbik et al., 2019).

All of our experiments were conducted for Spanish language datasets. We generated concept embeddings for several PLM of interest with different base architectures and domains. For the base architectures, we selected BERT, RoBERTa, and Flair. As for the domain, we chose, whenever possible, general, biomedical, and clinical models. As we did not find a publicly available BERT linguistic model for the clinical domain trained on Spanish text, we tuned a general domain model in Spanish (Cañete et al., 2020) with clinical text obtained from the Chilean Waiting List Corpus (Báez et al., 2020, 2022).

3.3 Implementation of intrinsic test

We build our intrinsic test as follows. First, we calculate the cosine similarity between concept embedding pairs. Then, we obtained the Spearman correlation between cosine similarity and path length, which we called ρ. This simple process allowed us to get our first metric. We expect that a greater path length between two concepts will result in a lower cosine similarity, given that they are farther semantically. Therefore, the Spearman correlation (ρ) between these two distances over all concepts pairs will be negative. If we compare embeddings generated by different PLM, we could expect that more domain-specific PLM will generate embeddings with more semantic differences between concepts within the domain, resulting in a more negative ρ. Thus, a more negative ρ indicates a PLM that can separate better semantically concepts within a domain.

As a part of our analysis, we calculated the average cosine similarity per path length. This step led us to obtain a complementary metric, the difference of mean cosine similarity for the shortest path length and the longest path length, that we called δ. The rationality behind this metric is similar to what we found in the previous one. However, in this case, a more positive δ indicates a PLM that can better separate concepts semantically within a domain.

3.4 Comparison with extrinsic test

Our intrinsic metrics were compared to extrinsic metrics using the F1 score in relevant biomedical and clinical NER datasets. The idea of incorporating extrinsic tests is to check if having better values of our intrinsic metrics will translate into better performance in downstream tasks for the selected PLM.

To build a reproducible extrinsic comparison for all PLM base architectures, we create a probing task for NER. In other words, we extracted contextualized embeddings from a PLM without fine-tuning for any downstream task, and those embeddings were input into a linear layer trained for NER.

The clinical and biomedical datasets in Spanish used for the NER probing task were:

- CANTEMIST[6] (Miranda-Escalada et al., 2020): annotated corpus with tumor morphology mentions in 1,301 oncological clinical case reports.
- PharmaCoNER[7] (Gonzalez-Agirre et al., 2020): annotated corpus with entities such as chemical compounds and drugs in 1,000 clinical case studies.
- CT-EBM-SP[8] (Campillos-Llanos et al., 2021): annotated corpus with UMLS entities in 1,200 texts about clinical trials studies and clinical trials announcements.
- NUBes[9] (Lopez et al., 2020): annotated corpus with negation and uncertainty entities in anonymised health records (29,682 sentences).

4 Results

We queried 20,000 pairs of random atoms to select UMLS concepts from the graph database. Figure 2 shows the histogram of those pairs by path length. We can see that pair frequency increases as path length increase until seven parent relationships of

[3]https://huggingface.co/models
[4]https://github.com/UKPLab/sentence-transformers
[5]https://github.com/flairNLP/flair
[6]https://zenodo.org/record/3978041
[7]https://zenodo.org/record/4270158
[8]http://www.lllf.uam.es/ESP/nlpmedterm_en
[9]https://github.com/Vicomtech/NUBes-negation-uncertainty-biomedical-corpus

distance. After that point, the frequency of pairs decreases until it reaches 14 relations of distance. We removed all path lengths containing less than 300 pairs of concepts to calculate the metrics ρ and δ.

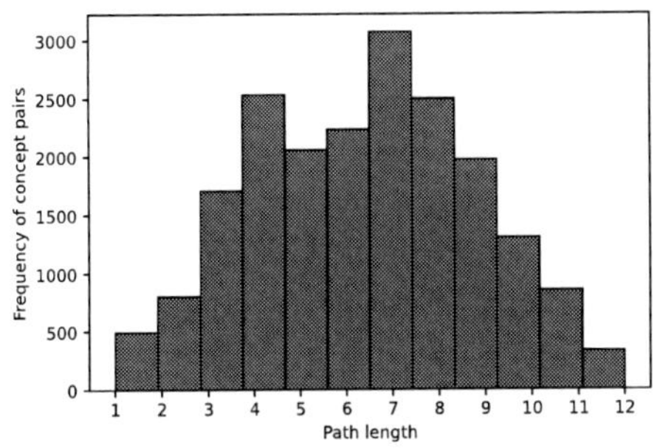

Figure 2: Histogram of UMLS concept pairs by path length

Then, we plot a boxplot of cosine similarity by path length for every PLM. Figure 3 shows such a boxplot for a general-domain BERT trained in Spanish text (Cañete et al., 2020)[10]. This plot allows us to understand how cosine similarity distributes along path length.

It is clear from the plot that average cosine similarity decreases as path length increases. However, the decline is near null or even negative from path length four onwards. Moreover, the average cosine similarity is not going near zero. We hypothesize this pattern is because all concepts are related to clinical and biomedical domains and also due to the anisotropic behavior of sentence embeddings obtained from PLM. As discussed in Ethayarajh (2019), contextualized embeddings obtained from PLM tend to distribute not evenly in the embedding space but in a small portion of it. Therefore, they still have a relatively high similarity when comparing dissimilar concepts.

To compare several PLM, we plot only average cosine similarity by path length for every language model, as shown in Figure 4. As we can see, average cosine similarity by path length varies for different base architectures and domains of PLM. However, they all repeat the same decline pattern as path length increases.

Similarly to Figure 3, Figure 4 does not show any average cosine similarity going near zero. However, the similarity level where each PLM stabilizes

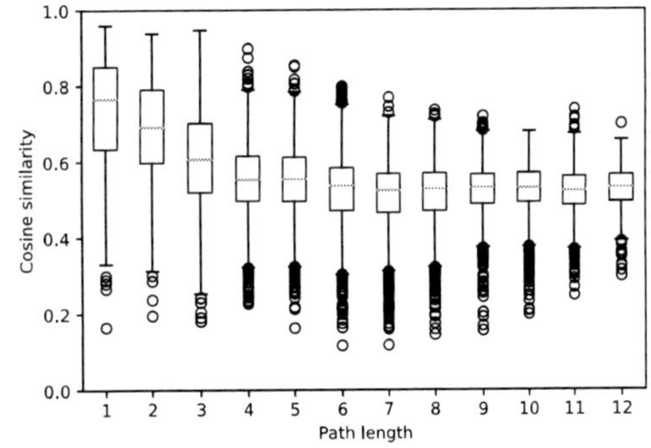

Figure 3: Boxplot of cosine similarity by path length for a general-domain BERT trained on Spanish text.

is different. Not surprisingly, language models trained on a similar corpus or being a fine-tuned version from another have comparable similarity levels. RoBERTa-es-clinical was trained with the same corpora as RoBERTa-es-biomedical plus a clinical corpus (Carrino et al., 2022), and BERT-es-clinical is a fine-tuned model from BERT-es-general over a clinical corpus.

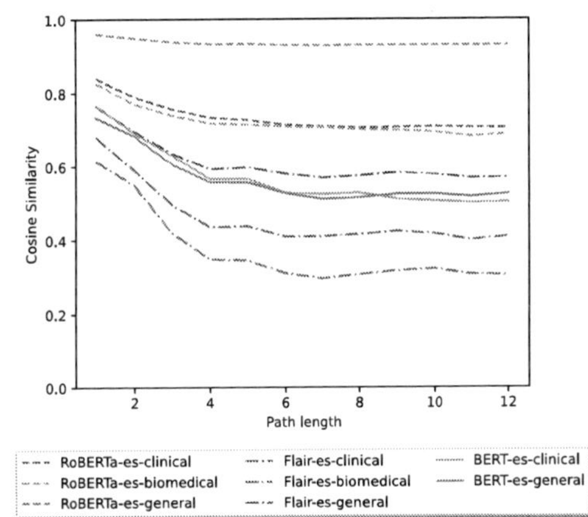

Figure 4: Average cosine similarity by path length for multiple language models

To measure the degree of the decline, we calculated the metrics ρ and δ for all the selected PLM, as shown in Table 1. We notice that ρ and δ are greater in absolute value for biomedical and clinical models than general ones within the same base architecture. This means that given a PLM base architecture, the degree of decline of the average cosine similarity is greater for domain-specific models than for general domain models. This finding suggests that domain-specific PLM and their concept embeddings better represent UMLS concepts;

[10]Other models' plots are included in the appendix

Reference	Architecture	Domain	ρ	δ
Ours	BERT	Clinical	-0.38	0.25
(Cañete et al., 2020)	BERT	General	-0.30	0.18
(Akhtyamova et al., 2020)	Flair	Biomedical	-0.24	0.27
(Rojas et al., 2022b)	Flair	Clinical	-0.23	0.27
(Akbik et al., 2018)	Flair	General	-0.20	0.11
(Carrino et al., 2021)	RoBERTa	Clinical	-0.31	0.09
(Carrino et al., 2021)	RoBERTa	Biomedical	-0.28	0.13
(Gutiérrez-Fandiño et al., 2022)	RoBERTa	General	-0.23	0.03

Table 1: Correlations and differences for each language representation. The table is sorted ascending by ρ and then by base architecture. Every ρ is statistically significant.

Architecture	Domain	CANTEMIST	PharmaCoNER	CT-EBM-SP	NUBes
BERT	Clinical	0.739 (0.018)	0.577 (0.013)	**0.742 (0.012)**	0.791 (0.009)
BERT	General	**0.757 (0.004)**	**0.582 (0.007)**	0.714 (0.006)	**0.797 (0.013)**
Flair	Biomedical	**0.784 (0.006)**	**0.615 (0.013)**	**0.725 (0.008)**	0.792 (0.003)
Flair	Clinical	0.771 (0.009)	0.580 (0.021)	0.694 (0.000)	**0.802 (0.003)**
Flair	General	0.714 (0.013)	0.558 (0.002)	0.633 (0.002)	0.780 (0.005)
RoBERTa	Clinical	**0.794 (0.009)**	**0.633 (0.010)**	0.792 (0.012)	0.820 (0.004)
RoBERTa	Biomedical	0.784 (0.006)	0.626 (0.009)	**0.794 (0.014)**	**0.821 (0.005)**
RoBERTa	General	0.767 (0.014)	0.584 (0.006)	0.734 (0.005)	0.804 (0.003)

Table 2: F1 scores and standard deviations for NER probing task over four datasets in Spanish. The table is sorted according the same criteria as Table 1

hence the similarity pattern displayed. However, it is important to note that we do not find this behavior when comparing different base architectures.

We can see F1 scores for every NER probing task by PLM in Table 2. As expected, we can see a tendency to obtain better F1 scores for clinical or biomedical PLM than general ones. However, in the case of BERT architecture, results are mixed. We believe this behavior could be due to the creation of the clinical BERT model. Instead of being trained from scratch with clinical and biomedical data, it is a fine-tuned version of a general BERT. On the other hand, clinical and biomedical Flair and RoBERTa models were trained from scratch with domain-specific data.

Interestingly, when ρ metric is greater for a clinical model compared to a biomedical one, F1 scores for NER probing tasks are also greater, as we can see in the case of RoBERTa architecture for CANTEMIST and PharmaCoNER datasets. In the case of CT-EBM-SP and NUBes, there are no such differences, but F1 scores for clinical and biomedical are almost the same. On the contrary, when ρ metric is greater for a biomedical model compared to a clinical one, then F1 scores present a similar behavior, as we can see in the case of Flair architecture for CANTEMIST, PharmaCoNER, and CT-EBM-SP datasets. And as same as the previous situation, F1 scores for another dataset (NUBes) are almost the same. We do not observe this pattern for δ metric.

This finding suggests that ρ metric could be applied as a useful intrinsic test for comparing PLM within the same base architecture. However, it is important to note when comparing ρ metric for different base architectures, we do not find a clear relation with F1 scores. Consequently, we present the ρ metric as an intrinsic test to measure improvements for PLM within the same base architecture.

5 Conclusion and future work

Using domain-specific PLMs for downstream tasks has allowed reaching the state-of-the-art in several benchmarks. However, since these models are trained in large corpora, fine-tuning them or training from scratch is time-consuming. Therefore, before using these models to solve downstream tasks, it is crucial to create intrinsic tests that validate whether a domain-specific PLM yields better results than its base version.

In this study, we build an intrinsic test for clinical and biomedical PLM using contextualized embeddings and the UMLS knowledge graph. We suggest that our intrinsic test can help compare domain-specific PLM performance within its base architecture, which could be used to evaluate improvements when building PLM. Our experimental results show that this intrinsic test can capture improvements in clinical and biomedical PLM over general ones. Also, it correlates with better results in a NER probing task over four datasets in Spanish.

In future work, we can implement this study for other languages. Additionally, we can compare our intrinsic test with other probing tasks such as POS-tagging or coreference or even other clinical downstream tasks such as patient mortality or unplanned readmission. On the other hand, since our experimental datasets contain nested entities, but for simplicity, they were ignored, we would like to explore the use of contextualized embeddings in models that can address them, such as those proposed in Rojas et al. (2022a). Finally, we can compare several experimental settings, such as multiple numbers of concept pairs.

Limitations

We can group the limitations of our study in the ones related to the graph knowledge, the selected PLM, comparison with other embedding techniques, and language. First, regarding graph knowledge, we could have chosen several random subsets of concept pairs of different lengths and types of relations to check if our findings are still present. Second, we selected three base architectures, and all of them were of encoder type. Third, we could have compared our results with static embeddings. And finally, we could have selected more languages for comparison.

Ethics Statement

We state that our work complies with the ACL Code of Ethics. We believe that our work could help the research community with a new tool for their work in clinical and biomedical PLM. Our study was based on publicly available and anonymized data to avoid the privacy issues that clinical data may raise.

Acknowledgements

This work was funded by ANID Chile: Basal Funds for Center of Excellence FB210005 (CMM) and FB210017 (CENIA); Millennium Science Initiative Program ICN17_002 (IMFD) and ICN2021_004 (iHealth), Fondecyt grant 11201250, and National Doctoral Scholarships 21211659 (C.A.) and 21220200 (F.V.). Regarding hardware, the research was partially supported by the supercomputing infrastructure of the NLHPC (ECM-02) and the Patagón supercomputer of Universidad Austral de Chile (FONDEQUIP EQM180042).

References

Alan Akbik, Tanja Bergmann, Duncan Blythe, Kashif Rasul, Stefan Schweter, and Roland Vollgraf. 2019. FLAIR: An easy-to-use framework for state-of-the-art NLP. In *Proceedings of the 2019 Conference of the North American Chapter of the Association for Computational Linguistics (Demonstrations)*, pages 54–59, Minneapolis, Minnesota. Association for Computational Linguistics.

Alan Akbik, Duncan Blythe, and Roland Vollgraf. 2018. Contextual string embeddings for sequence labeling. In *Proceedings of the 27th International Conference on Computational Linguistics*, pages 1638–1649, Santa Fe, New Mexico, USA. Association for Computational Linguistics.

Liliya Akhtyamova, Paloma Martínez, Karin Verspoor, and John Cardiff. 2020. Testing Contextualized Word Embeddings to Improve NER in Spanish Clinical Case Narratives. *IEEE Access*, 8:164717–164726.

Emily Alsentzer, John Murphy, William Boag, Wei-Hung Weng, Di Jindi, Tristan Naumann, and Matthew McDermott. 2019. Publicly available clinical BERT embeddings. In *Proceedings of the 2nd Clinical Natural Language Processing Workshop*, pages 72–78, Minneapolis, Minnesota, USA. Association for Computational Linguistics.

Pablo Báez, Felipe Bravo-Marquez, Jocelyn Dunstan, Matías Rojas, and Fabián Villena. 2022. Automatic extraction of nested entities in clinical referrals in spanish. *ACM Transactions on Computing for Healthcare (HEALTH)*, 3(3):1–22.

Pablo Báez, Fabián Villena, Matías Rojas, Manuel Durán, and Jocelyn Dunstan. 2020. The Chilean waiting list corpus: a new resource for clinical named entity recognition in Spanish. In *Proceedings of the 3rd Clinical Natural Language Processing Workshop*, pages 291–300, Online. Association for Computational Linguistics.

Amir Bakarov. 2018. A survey of word embeddings evaluation methods. *arXiv preprint arXiv:1801.09536*.

Iz Beltagy, Kyle Lo, and Arman Cohan. 2019. SciBERT: A pretrained language model for scientific text. In *Proceedings of the 2019 Conference on Empirical Methods in Natural Language Processing and the*

9th International Joint Conference on Natural Language Processing (EMNLP-IJCNLP)*, pages 3615–3620, Hong Kong, China. Association for Computational Linguistics.

Olivier Bodenreider. 2004. The Unified Medical Language System (UMLS): integrating biomedical terminology. *Nucleic Acids Research*, 32(Database issue):D267–D270.

Leonardo Campillos-Llanos, Ana Valverde-Mateos, Adrián Capllonch-Carrión, and Antonio Moreno-Sandoval. 2021. A clinical trials corpus annotated with UMLS entities to enhance the access to evidence-based medicine. *BMC Medical Informatics and Decision Making*.

Casimiro Pio Carrino, Jordi Armengol-Estapé, Asier Gutiérrez-Fandiño, Joan Llop-Palao, Marc Pàmies, Aitor Gonzalez-Agirre, and Marta Villegas. 2021. Biomedical and clinical language models for spanish: On the benefits of domain-specific pretraining in a mid-resource scenario. *CoRR*, abs/2109.03570.

Casimiro Pio Carrino, Joan Llop, Marc Pàmies, Asier Gutiérrez-Fandiño, Jordi Armengol-Estapé, Joaquín Silveira-Ocampo, Alfonso Valencia, Aitor Gonzalez-Agirre, and Marta Villegas. 2022. Pretrained biomedical language models for clinical NLP in Spanish. In *Proceedings of the 21st Workshop on Biomedical Language Processing*, pages 193–199, Dublin, Ireland. Association for Computational Linguistics.

José Cañete, Gabriel Chaperon, Rodrigo Fuentes, Jou-Hui Ho, Hojin Kang, and Jorge Pérez. 2020. Spanish pre-trained bert model and evaluation data. In *PML4DC at ICLR 2020*.

Hercules Dalianis. 2018. Characteristics of Patient Records and Clinical Corpora. In *Clinical Text Mining*, pages 21–34. Springer International Publishing, Cham.

Jacob Devlin, Ming-Wei Chang, Kenton Lee, and Kristina Toutanova. 2019. BERT: Pre-training of deep bidirectional transformers for language understanding. In *Proceedings of the 2019 Conference of the North American Chapter of the Association for Computational Linguistics: Human Language Technologies, Volume 1 (Long and Short Papers)*, pages 4171–4186, Minneapolis, Minnesota. Association for Computational Linguistics.

Lisa Ehrlinger and Wolfram Wöß. 2016. Towards a definition of knowledge graphs. In *SEMANTiCS*.

Kawin Ethayarajh. 2019. How contextual are contextualized word representations? Comparing the geometry of BERT, ELMo, and GPT-2 embeddings. In *Proceedings of the 2019 Conference on Empirical Methods in Natural Language Processing and the 9th International Joint Conference on Natural Language Processing (EMNLP-IJCNLP)*, pages 55–65, Hong Kong, China. Association for Computational Linguistics.

Aitor Gonzalez-Agirre, Antonio Miranda-Escalada, Obdulia Rabal, and Martin Krallinger. 2020. PharmaCoNER corpus: gold standard annotations of Pharmacological Substances, Compounds and proteins in Spanish clinical case reports. *Zenodo*. Funded by the Plan de Impulso de las Tecnologías del Lenguaje (Plan TL).

Yu Gu, Robert Tinn, Hao Cheng, Michael Lucas, Naoto Usuyama, Xiaodong Liu, Tristan Naumann, Jianfeng Gao, and Hoifung Poon. 2021. Domain-specific language model pretraining for biomedical natural language processing. *ACM Trans. Comput. Healthcare*, 3(1).

Asier Gutiérrez-Fandiño, Jordi Armengol-Estapé, Marc Pàmies, Joan Llop-Palao, Joaquin Silveira-Ocampo, Casimiro Pio Carrino, Carme Armentano-Oller, Carlos Rodriguez-Penagos, Aitor Gonzalez-Agirre, and Marta Villegas. 2022. MarIA: Spanish language models. *Procesamiento del Lenguaje Natural*, 68(0):39–60.

Qiao Jin, Bhuwan Dhingra, William Cohen, and Xinghua Lu. 2019. Probing biomedical embeddings from language models. In *Proceedings of the 3rd Workshop on Evaluating Vector Space Representations for NLP*, pages 82–89, Minneapolis, USA. Association for Computational Linguistics.

Daniel Jurafsky and James H. Martin. 2000. *Speech and Language Processing: An Introduction to Natural Language Processing, Computational Linguistics, and Speech Recognition*, 1st edition. Prentice Hall PTR, USA.

Guillaume Lample, Miguel Ballesteros, Sandeep Subramanian, Kazuya Kawakami, and Chris Dyer. 2016. Neural architectures for named entity recognition. In *Proceedings of the 2016 Conference of the North American Chapter of the Association for Computational Linguistics: Human Language Technologies*, pages 260–270, San Diego, California. Association for Computational Linguistics.

Jinhyuk Lee, Wonjin Yoon, Sungdong Kim, Donghyeon Kim, Sunkyu Kim, Chan Ho So, and Jaewoo Kang. 2020. BioBERT: a pre-trained biomedical language representation model for biomedical text mining. *Bioinformatics*, 36:1234 – 1240.

Bohan Li, Hao Zhou, Junxian He, Mingxuan Wang, Yiming Yang, and Lei Li. 2020. On the sentence embeddings from pre-trained language models. In *Proceedings of the 2020 Conference on Empirical Methods in Natural Language Processing (EMNLP)*, pages 9119–9130, Online. Association for Computational Linguistics.

Nelson F. Liu, Matt Gardner, Yonatan Belinkov, Matthew E. Peters, and Noah A. Smith. 2019. Linguistic knowledge and transferability of contextual representations. In *Proceedings of the 2019 Conference of the North American Chapter of the Association for Computational Linguistics: Human Lan-*

guage Technologies, Volume 1 (Long and Short Papers), pages 1073–1094, Minneapolis, Minnesota. Association for Computational Linguistics.

Salvador Lima Lopez, Naiara Perez, Montse Cuadros, and German Rigau. 2020. NUBes: A Corpus of Negation and Uncertainty in Spanish Clinical Texts. In *Proceedings of The 12th Language Resources and Evaluation Conference (LREC2020)*, pages 5772–5781, Marseille, France. European Language Resources Association.

Antonio Miranda-Escalada, Eulalia Farré, and Martin Krallinger. 2020. Named entity recognition, concept normalization and clinical coding: Overview of the cantemist track for cancer text mining in Spanish, corpus, guidelines, methods and results. In *Proceedings of the Iberian Languages Evaluation Forum (IberLEF 2020), CEUR Workshop Proceedings*.

Yifan Peng, Shankai Yan, and Zhiyong Lu. 2019. Transfer learning in biomedical natural language processing: An evaluation of BERT and ELMo on ten benchmarking datasets. In *Proceedings of the 18th BioNLP Workshop and Shared Task*, pages 58–65, Florence, Italy. Association for Computational Linguistics.

Matthew E. Peters, Mark Neumann, Mohit Iyyer, Matt Gardner, Christopher Clark, Kenton Lee, and Luke Zettlemoyer. 2018. Deep contextualized word representations. In *Proceedings of the 2018 Conference of the North American Chapter of the Association for Computational Linguistics: Human Language Technologies, Volume 1 (Long Papers)*, pages 2227–2237, New Orleans, Louisiana. Association for Computational Linguistics.

Alec Radford, Jeffrey Wu, Rewon Child, David Luan, Dario Amodei, Ilya Sutskever, et al. 2019. Language models are unsupervised multitask learners. *OpenAI blog*, 1(8):9.

Nils Reimers and Iryna Gurevych. 2019. Sentence-BERT: Sentence embeddings using Siamese BERT-networks. In *Proceedings of the 2019 Conference on Empirical Methods in Natural Language Processing and the 9th International Joint Conference on Natural Language Processing (EMNLP-IJCNLP)*, pages 3982–3992, Hong Kong, China. Association for Computational Linguistics.

Matias Rojas, Felipe Bravo-Marquez, and Jocelyn Dunstan. 2022a. Simple yet powerful: An overlooked architecture for nested named entity recognition. In *Proceedings of the 29th International Conference on Computational Linguistics*, pages 2108–2117, Gyeongju, Republic of Korea. International Committee on Computational Linguistics.

Matías Rojas, Jocelyn Dunstan, and Fabián Villena. 2022b. Clinical flair: A pre-trained language model for Spanish clinical natural language processing. In *Proceedings of the 4th Clinical Natural Language Processing Workshop*, pages 87–92, Seattle, WA. Association for Computational Linguistics.

Bin Wang, Angela Wang, Fenxiao Chen, Yuncheng Wang, and C.-C. Jay Kuo. 2019. Evaluating word embedding models: methods and experimental results. *APSIPA Transactions on Signal and Information Processing*, 8:e19. Publisher: Cambridge University Press.

Michael Zhai, Johnny Tan, and Jinho Choi. 2016. Intrinsic and Extrinsic Evaluations of Word Embeddings. *Proceedings of the AAAI Conference on Artificial Intelligence*, 30(1).

A Appendix

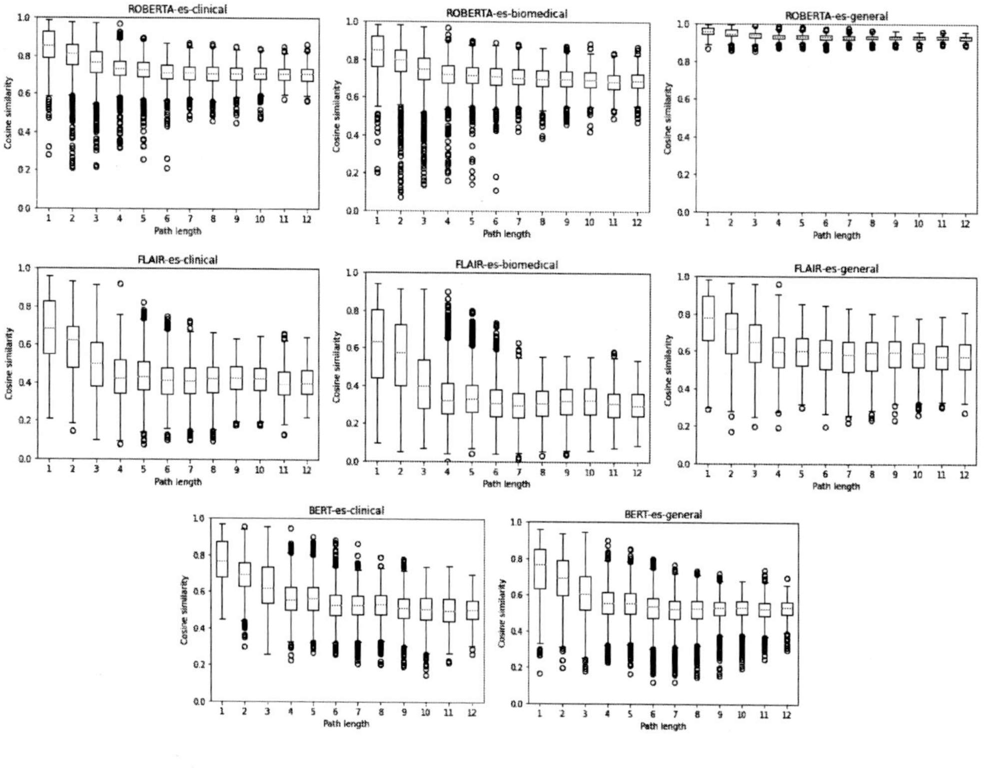

Figure 5: Boxplots of cosine similarity by path length for selected PLM trained in Spanish text

Enriching Deep Learning with Frame Semantics for Empathy Classification in Medical Narrative Essays

Priyanka Dey
Computer Science Department
University of Illinois, Urbana-Champaign
pdey3@illinois.edu

Roxana Girju
Department of Linguistics,
Computer Science Department,
Beckman Institute,
University of Illinois, Urbana-Champaign
girju@illinois.edu

Abstract

Empathy is a vital component of health care and plays a key role in the training of future doctors. Paying attention to medical students' self-reflective stories of their interactions with patients can encourage empathy and the formation of professional identities that embody desirable values such as integrity and respect. We present a computational approach and linguistic analysis of empathic language in a large corpus of 440 essays written by pre-med students as narrated simulated patient – doctor interactions. We analyze the discourse of three kinds of empathy: cognitive, affective, and prosocial as highlighted by expert annotators. We also present various experiments with state-of-the-art recurrent neural networks and transformer models for classifying these forms of empathy. To further improve over these results, we develop a novel system architecture that makes use of frame semantics to enrich our state-of-the-art models. We show that this novel framework leads to significant improvement on the empathy classification task for this dataset.

1 Introduction

Empathy is a complex phenomenon concerning how we seek to understand and experience, to some extent, the experiences of others (Ratcliffe, 2017) – i.e., having a sense of the other's story and the context in which it takes place. One way to get to an appreciation of one's complex situation (i.e., the embodied actions and the contexts within which they act) is through narratives of lived experience (McIntyre, 1981; Gallagher, 2012). Self-reflective (i.e., first person) narratives, for instance, offer a wide range of resources for empathy, as they bring together one's inner and outer worlds, thus giving meaning to experience (Mattingly, 2000). In this respect, narratives seem necessary for empathy, as our first-person experience is grounded in the contextualized content of the narrative. They also provide a form or structure that allows us to frame an understanding of others, together with a learned set of skills and practical knowledge that shapes our understanding of what we and others are experiencing.

Reflective writing is a dynamic process that allows for an active engagement with knowledge and experience, being widely used in clinical practice (Jasper et al., 2013; Burkhardt et al., 2019; Artioli et al., 2021). Putting into words the focused inspection of their thoughts, feelings, and events enables one to reprocess the experience, build new insights, and new ways to conceive reality (Artioli et al., 2021). Thus, narrative exercises like self-reflective stories can help medical students recognise and derive meaning from key experiences, which in turn can support critical thinking, self-consciousness, and the development of personal skills, communication and empathy skills, self-knowledge, professional identify development, and instill behavior change (Craft, 2005; Borgstrom et al., 2016; Mintz-Binder et al., 2019; Allan and Driscoll, 2014; Peterson et al., 2018; Liu et al., 2016; Bekker et al., 2013). Such writing can lead to an increase in experience-taking skills (Kaufman and Libby, 2012) and can decrease stereotyping, prejudice, and racial bias in healthcare (Williams and Wyatt, 2015).

In this research, we take a narrative approach to empathy and explore the experiences of premed students at a large university by analysing their self-reflective writing portfolios (a large corpus of first-person essays written by premed students in narrated simulated patient-doctor interactions). Specifically, we introduce an exploratory study of empathy in clinical encounters paying attention to the discourse of three types of empathy: *cognitive* (the drive and ability to identify and understand another's emotional or mental states), *affective* (the capacity to experience an appropriate emotion in response to another's emotional or mental state), and *prosocial behavior* (a response to having identified

Proceedings of the 13th International Workshop on Health Text Mining and Information Analysis (LOUHI), pages 207 - 217
December 7, 2022 ©2022 Association for Computational Linguistics

the perspective of another with the intention of acting upon the other's mental and/or emotional state), following established practices in psychology (Cuff et al., 2016; Eisenberg et al., 2006; Rameson et al., 2012). We introduce a set of informative baseline experiments using state-of-the-art recurrent neural networks and transformer models for classifying the various forms of empathy. As initial experiments show relatively low scores, we explore a novel FrameNet-based system architecture where we use sentence frames to extract additional semantic features. We apply this framework to state-of-the-art and representative neural network models and show significant improvement in the empathy classification task for this dataset. Although previous research suggests that narrative-based interventions tend to be effective education-based methods, it is less clear what are some of the mechanisms through which narratives achieve such an effect, which is another contribution of this research.

2 Related Work

In spite of its increasing theoretical and practical interest, empathy research in computational linguistics has been relatively sparse and lacks cohesion. Even more so, investigations of empathy as it relates to clinical practice have received even less attention mainly due to data and privacy concerns.

Most of the research on empathy detection has focused on conversations or interactions, as dialogue systems (Zhong et al., 2020; Chen et al., 2022a; Samad et al., 2022), or in online platforms (e.g. (Pérez-Rosas et al., 2017; Khanpour et al., 2017; Otterbacher et al., 2017; Sharma et al., 2020; Lahnala et al., 2022; Sharma et al., 2021; Hosseini and Caragea, 2021), a few on news stories and other narratives (Buechel et al., 2018; Wambsganss et al., 2021b; Sedoc et al., 2020; Mundra et al., 2021; Guda et al., 2021), and even less on empathy in clinical settings (Zhou et al., 2021; Shi et al., 2021). Buechel et al. (2018) used crowd-sourced workers to self-report their empathy and distress levels and to write empathic reactions to news stories. Wambsganss et al. (2021b) built a text corpus of student peer reviews collected from a German business innovation class annotated for cognitive and affective empathy levels. Furthermore, using Batson's Empathic Concern-Personal Distress Scale (Batson et al., 1987), Buechel et al. (2018) have focused only on negative empathy instances (i.e., pain and sadness "by witnessing another person's suffering"). This year, the WASSA shared task focused on predicting empathy, emotion, and personality in reaction to news stories (Barriere et al., 2022; Vasava et al., 2022). The dataset is an extension of Buechel et al. (2018)'s dataset – i.e., it includes news articles that express harm to an entity (e.g. individual, group of people, nature). Each article comes with reaction essays in which authors expressed their empathy and distress toward these news articles. Each essay is annotated for empathy and distress, and with authors' personality traits and demographic information (age, gender, ethnicity, income, and education level). Here, we could not compare our models with the WASSA results – our dataset does not capture the meta-data in WASSA. Moreover, our empathy instances are not always negative (Fan et al., 2011): a dataset reflecting empathetic language should ideally allow for expressions of empathy that encompass a variety of positive and negative emotions. We could not compare against its best performing system due to limited reproducibility (Chen et al., 2022b).

In multimodal research, R. M. Frankel (2000) and Cordella and Musgrave (2009) identify sequential patterns of empathy frequently expressed in video-recorded exchanges by medical graduates interacting with a cancer patient. Sharma et al. (2020) analyzed the discourse of conversations in online peer-to-peer support platforms. They successfully trained novice writers to improve low-empathy responses by giving the writers feedback with examples of sentences that are typical of recognition and interpretation of others' feelings or experiences. In a subsequent set of experiments (Sharma et al., 2021), they suggested that empathic written discourse should be coherent, specific to the conversation at hand, and lexically diverse.

To our knowledge, no self-reflective narrative text corpora have been developed for computational linguistics investigations of clinical student training. Adding to the scarcity of empathy-dedicated resources, there is also a lack of understanding of which linguistic features might contribute to the various types of empathy, like cognitive, affective, and prosocial behavior.

3 Self-reflective Narrative Essays in Medical Training

In this research, we focus on self-reflective narratives written by premed students given a simulated scenario. Simulation is strongly set on our first-person experiences, relying on resources that are available to the simulator. In a simulation process, the writer puts themselves in the other's situation and asks what "I would do if I were in that situation." Perspective taking (i.e., cognitive empathy) is crucial for fostering affective abilities, enabling writers to imagine and learn about the emotions of others and to share them, too. As empathy is other-directed (De Vignemont and Jacob, 2012; Gallagher, 2012), this means that we, as narrators, are open to the experience and the life of the other, in their context, as we can understand it.

This study's intervention was designed as a written assignment in which premed students were asked to consider a hypothetical scenario where they took the role of a physician breaking the news of an unfavorable diagnosis of high blood cholesterol to a middle-aged patient[1]. They were instructed to recount (in first person voice) the hypothetical doctor-patient interaction where they explained the diagnosis and prescribed medical treatment using layman terms and language they believed would comfort as well as persuade the hypothetical patient to adhere to their prescription.

With the students' consent, we collected a corpus of 774 essays over a period of one academic year (Shi et al., 2021). Following a thorough annotation process, annotators (undergraduate and graduate students in psychology and social work)[2] labeled a subset of 440 randomly selected essays (henceforth, "the corpus"). Using a rich color code schema, each sentence in every essay was labeled as either cognitive empathy (green; e.g., "She looked tired"), affective empathy (yellow; e.g.: "I felt the pain"), or prosocial behavior (cyan; e.g.: "I reassured her this was the best way") (everything else was "no empathy") (Cuff et al., 2016; Eisenberg et al., 2006; Rameson et al., 2012). The six paid undergraduate students were trained on the task and instructed to annotate the data. Two meta-annotators, paid graduate students with prior experience with the task, reviewed the work of the annotators and updated the annotation guidelines at regular intervals, in an iterative loop process after each batch of essays[3]. The meta-annotators reached a Cohen's kappa of 0.82, a good level of agreement. Disagreed cases were discussed and mitigated. At the end, all the essays were re-annotated per the most up-to-date guidelines. The resulting annotated data shows an uneven label distribution in the annotated corpus (11,763 total): 667 (cognitive), 1,659 (affective), and 723 (prosocial) sentences (and 8,714 non-empathy sentences).

[1]The patient was referred to as Betty or John.

[2]The students were hired based on previous experience with similar projects in social work and psychology.

[3]10 essays per week

4 Empathy Classification Task

In this research, our goal is to explore machine learning models of empathy classification in narrative essays to better our understanding of the mechanisms through which empathy can be expressed. Since we are interested in the linguistic expressions of empathy, we zoom in to the sentence level. Given such a corpus of essay sentences, we first build a binary classifier which can be useful in applications requiring a general linguistic understanding of the presence of empathy. In some cases such as medical communication training of pre-med students, a more fine-grained understanding of different kinds of empathy is useful. Thus, we also build a classifier that can identify each type of empathy: cognitive, affective, and prosocial.

For both types of classifiers, we first experiment with several state-of-the-art statistical and machine learning models. As our research is focused on the subcategorization of empathy, we seek to improve our multi-label classifier. Thus, we introduce a new and better performing system architecture by employing FrameNet (Baker et al., 1998), the research and development project which builds on the theory of frame semantics. Using a state-of-the-art FrameNet sentence parser (Swayamdipta et al., 2017), we extract semantic frames from each sentence in our corpus and use this resource to enhance our original (baseline) models with these additional knowledge. As we will show in Subsection 4.4, incorporating FrameNet semantics into state-of-the-art deep learning models leads to an increase in empathy classification results.

4.1 Baseline Models

We started with the following representative baseline models: Naive Bayes (NB), support vector machines (SVM), and logistic regression (logR). We are also interested in observing the performance of deep learning methods and, among them, we experiment with long-short term memory (LSTM) (Hochreiter and Schmidhuber, 1997) and bidirectional long-short term memory (bi-LSTM) (Graves and Schmidhuber, 2005) models; additionally, we use the transformer neural network models BERT

(Devlin et al., 2018) and RoBERTa (Liu et al., 2019). We used unigrams as our features. We also initialized the embedding layers in our neural models (LSTM and bi-LSTM) with GloVe embeddings since the expression of empathy involves larger units than words, and embeddings are known to better capture contextual information. For the transformer models, we use the default BERT embeddings. Since our dataset is imbalanced, we report the precision, recall, and F1-score (harmonic mean of the precision and recall).

We identify sentences with empathy by using the annotator's highlights – e.g., a sentence containing cyan and green highlights is considered a prosocial and cognitive empathy sentence. For our binary empathy classification, we use colored sentences as empathy sentences. We consider sentences with no highlights as no empathy sentences.

For the NB, logistic regression, and SVM models, we generate binary classifiers for each type of empathy. For all the neural network models, we generate multi-label classifiers. For each type of empathy highlighted sentences, we reserve 80/20 training/test ratio, with 5-fold cross validation. For the logistic regression models, we use a L2 regularization and for the SVM models, a linear kernel function. We decided to apply an attention layer for the LSTM and bi-LSTM models to learn patterns that may improve the classification. For our final output layer, we use the sigmoid activation function, as we are dealing with a multi-label classification task. For the BERT and RoBERTa models, we apply a dropout layer with probability 0.4 which helps to regularize the model; we use a linear output layer and apply a sigmoid on the outputs.

For our binary empathy classification task, we find that the imbalanced dataset greatly affects the performance of most models; the best performing model: BERT achieves an F1-score of 0.56 for empathy sentences and 0.79 for no empathy sentences. To combat this imbalance, we randomly downsampled the no empathy sentence dataset (to get an equal number of empathy and no-empathy sentences). This resulted in an improved BERT model (0.72 F1 for empathy and 0.79 F1 for no empathy sentences). For our second empathy classification task, we again downsample the total number of no empathy sentences, resulting in a final dataset of 1,659 affective empathy sentences, 723 prosocial sentences, 667 cognitive sentences, and 1,659 no empathy sentences. Table 1 shows the precision, recall, and F1-measure scores for these baseline experiments. As only 5.81% of our sentences contain multiple types of empathy, we only present collapsed results for each category. We leave the study of these sentences for future research.

The Naive Bayes, SVM, and logistic regression models all overfit the training data and, in general, do not handle the imbalanced dataset well. The neural network models provide more promising results, with affective empathy even reaching 0.81 F1 scores. Prosocial empathy seems to be the most difficult to identify, with the highest F1 of 0.73 as obtained by the BERT model. Overall, the transformer models, BERT and RoBERTa, achieve the best performance across all three types of empathy.

4.2 Incorporating FrameNet to Improve Empathy Classification

In our attempt to improve the classification of our empathic narrative sentences, we decided to explore feature generation to further enhance these models. Since empathy is a highly complex semantic-pragmatic phenomenon, one intuition is that semantic knowledge should help the classifiers. One linguistic theory called frame semantics deconstructs a sentence into predicate-argument structures that describe meaning not at the level of individual words, but is instead based on the concept of a scenario, scene, or event called a frame. Frames are defined by the group of words that evoke the scene (frame-evoking elements or FEEs), as well as by their expected semantic arguments (frame elements). A JUDGMENT frame, for instance, has FEEs like *praise.v, criticize.v*, and *disapprove.v*, and frame elements such as Cognizer, Evalueee, Expressor, Reason. The Berkeley FrameNet project (Baker et al., 1998; Ruppenhofer et al., 2016) is the most well-known lexical resource of frame semantics, with definitions for over 1200 frames.[4]

To generate new features, we leverage frame semantics to identify all the frames that occur in a sentence. Each sentence in our essay corpus is parsed with the Frame-Semantic Parser (Swayamdipta et al., 2017), which is based on a softmax margin segmental recurrent neural network model. Specifically, we use the FrameNet 1.7 pretrained models to predict frames for each of our sentences. For instance, for "He played an important role in preventing her from becoming depressed", the frame

[4]We used the release 1.7 which has 1,222 frame annotations (http://framenet.icsi.berkeley.edu).

Classifier	Cognitive			Affective			Prosocial			None		
	Prec.	Rec.	F1	Prec.	Rec.	F1	Prec.	Rec.	F1	Prec.	Rec.	F1
NB	0.03	0.18	0.05	0.05	0.38	0.09	0.14	0.05	0.07	1.0	0.72	0.84
SVM	0.30	0.19	0.23	0.46	0.50	0.48	0.44	0.37	0.40	0.80	0.71	0.75
LogR	0.44	0.38	0.40	0.74	0.58	0.65	0.20	0.25	0.22	0.77	0.71	0.74
LSTM	0.62	0.72	0.67	0.63	0.61	0.62	0.51	0.59	0.55	0.71	0.76	0.73
biLSTM	0.64	0.71	0.67	0.79	0.62	0.69	0.59	0.62	0.60	0.78	0.74	0.76
BERT	0.74	0.78	0.76	0.92	0.73	0.81	0.72	0.75	0.73	0.75	0.84	0.79
RoBERTa	0.74	0.83	0.78	0.77	0.78	0.77	0.69	0.68	0.68	0.77	0.80	0.78
FN-LSTM	0.73	0.73	0.73	0. 83	0.68	0.75	0.66	0.78	0.72	**0.79**	0.77	**0.78**
FN-biLSTM	0.71	0.88	0.79	0.85	0.78	0.81	0.72	0.86	0.78	0.73	0.75	0.74
FN-BERT	**0.78**	**0.89**	**0.83**	**0.88**	**0.79**	**0.83**	**0.82**	**0.88**	**0.85**	0.71	**0.80**	0.75
FN-RoBERTa	0.73	0.88	0.80	0.85	**0.79**	0.82	**0.82**	0.86	0.84	0.71	**0.80**	0.75

Table 1: Precision, recall and F1 scores of all baseline and FrameNet-incorporated classifiers on the test dataset: 133 cognitive, 332 affective, 145 prosocial, and 332 no-empathy sentences. Bolded numbers indicate best performance.

semantic parser identifies four frames: PERFORMERS_AND_ROLES (i.e., he played a role), IMPORTANCE (i.e., important role), THWARTING (i.e., preventing her), EMOTIONS_BY_POSSIBILITY (i.e., becoming depressed).

Given the parser's extraction of 669 unique frames from our entire sentence dataset, we explore the most common frames present in sentences containing each type of empathy (Table 2). Many of the frames exhibited in cognitive empathy sentences focus on *speaking*, *supporting*, and *seeing*, while affective empathy sentences contain frames related to *responses*, *stimulating emotions*, and *perceiving emotions/states*. Many of the prosocial empathy sentences include frames that discuss a form of action e.g. *trying to [perform an action], reassure, seek to achieve*, etc.

To use the frame identification as a feature in our models, we generate a frequency vector to encode the occurrences of a frame in a sentence. For example, if we had a total of 3 frames *Fa*, *Fb*, *Fc*, and sentence x contained one mention of frame *Fa*, 2 mentions of *Fb*, and no *Fc*, our encoding vector would be: [1, 2, 0], representing their frequencies. Thus, we generate a vector of size 669 for each of our sentences in the whole essay dataset.

In our quest for improved empathy classification, we focus on our neural network (LSTM, bi-LSTM, BERT, and RoBERTa) models as these proved to perform best in our baseline experiments. For our LSTM and bi-LSTM models, we use GloVe embeddings to encode the processed sentence, and then add the FrameNet encoding vector to the end of the embedding vector. We then apply the LSTM or bi-LSTM layer followed by the attention layer and transform outputs using the sigmoid activation to get class probabilities (Figure 1 shows the system architecture for this framework).

For the BERT and RoBERTa models, we first input the processed sentence and extract the textual embeddings, and append the FrameNet encoding vector to the embedding vector. We then apply a feedforward neural network – i.e., a multi layer perceptron (MLP) with a sigmoid activation function – to get predictions (Figure 2 shows the system architecture for this framework).

4.3 Constructing a Frame Lattice

An initial exploration of the FN parser shows that our training dataset contained a total of 616 unique frames, roughly 50% of them appearing only in at most 5 sentences. To optimize learning in the neural network models, we identify a lattice of frames from our training corpus that most improves the classification performance. To do this, we iterate through each combination of subsets of size K of the identified frames in our training dataset. We then compute weighted average accuracy scores for empathy classification using the training dataset and identify the set of frames most influential in each of the four models considered. An initial set of exploratory experiments has shown that lattices of sizes between 5 and 20 yield the highest improvement. Frame lattices of size 2, 3, and 4 did not show any significant improvement (i.e., no increase in score above 0.01). Latices larger than 20 become very noisy and, thus, negatively impact performance. Thus, we decided to further explore this

Cognitive	Affective	Prosocial
JUDGMENT (159)	RESPONSE (831)	GESTURE (458)
MAKE_COGNITIVE_CONNECTION (143)	COMMUNICATION_RESPONSE (368)	DESIRABILITY (290)
SPEAK_ON_TOPIC (134)	PERCEPTION (293)	REASSURING (274)
SUPPORTING (108)	STIMULATE_EMOTION (209)	FACIAL_EXPRESSION (231)
SEE_THROUGH (96)	SOCIABILITY (148)	AWARENESS (173)

Table 2: Most common frame classes for each empathy class

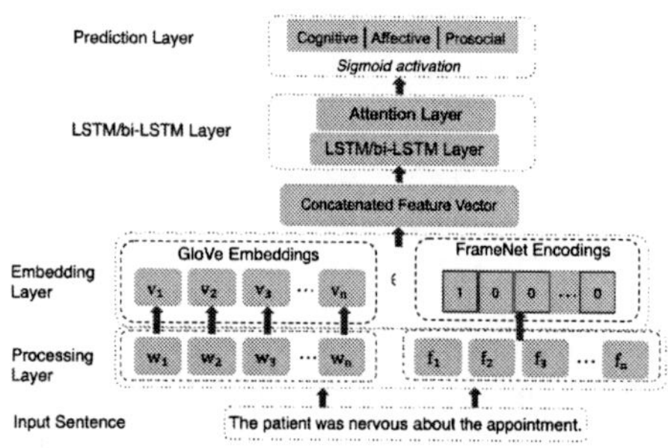

Figure 1: Architecture for LSTM & bi-LSTM models

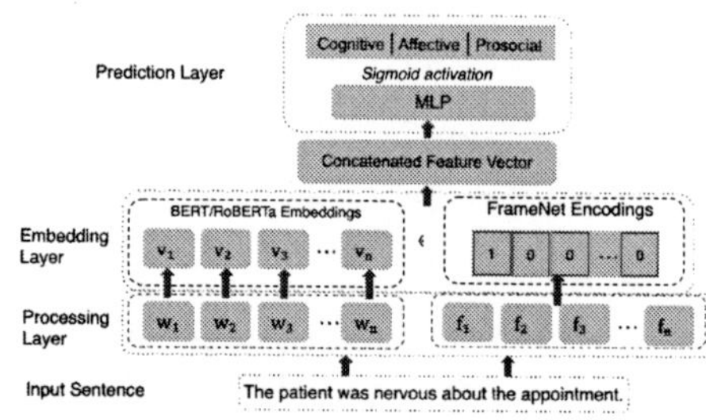

Figure 2: Architecture for BERT & RoBERTa models

5-20 range. Specifically, we iterate through all possible combinations of 5 frames that appeared in the training corpus. We then increment the frame size by 1 in each iteration, and recompute performance. Results on test data are shown in Fig. 3.

Since we wanted to use a metric that would measure the performance for all three empathy types together, we did not use the individual F1 scores for our categories. The closest measurement was the macro-F1 score, but this is still an unweighted average (since we have already had good performance for affective empathy, using this metric, the results would not increase by much). Thus, the weighted average made more sense to identify the best lattice.

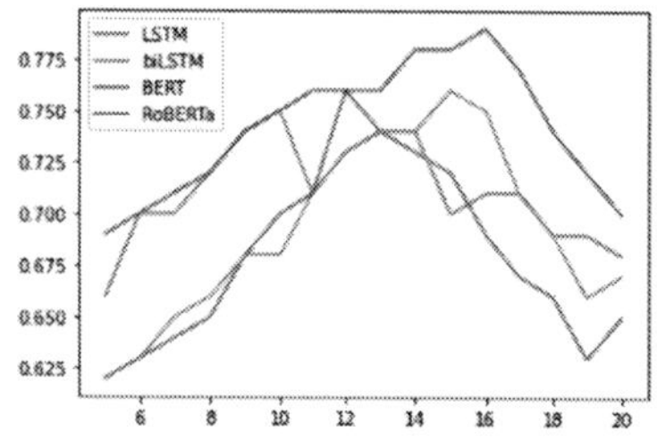

Figure 3: Weighted average scores for varying lattice sizes: 5 to 20

4.4 FrameNet Experiments' Results

To improve classification, we thus incorporate each neural network's best performing lattice and build a frame encoding vector for each sentence in our dataset. We then follow the system architectures in Figures 1 and 2 and compute the performance for each model (See Table 1).

The experimental results show that the inclusion of the FrameNet lattices improves performance considerably. The best models are FrameNet-BERT and FrameNet-RoBERTa, for which all the metric scores significantly improve with this additional feature. We also notice that the classification performance for prosocial empathy significantly improved over the baseline models (0.85 vs. previous score: 0.73). The enhanced BERT model yields the highest F1 scores for all three empathy types, with all empathy categories scoring above 0.8; the no-empathy category however does drop in performance (0.75 vs. previous score: 0.79).

These experiments indicate that our system learns best from a lattice of different sizes for each learning model. Table 3 shows the specific frames per model. Many of the learning architectures choose the same frames in their lattices, e.g. INTENTIONALLY_ACT, GESTURE, SOCIABILITY, PERCEPTION, SENSATION. Interestingly, the transformer models select some additional frames directly linked to certain types of empathy: cognitive (MAKE_COGNITIVE_CONNECTION, MENTION), affective (PERCEPTION, RESPONSE), and prosocial (SEEKING_TO_ACHIEVE). These frames are possibly somewhat tied to our specific dataset and narrative genre, issue we leave for future research.

FrameNet-BERT vs. BERT

We also examined a bit closer the results to get

LSTM (latice size = 13)	**bi-LSTM** (latice size = 14)
CAUSE_EMOTIONS, INTENTIONALLY_ACT, GESTURE, JUDGMENT, DESIRABILITY, PERCEPTION, COMMUNICATION_RESPONSE, SEEKING_TO_ACHIEVE, SENSATION, SOCIABILITY, TELLING, WORRY)	EMOTIONS, EMOTIONS_BY_POSSIBILITY, EVOKING, GESTURE, JUDGMENT, MENTION, OPINION, INTENTIONALLY_ACT, PERCEPTION, RESPONSE, RESPOND_TO_PROPOSAL, COMMUNICATION_RESPONSE_SCENARIO, SENSATION, SOCIABILITY, STIMULATE_EMOTION, SEEKING_TO_ACHIEVE
BERT (latice size = 14)	**RoBERTa** (latice size = 12)
CAUSE_EMOTIONS , EMOTIONS_BY_POSSIBILITY, EVOKING, GESTURE, INTENTIONALLY_ACT, MAKE_COGNITIVE_CONNECTION, MENTION, OPINION, PERCEPTION, RESPONSE, SENSATION, SOCIABILITY, SUPPORTING, WORRY	CAUSE_EMOTIONS, EMOTIONS_BY_POSSIBILITY, GESTURE, FACIAL_EXPRESSION, INTENTIONALLY_ACT, JUDGMENT, MAKE_COGNITIVE_CONNECTION, MENTION, PERCEPTION, RESPONSE, SEEKING_TO_ACHIEVE, SPEAK_ON_TOPIC

Table 3: Best frame latices for each learning model

more insights into the contribution of the FrameNet external semantic knowledge to the task of empathy classification. Specifically, we wanted to see what kinds of examples FrameNet-BERT classifies correctly over the baseline transformer BERT.

Overall, there was a total of 197 instances (affective: 76; cognitive: 59; prosocial: 62) that FrameNet-BERT classified correctly and BERT incorrectly. A look at these sentences shows a balanced combination of frames like MEDICAL_CONDITIONS, DIFFICULTY, QUESTIONING, BIOLOGICAL_CLASSIFICATION, EXPLAINING_THE_FACTS, CURE, as well as AWARENESS, EMOTION_DIRECTED, COMING_TO_BELIEVE, EXPERIENCER_FOCUS, EXPERIENCER_OBJ, FEAR. These empirical results support new evidence in medical education (Warmington, 2019; Warmington et al., 2022) – meaning, they highlight how important it is for future doctors to focus and reflect not only on how to diagnose and provide proper treatment to the patient, but also to develop an awareness of how patients experience their illness and focus on how patients need their experience of illness acknowledged.

In addition to these frames, a specific subset deserves particular attention and discussion, subset which works best in combination with those mentioned above. Table 4 lists the most frequent frames of non-verbal communication that tend to occur in true positive test instances as identified by FrameNet-BERT. These results indicate that, even in self-reflective narratives, both verbal and non-verbal aspects of interaction play an important role. What we wear and the way we physically interact with others communicate a great deal about who we are (Iedema and Caldas-Coulthard, 2008). Such narratives include information about non-verbal communication and impressions of other aspects of the context. For instance, the importance of the senses of sight and sound in building up a rich description of both the setting and events is well recognised (i.e., laughter, cry, the tone or volume of voices). These empirical results indicate that cognitive and sensory self-awareness are critical to the clinical encounter process. Doctors paying close attention to their patients' as well as to their own sensations, perceptions and emotional responses picture a process that emphasizes the importance of self-awareness and awareness of others, both indispensable in effective empathic communication.

5 Discussion and Conclusions

Medical education should and can incorporate guided self-reflective practices that show how important it is for the students to develop an awareness of the emotional and relational aspects of the clinical encounter with their patients (Warmington, 2019). The way people identify themselves and perform in particular roles and in relations to others brings together a specific set of values, attitudes, and competencies that can be supported through ongoing self-reflection. Thus, students learn not only how to diagnose and treat patients' medical conditions, but also how to witness the patient's illness experience. In practice, they often switch between these positions: witnessing what it is like for the patient, as well as understanding what they need medically.

Often, clinical encounters can be highly charged emotionally especially for patients in case of serious illness. Unfortunately, medicine lags behind other health professions (like nursing, social work, psychology) which learn from reflective practice and respect it from the beginning. Yet, acknowledging the patient's situation, who they are and their experience can make a huge impact on the quality

Frame	Examples	Count
BODY_PARTS	I noticed Betty fidgeting and clasping her hands, and so I tried to reassure her we would work together and develop a recovery plan.	59
SENSATION	After he left the meeting room, I began feeling very helpless.	71
BODY_MOVEMENT	He seemed to almost roll his eyes at that moment which I don't blame him for.	41
CHANGE_POSTURE	He quietly sat down with his hands folded without responding to my remark.	19
FACIAL_EXPRESSION	I noticed after I told her the news, her mouth forming into a frown and she seemed very depressed.	38
GESTURE	I proceeded with the diagnosis to explain the severity of elevated levels but stopped as she waved her hand.	83
BREATHING	Betty and her family both sighed a breath of relief.	40
SOUND_LEVEL	After I told him the bad news, my patient became silent.	16

Table 4: Examples of empathy sentences with non-verbal communication frames

of that relationship and the trust that is built up for the patient. Narrative-based interventions and activities can facilitate self-reflection and enrich medical students' professional identity formation.

Computational approaches to empathy can be very valuable, but it is clear that such AI initiatives must be multidisciplinary, using and developing a variety of core sets of requirements and expertise and engaging many participants, e.g. AI designers, developers, frontline clinical teams, ethicists, humanists, patients, caregivers (Matheny et al., 2019).

The research experiments and findings summarized in this paper are part of a larger interdisciplinary and highly collaborative project where we analyze both self-reflective narratives of simulated interactions, as well as multimodal patient-doctor encounters in real clinical settings (Girju, 2021; Girju and Girju, 2022). In this paper, we presented a computational approach and linguistic analysis of empathic language in a large corpus of premed student essays of narrated simulated patient-doctor interactions. Specifically, we showed that semantic information at the sentence level can be very useful not only in empathy identification but provides details on the differences among the three main types of empathy: cognitive, affective, and prosocial. We presented novel and performant FrameNet-based transformer models for empathy classification. In future work, we will expand this analysis by considering discourse-level context. We will also integrate other resources like WordNet (Miller, 1995), VerbNet (Kipper et al., 2000), and take advantage of larger discourse.

6 Ethical Considerations

Despite the clear benefits that such empathy detection systems can bring, there are also ethical issues that arise from their use. First, machine learning models are susceptible to design biases that may result in systematic errors, in addition to lower transparency, loss of control, and potential lack of trust by human users (Wambsganss et al., 2021a). Moreover, such models are data-driven – and most of the time such data is potentially biased, highly sensitive, where user privacy becomes an even more important concern. For instance, although we followed the ethical protocols put forward in academia for data collection and annotation, our data is imbalanced demographically (for both pre-med students and the hypothetical patient) and limited to only one clinical scenario (i.e., breaking bad news). Furthermore, special attention should be given to models designed to empathize with vulnerable population like children and people of various abilities. Moreover, focusing only on one hypothetical medical scenario, resulted in a dataset with limited diversity. Another aspect to consider in future research is the use of self-assessment vs. third-party empathy reports. Although most of our pre-med students were highly confident in their empathetic abilities, more thorough research is needed in this direction. AI research on empathy should compare against and even integrate qualitative metrics like the Jefferson Scale of Physician Empathy (Hojat et al., 2001) or the Consultation and Relational Empathy (CARE) Measure (Mercer et al., 2004).

Obviously, we are currently far from being able to deploy such models to help in medical student training. However, our annotated corpus and experiments help shed new light on the empathy classification task and show what kind of linguistic (semantic) knowledge can contribute to it. We also hope such work will encourage future research and collaboration between AI practitioners and clinicians. Overall, developers and providers alike need to increasingly follow ethical considerations in the human-value sensitive design of these systems to ensure the well-being of their users.

References

Elizabeth G Allan and Dana Lynn Driscoll. 2014. The three-fold benefit of reflective writing: Improving program assessment, student learning, and faculty professional development. *Assessing Writing*, 21:37–55.

Giovanna Artioli, Laura Deiana, Francesco De Vincenzo, Margherita Raucci, Giovanna Amaducci, Maria Chiara Bassi, Silvia Di Leo, Mark Hayter, and Luca Ghirotto. 2021. Health professionals and students' experiences of reflective writing in learning: A qualitative meta-synthesis. *BMC medical education*, 21(1):1–14.

Collin F Baker, Charles J Fillmore, and John B Lowe. 1998. The berkeley framenet project. In *COLING 1998 Volume 1: The 17th International Conference on Computational Linguistics*.

Valentin Barriere, Shabnam Tafreshi, João Sedoc, and Sawsan Alqahtani. 2022. WASSA 2022 shared task: Predicting empathy, emotion and personality in reaction to news stories. In *Proceedings of the 12th Workshop on Computational Approaches to Subjectivity, Sentiment & Social Media Analysis*, pages 214–227, Dublin, Ireland. Association for Computational Linguistics.

C Daniel Batson, Jim Fultz, and Patricia A Schoenrade. 1987. Distress and empathy: Two qualitatively distinct vicarious emotions with different motivational consequences. *Journal of personality*, 55(1):19–39.

Hilary L Bekker, Anna E Winterbottom, Phyllis Butow, Amanda J Dillard, Deb Feldman-Stewart, Floyd J Fowler, Maria L Jibaja-Weiss, Victoria A Shaffer, and Robert J Volk. 2013. Using personal stories. *BMC Medical Informatics and Decision Making*, 13(59).

Erica Borgstrom, Rachel Morris, Diana Wood, Simon Cohn, and Stephen Barclay. 2016. Learning to care: medical students' reported value and evaluation of palliative care teaching involving meeting patients and reflective writing. *BMC medical education*, 16(1):1–9.

Sven Buechel, Anneke Buffone, Barry Slaff, Lyle Ungar, and Joao Sedoc. 2018. Modeling empathy and distress in reaction to news stories. *arXiv preprint arXiv:1808.10399*.

Crystal Burkhardt, Ashley Crowl, Margaret Ramirez, Brianna Long, and Sarah Shrader. 2019. A reflective assignment assessing pharmacy students' interprofessional collaborative practice exposure during introductory pharmacy practice experiences. *American Journal of Pharmaceutical Education*, 83(6).

Mao Yan Chen, Siheng Li, and Yujiu Yang. 2022a. EmpHi: Generating empathetic responses with human-like intents. In *Proceedings of the 2022 Conference of the North American Chapter of the Association for Computational Linguistics: Human Language Technologies*, pages 1063–1074, Seattle, United States. Association for Computational Linguistics.

Yue Chen, Yingnan Ju, and Sandra Kübler. 2022b. Iucl at wassa 2022 shared task: A text-only approach to empathy and emotion detection. In *Proceedings of the 12th Workshop on Computational Approaches to Subjectivity, Sentiment & Social Media Analysis*, pages 228–232.

M. Cordella and S. Musgrave. 2009. Oral communication skills of international medical graduates: Assessing empathy in discourse. *Communication and Medicine*, 6(2):129–142.

Melissa Craft. 2005. Reflective writing and nursing education. *Journal of nursing education*, 44(2):53–57.

Benjamin MP Cuff, Sarah J Brown, Laura Taylor, and Douglas J Howat. 2016. Empathy: A review of the concept. *Emotion review*, 8(2):144–153.

Frédérique De Vignemont and Pierre Jacob. 2012. What is it like to feel another's pain? *Philosophy of science*, 79(2):295–316.

Jacob Devlin, Ming-Wei Chang, Kenton Lee, and Kristina Toutanova. 2018. Bert: Pre-training of deep bidirectional transformers for language understanding. *arXiv preprint arXiv:1810.04805*.

Nancy Eisenberg, Richard A Fabes, and Tracy L Spinrad. 2006. Prosocial development. In *Volume III. Social, Emotional, and Personality Development*. John Wiley & Sons, Inc.

Y. Fan, Duncan NW, de Greck M, and Northoff G. 2011. Is there a core neural network in empathy? an fmri based quantitative meta-analysis. *Neuroscience Biobehavioral Review*, 35(3):903–911.

Shaun Gallagher. 2012. Empathy, simulation, and narrative. *Science in Context*, 25(3):355–381.

Roxana Girju. 2021. Adaptive multimodal and multisensory empathic technologies for enhanced human communication. In *Rethinking the Senses: A Workshop on Multisensory Embodied Experiences and Disability Interactions, the ACM CHI Conference on Human Factors in Computing Systems*. arXiv preprint arXiv:2110.15054.

Roxana Girju and Marina Girju. 2022. Design considerations for an NLP-driven empathy and emotion interface for clinician training via telemedicine. In *Proceedings of the Second Workshop on Bridging Human–Computer Interaction and Natural Language Processing*, pages 21–27, Seattle, Washington. Association for Computational Linguistics.

Alex Graves and Jürgen Schmidhuber. 2005. Framewise phoneme classification with bidirectional lstm and other neural network architectures. *Neural networks*, 18(5-6):602–610.

Bhanu Prakash Reddy Guda, Aparna Garimella, and Niyati Chhaya. 2021. Empathbert: A bert-based framework for demographic-aware empathy prediction. In *EACL*.

Sepp Hochreiter and Jürgen Schmidhuber. 1997. Long short-term memory. *Neural computation*, 9(8):1735–1780.

Mohammadreza Hojat, Salvatore Mangione, Thomas J Nasca, Mitchell JM Cohen, Joseph S Gonnella, James B Erdmann, Jon Veloski, and Mike Magee. 2001. The jefferson scale of physician empathy: development and preliminary psychometric data. *Educational and psychological measurement*, 61(2):349–365.

Mahshid Hosseini and Cornelia Caragea. 2021. Distilling knowledge for empathy detection. In *Findings of the Association for Computational Linguistics: EMNLP 2021*, pages 3713–3724, Punta Cana, Dominican Republic. Association for Computational Linguistics.

Rick Iedema and Carmen Rosa Caldas-Coulthard. 2008. Introduction: Identity trouble: Critical discourse and contested identities. In *Identity trouble*, pages 1–14. Springer.

Melanie Jasper, Megan Rosser, and Gail Mooney. 2013. *Professional development, reflection and decision-making in nursing and healthcare*. John Wiley & Sons.

Geoff F Kaufman and Lisa K Libby. 2012. Changing beliefs and behavior through experience-taking. *Journal of personality and social psychology*, 103(1):1—19.

Hamed Khanpour, Cornelia Caragea, and Prakhar Biyani. 2017. Identifying empathetic messages in online health communities. In *Proceedings of the Eighth International Joint Conference on Natural Language Processing (Volume 2: Short Papers)*, pages 246–251.

Karin Kipper, Hoa Trang Dang, Martha Palmer, et al. 2000. Class-based construction of a verb lexicon. *AAAI/IAAI*, 691:696.

Allison Lahnala, Charles Welch, Béla Neuendorf, and Lucie Flek. 2022. Mitigating toxic degeneration with empathetic data: Exploring the relationship between toxicity and empathy. In *Proceedings of the 2022 Conference of the North American Chapter of the Association for Computational Linguistics: Human Language Technologies*, pages 4926–4938, Seattle, United States. Association for Computational Linguistics.

Geoffrey Z Liu, Oliver K Jawitz, Daniel Zheng, Richard J Gusberg, and Anthony W Kim. 2016. Reflective writing for medical students on the surgical clerkship: oxymoron or antidote? *Journal of surgical education*, 73(2):296–304.

Yinhan Liu, Myle Ott, Naman Goyal, Jingfei Du, Mandar Joshi, Danqi Chen, Omer Levy, Mike Lewis, Luke Zettlemoyer, and Veselin Stoyanov. 2019. Roberta: A robustly optimized bert pretraining approach. *arXiv preprint arXiv:1907.11692*.

M. Matheny, S. Thadaney Israni, M. Ahmed, and D. Whicher (editors). 2019. *Artificial Intelligence in Health Care: The Hope, the Hype, the Promise, the Peril*. National Academy of Medicine, Washington, DC.

Linda C Garro; Cheryl Mattingly. 2000. *Narrative and the cultural construction of illness and healing*. Univ. of California Press, Berkeley, California.

Alestair McIntyre. 1981. *After Virtue*. South Bend: University of Notre Dame Press, Notre Dame, IN.

Stewart W Mercer, Margaret Maxwell, David Heaney, and Graham Watt. 2004. The consultation and relational empathy (care) measure: development and preliminary validation and reliability of an empathy-based consultation process measure. *Family practice*, 21(6):699–705.

George A Miller. 1995. Wordnet: a lexical database for English. *Communications of the ACM*, 38(11):39–41.

R Mintz-Binder, MM Jones, et al. 2019. When a clinical crisis strikes: Lessons learned from the reflective writings of nursing students. In *Nursing Forum*, volume 54, pages 345–351.

Jay Mundra, Rohan Gupta, and Sagnik Mukherjee. 2021. WASSA@IITK at WASSA 2021: Multi-task learning and transformer finetuning for emotion classification and empathy prediction. In *Proceedings of the Eleventh Workshop on Computational Approaches to Subjectivity, Sentiment and Social Media Analysis*, pages 112–116, Online. Association for Computational Linguistics.

Jahna Otterbacher, Chee Siang Ang, Marina Litvak, and David Atkins. 2017. Show me you care: Trait empathy, linguistic style, and mimicry on facebook. *ACM Transactions on Internet Technology (TOIT)*, 17(1):1–22.

Verónica Pérez-Rosas, Rada Mihalcea, Kenneth Resnicow, Satinder Singh, and Lawrence An. 2017. Understanding and predicting empathic behavior in counseling therapy. In *Proceedings of the 55th Annual Meeting of the Association for Computational Linguistics (Volume 1: Long Papers)*, pages 1426–1435.

William J Peterson, Joseph B House, Cemal B Sozener, and Sally A Santen. 2018. Understanding the struggles to be a medical provider: view through medical student essays. *The Journal of Emergency Medicine*, 54(1):102–108.

R. M. Frankel. 2000. *The (socio) linguistic turn in physician-patient communication research*. Georgetown University Press, Boston, MA.

Lian T Rameson, Sylvia A Morelli, and Matthew D Lieberman. 2012. The neural correlates of empathy: experience, automaticity, and prosocial behavior. *Journal of cognitive neuroscience*, 24(1):235–245.

Matthew Ratcliffe. 2017. Empathy without simulation. In *Imagination and Social Perspectives*, page 22. Routledge.

Josef Ruppenhofer, Michael Ellsworth, Myriam Schwarzer-Petruck, Christopher R Johnson, and Jan Scheffczyk. 2016. Framenet ii: Extended theory and practice. Technical report, International Computer Science Institute.

Azlaan Mustafa Samad, Kshitij Mishra, Mauajama Firdaus, and Asif Ekbal. 2022. Empathetic persuasion: Reinforcing empathy and persuasiveness in dialogue systems. In *Findings of the Association for Computational Linguistics: NAACL 2022*, pages 844–856, Seattle, United States. Association for Computational Linguistics.

João Sedoc, Sven Buechel, Yehonathan Nachmany, Anneke Buffone, and Lyle Ungar. 2020. Learning word ratings for empathy and distress from document-level user responses. In *Proceedings of the 12th Language Resources and Evaluation Conference*, pages 1664–1673, Marseille, France. European Language Resources Association.

Ashish Sharma, Inna W Lin, Adam S Miner, David C Atkins, and Tim Althoff. 2021. Towards facilitating empathic conversations in online mental health support: A reinforcement learning approach. In *Proceedings of the Web Conference 2021*, pages 194–205.

Ashish Sharma, Adam Miner, David Atkins, and Tim Althoff. 2020. A computational approach to understanding empathy expressed in text-based mental health support. In *Proceedings of the 2020 Conference on Empirical Methods in Natural Language Processing (EMNLP)*, pages 5263–5276, Online. Association for Computational Linguistics.

Shuju Shi, Yinglun Sun, Jose Zavala, Jeffrey Moore, and Roxana Girju. 2021. Modeling clinical empathy in narrative essays. In *2021 IEEE 15th International Conference on Semantic Computing (ICSC)*, pages 215–220.

Swabha Swayamdipta, Sam Thomson, Chris Dyer, and Noah A. Smith. 2017. Frame-semantic parsing with softmax-margin segmental rnns and a syntactic scaffold. *CoRR*, abs/1706.09528.

Himil Vasava, Pramegh Uikey, Gaurav Wasnik, and Raksha Sharma. 2022. Transformer-based architecture for empathy prediction and emotion classification. In *Proceedings of the 12th Workshop on Computational Approaches to Subjectivity, Sentiment & Social Media Analysis*, pages 261–264, Dublin, Ireland. Association for Computational Linguistics.

Thiemo Wambsganss, Anne Höch, Naim Zierau, and Matthias Söllner. 2021a. Ethical design of conversational agents: towards principles for a value-sensitive design. In *International Conference on Wirtschaftsinformatik*, pages 539–557. Springer.

Thiemo Wambsganss, Christina Niklaus, Matthias Söllner, Siegfried Handschuh, and Jan Marco Leimeister. 2021b. Supporting cognitive and emotional empathic writing of students. *arXiv preprint arXiv:2105.14815*.

Sally G Warmington. 2019. *Storytelling encounters as medical education: crafting relational identity*. Routledge.

Sally G. Warmington, May-Lill Johansen, and Hamish Wilson. 2022. Identity construction in medical student stories about experiences of disgust in early nursing home placements: a dialogical narrative analysis. *BMJ open*, 12(2):e051900.

David R Williams and Ronald Wyatt. 2015. Racial bias in health care and health: challenges and opportunities. *Jama*, 314(6):555–556.

Peixiang Zhong, Chen Zhang, Hao Wang, Yong Liu, and Chunyan Miao. 2020. Towards persona-based empathetic conversational models. In *Proceedings of the 2020 Conference on Empirical Methods in Natural Language Processing (EMNLP)*, pages 6556–6566, Online. Association for Computational Linguistics.

Yanmengqian Zhou, Michelle L Acevedo Callejas, Yuwei Li, and Erina L MacGeorge. 2021. What does patient-centered communication look like?: Linguistic markers of provider compassionate care and shared decision-making and their impacts on patient outcomes. *Health Communication*, pages 1–11.

Document-level Condition-Treatment Relation Extraction on Disease-related Social Media Forums

Sichang Tu
Computer Science
Emory University
Atlanta GA 30322 USA
sichang.tu@emory.edu

Stephen Doogan
Real Life Sciences
Wayne PA 19087 USA
sdoogan@rlsciences.com

Jinho D. Choi
Computer Science
Emory University
Atlanta GA 30322 USA
jinho.choi@emory.edu

Abstract

Social media has become a popular platform where people share information about personal healthcare conditions, diagnostic histories, and medical plans. Analyzing posts on social media depicting such realistic information can help improve quality and clinical decision-making; however, the lack of structured resources in this genre limits us to build robust NLP models for meaningful analysis. This paper presents a new corpus annotating relations among many types of conditions, treatments, and their attributes illustrated in social media posts by patients and caregivers. For experiments, a transformer encoder is pretrained on 1M raw posts and used to train several document-level relation extraction models using our corpus. Our best-performing model achieves the F1 scores of 70.9 and 51.7 for Entity Recognition and Relation Extraction, respectively. These results are encouraging as it is the first neural model extracting complex relations of this kind on social media data.

1 Introduction

There is an increasing number of disease-related posts published online every day. On social media platforms such as Reddit and Twitter, people discuss medical conditions and treatments they use to obtain insights from one another. Capturing medical entities and their relations in these real-world data may significantly benefit tasks such as disease detection (Amin et al., 2020), adverse drug event (O'Connor et al., 2014), and pharmacovigilance (Nikfarjam et al., 2015).

Previous studies have established guidelines and corpora focusing on medical mention, chemical-disease relations, and drug-drug interactions (Uzuner et al., 2011; Patel et al., 2018; Schulz et al., 2020). One limitation of most existing corpora is that their data are collected from well-structured medical text, including electronic health records (EHRs), medical discharges, and clinical notes. Models trained on the corpora of formal medical texts may not perform well on the social media data because social media data are noisy (Baldwin et al., 2013) with poor sentence structures and spelling mistakes. An annotated corpus with carefully designed guidelines is necessary to take full advantage of the large-scale disease-related social media data. However, only a few research works contribute to medical text mining in the social media context (Nikfarjam et al., 2015; Jimeno-Yepes et al., 2015; Basaldella et al., 2020), and no work has directly investigated the condition-treatment relation extraction (RE) on social media data.

To bridge the research gap mentioned above, we develop annotation guidelines and address the automatic extraction of medical entities and condition-treatment relations on social media data (Section 2). Our annotation scheme and the new corpus are illustrated in Section 4. We then experiment with joint models between NER and RE using our corpus (Section 5). Finally, a detailed error analysis of the experiment results is provided in Section 6. The contributions of this paper are as follows:

1. We present annotation guidelines that do not require prior medical knowledge. Unlike many existing medical annotation schemes, our guidelines are not restricted to specific conditions or drugs.
2. We introduce an open-access corpus of 1,150 annotated social media posts in terms of 14 entity types and 2 relation types. To the best of our knowledge, this is the first English condition-treatment RE corpus targeting social media posts.
3. We conduct pilot experiments on automatic entity detection and relation extraction, using a state-of-art document-level joint model. With the pre-trained language model on one million medical social media posts, the best F1 scores for entity detection and relation extraction are 70.9 and 51.7.

Proceedings of the 13th International Workshop on Health Text Mining and Information Analysis (LOUHI), pages 218 - 228
December 7, 2022

2 Related Work

2.1 Medical Datasets

Annotated corpora are essential resources for supervised machine learning. With the advance of NLP in the medical domain, there is increasing research on developing reliable medical corpora for various tasks. For Named Entity Recognition (NER), many datasets are restricted to specific tasks (Uzuner et al., 2008; Uzuner, 2009; Uzuner et al., 2010). For example, in n2c2 datasets[1] (originally known as i2b2), one of their subsets, i2b2 medication dataset (Uzuner et al., 2010) only annotates *Medications* and related entities such as *Dosage*, *Frequency*, and *Duration* in discharge summaries. Moreover, the sources of most datasets are discharge summaries, clinical reports, electronic healthcare records, and biomedical literature.

Very few datasets aim to capture medical entities on social media. Karimi et al. (2015) presented CADEC, the first open-access corpus of medical forum posts. Their corpus comprises 1,321 posts, with annotated entities that are linked to medical terms in controlled vocabularies, such as drug names, adverse drug event, disease, and symptoms. However, one limitation of CADEC is that the corpus only covers 12 drugs and their adverse events. Jimeno-Yepes et al. (2015) introduced a corpus of 1300 posts collected on Twitter, with 3 types of entities: *disease*, *pharmacologic substance*, and *symptom*. Furthermore, they experimented with automatic NER and achieved an F1 score ranging from 55% to 66%. Alvaro et al. (2017) collected 2,000 posts from Twitter and PubMed articles by searching 30 drugs. Annotated entities include *drug* in SIDER database, *disease* and *symptom* in the MedDRA ontology. Scepanovic et al. (2020) obtained 1,980 posts from 18 disease-specific subreddits and annotated *symptom/disease* and *drug names*. They further adopted the BiLSTM-CRF model to extract entities and trained a classifier to categorize the Reddit posts on a large scale.

As for relation extraction, even fewer datasets are available. Uzuner et al. (2011) published the i2b2 clinical relation corpus with 871 annotated clinical records. Their corpus captures the relations in terms of the medical problem–treatment, medical problem–test, and medical problem–medical problem. Segura-Bedmar et al. (2013) provided the DDI Corpus, which annotates the drug-drug interaction in 1,017 documents from the DrugBank database and MedLine abstracts. Focusing on radiology reports, Jain et al. (2021) created the RadGraph dataset, which consists of 4 entity types and 4 relation labels. In addition, the authors developed a benchmark model for relation extraction, with a micro F1 score of 82.3/72.9 on two test datasets.

[1]https://portal.dbmi.hms.harvard.edu/projects/n2c2-nlp/

2.2 Medical Text Mining in Social Media

In the past few years, there has been a surge of interest in social media medical text mining, including tasks such as mental illness detection (Jimeno Yepes et al., 2015; Benton et al., 2017; Gkotsis et al., 2017), pharmacovigilance (MacKinlay et al., 2015; Sarker et al., 2016; Correia et al., 2020), and monitoring epidemic (Drinkall et al., 2022). For medical entity extraction in social media, recent studies show that neural network models (Yepes and MacKinlay, 2016; Scepanovic et al., 2020) outperform traditional approaches using conditional random fields or support vector machine.

3 Data

Our data is collected from various social media forums, using the keyword-based method to filter out disease-related posts. The source sites include online support groups, disease forums, message boards, etc. We obtain approximately one million unlabeled social media posts. Table 1 describes the statistics and source site distributions of the data.

4 Annotation Scheme

4.1 Annotation Environment

The annotation platform used for this project is INCEpTION (Klie et al., 2018), a web-based text annotation environment that allows users to create customized annotation layers and import/export documents in various formats. We created one span layer for entities and one relation layer for relations between entities. Each layer is assigned a tagset that controls the possible values for annotation labels (see Section 4.2 and Section 4.3 for details). Figure 3 shows how the post is annotated in INCEpTION. The dataset is exported in the format of WebAnno TSV 3.3 since it supports custom layers. The format captures document properties, including full text, token positions, token offsets, and annotations on custom layers with disambiguation IDs to identify stacked and multi-unit annotations. Appendix A.1 provides a detailed example of exported annotation in TSV format.

Total Posts	1,068,330
Average Word Count	307.9
Source Sites	*Proportion(%)*
dailystrength	9.90
healthboards.com	9.11
mdjunction.com	4.84
cancercompass.com	4.38
netmums.com	4.01
csn.cancer.org	3.91
alzheimers.org.uk	3.59
celiac.com	3.30
psychforums.com	3.05
experienceproject.com	2.87
addforums.com	2.76
forum.childrenwithdiabetes.com	2.68
alzconnected.org	2.56
ehealthforum.com	2.29
inspire.com	2.06
neurotalk.psychcentral.com	1.98
ibsgroup.org	1.79
crohnsforum.com	1.64
diabetes.co.uk	1.55
cancerforums.net	1.53
depresionforums.org	1.49
exchanges.webmd.com	1.32
ourhealth	1.29
diabetesdaily	1.28
reddit.api	1.28
Other (101)	23.5

Table 1: Data statistics. **Source sites**: the data distribution of the top 25 sites and the remaining 101 sites.

4.2 Entity Types

The *Entity* layer tagset contains 14 labels in total, which are further divided into 4 subcategories: *Condition*, *Treatment*, *Attribute*, and *Miscellaneous*.

Condition Generally, condition labels capture the disease and any related symptoms, side effects, or impairment caused by the disease or medication. Depending on whom the sufferer is, we annotate the condition as follows:

- PATIENT CONDITION refers to the condition from which the writer of the passage suffers. 'lupus' in Fig 1a is labeled as PATITENT CONDITION since the sufferer is the writer of the post.
- CAREGIVER CONDITION marks the condition affecting someone the writer of the passage cares for (e.g., family members or friends). We annotate 'tourette's' in Fig 1b as CAREGIVER CONDITION, since the patient is the son of the writer.
- UNSPECIFIED CONDITION appears in the context where the sufferer of the condition is unknown or unclear. Another case of UNSPECIFIED CONDITION happens when the condition is assumed or deduced. In Fig 1c, the sufferer is another user in the previous post threads. Hence, 'PND' is labeled as UNSPECIFIED CONDITION.

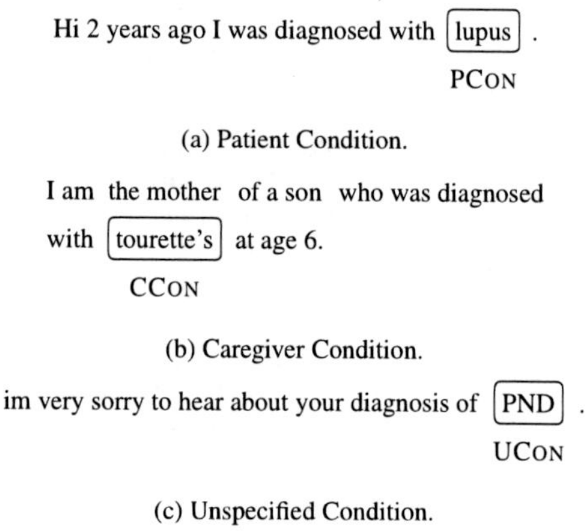

Figure 1: Examples for *Condition* labels.

Treatment Treatment labels annotate medical treatments (e.g., medicine, surgery, or even counseling) performed to deliver healthcare.

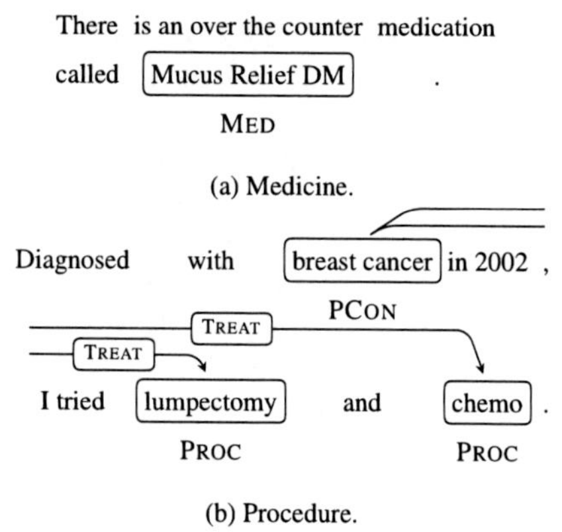

Figure 2: Examples for *Treatment* labels.

- MEDICINE refers to any substance used in treating disease and illness. It could be a drug name, a brand name, or a type of medication. Example is shown in Fig 2a.

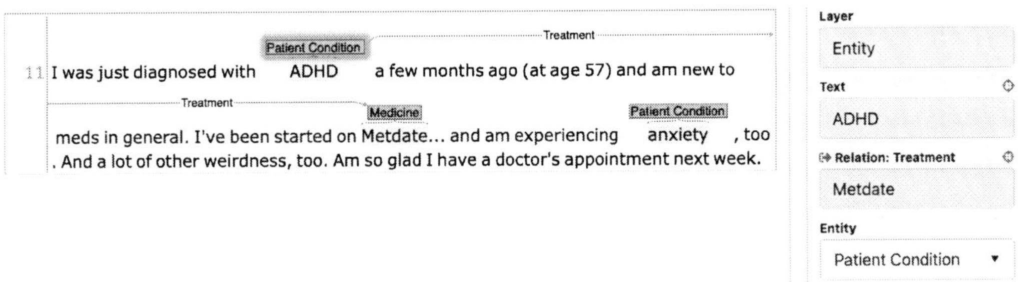

Figure 3: Annotation Interface.

- PROCEDURE marks any medical procedure except for the diagnostic procedure. Common kinds of procedures include surgical procedures (e.g., 'lumpectomy' in Fig 2b) and medical therapy (e.g., 'chemo' in Fig 2b).

Attribute A condition or treatment may have modifiers (usually adjectives or nouns) used attributively to describe them. After carefully examining possible modifier types in the dataset, we conclude 8 attribute labels as follows:

- LOCATION describes where the condition is located or where the treatment happens, such as body parts, anatomical structures, and organs. 'lip' in Fig 4a gives an example for the LOCATION label.
- OBJECT annotates the object to which the treatment is directed. Sometimes it is difficult to distinguish from LOCATION. For example, the 'thyroid' in the second sentence of Fig 4c is labeled as OBJECT since it is the object that was removed. However, the 'thyroid' in the first sentence specifies where the pain occurs and thus is annotated as LOCATION.
- QUANTITY marks the quantity determiner used to specify the condition. It could be concrete numbers (e.g., 'a' in Fig 4a) or quantifiers (e.g., 'several' and 'some').
- COLOR refers to the modifiers that describe the color of the condition. 'yellowish' in Fig 4a gives an example of this label.
- SIZE marks the magnitude and dimension of the condition. It could be linear dimensions (e.g., '2mm' in Fig 4a) or size adjectives (e.g., 'large' and 'small').
- DEGREE shows how severe the condition is, such as disease stages that provides important information on disease development. We label both disease staging (e.g., 'stage IV' in Fig 4b) and adjectives like 'severe' and 'bad' as DEGREE.
- TYPE annotates the specific types of the condition. For instance, 'diabetes' in Fig 4b has three main types, each of which has different symptoms. And the patient is suffering from 'type 2' in the post.
- PROPERTY captures other modifiers that do not fit into the previous attribute labels but provide important properties or characteristics for the condition (e.g., 'shooting' in Fig 4c).

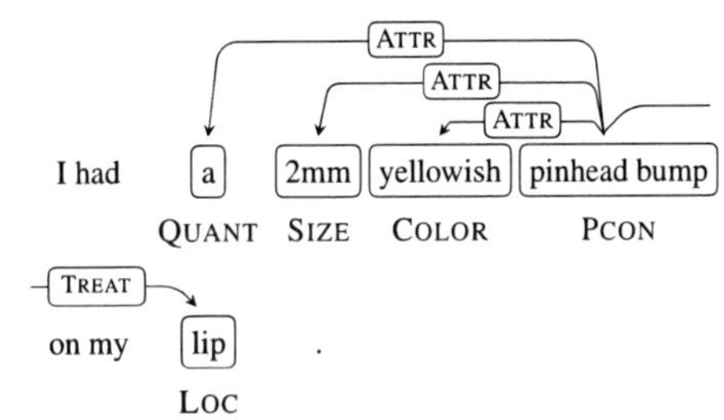

(a) Example for Quantity, Size, Color, and Location.

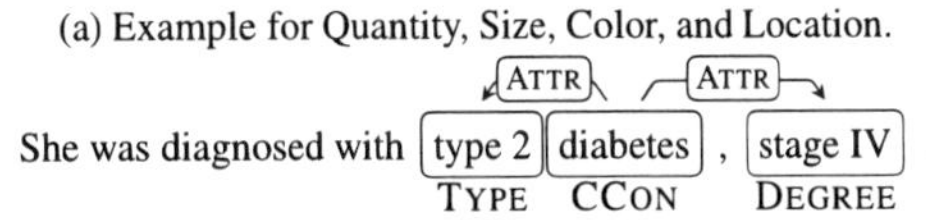

(b) Example for Type and Degree.

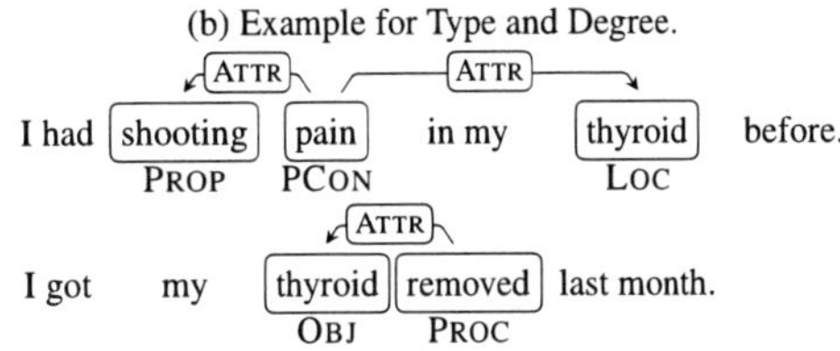

(c) Example for Location, Object, and Property.

Figure 4: Examples for *Attribute* labels.

Note that attribute labels are always attached to corresponding conditions or treatments. Normally, attribute labels would not appear without condition/treatment entities.

Miscellaneous *Miscellaneous* covers entities that do not fit in any of the previous categories, and that may be useful for condition-treatment extraction. Currently, we have one label, PROFILE, under this subcategory. Social media posts on some forums may follow specific conventions, providing additional information after the post content. As shown in Fig 5, the user adds personal information, including username, their relation to the patient, and the patient's medical history to the end of the post. Since it is not a grammatical or complete sentence, we label it as PROFILE separately.

[...post...] Rella. mom to Bredan – 15-yrs-old, dx'd March '08 at 8 years old Navigator CGM since 2/11

PROFILE

Figure 5: Example for PROFILE.

4.3 Relation Types

Apart from entities, we also annotate directed relations between entities, where applicable. The direction of the relationship is always from the governor to the dependent.

- ATTRIBUTE captures relations between condition/treatment labels and their attribute labels. As shown in Fig 4, all ATTRIBUTE relations go from conditions to attributes. Note that ATTRIBUTE relations are usually intra-sentence relations.
- TREATMENT annotates relations between condition labels and their corresponding treatments. The treatment should be attached to the closest condition with an in-going arc. Fig 2b gives example annotations.

4.4 Corpus Analytics

Since the annotation guidelines we developed require no prior medical knowledge, we recruited undergraduates from Computer Science and Linguistics departments. All annotators went through at least three rounds of annotation training before starting annotation. Initially, 2 annotators were invited and asked to test the guidelines on 6 batches of annotation (10~15 posts per batch). We discussed the issues reported and revised the guidelines accordingly. After this pilot phase, another 2 annotators were recruited to expedite the annotation process. All annotations have been examined and curated by one of the authors.

Table 2 displays the Inter-Annotator Agreement (IAA) scores on the final 3 training batches before the single annotation. Previous study on interrater reliability (Hripcsak and Rothschild, 2005) proves that F1 score is preferable for tasks where the negative case count is unknown or undefined. Our annotation task requires annotators to identify entity boundaries, choose entity labels, and connect relations if applicable. In this case, the annotated entities and relations do not contain any negative cases, which makes traditional metrics such as Cohen's Kappa score inapplicable. Furthermore, calculating the Kappa score on the token level may yield either an unfairly high score if including unannotated tokens or an extremely low score if ignoring unannotated tokens(Brandsen et al., 2020). Hence, F-measure is adopted as the evaluation metric for IAA scores. The F1 score is measured between annotations labeled by annotators and ground-truth annotations we created for the training purpose.

	Round 1 (45)		**Round 2 (50)**		**Round 3 (50)**	
	Ent	Rel	Ent	Rel	Ent	Rel
Annotator 1	42.5	15.9	67.0	44.9	75.5	60.0
Annotator 2	44.5	15.6	69.5	57.9	79.6	76.8
Annotator 3	67.5	55.6	66.2	39.1	78.0	53.5
Annotator 4	73.9	55.9	64.1	44.4	76.5	53.9

Table 2: Inter-Annotator Agreement results measured by F1 score. The number of posts annotated in each round is given in the parenthesis.

On average, we reach an IAA score ~77 for *Entity* and ~60 for *Relation*. Though the IAA scores of *Relations* are lower than *Entity*, note that the relation is correct only if the boundaries and labels of two entities and the relation label are exactly the same. It is noticeable that Annotator 1 and 2 obtained F1 scores ~16 for *Relation* in Round 1. It could be explained by the fact that the guidelines were updated after the two annotators finished the pilot phase, and the agreement scores were measured against the improved ground-truth annotations.

To further analyze the results, we examined annotation disagreements. Disease-related social media data poses certain challenges to the annotation process. First, different from discharge notes or electronic health records, the texts in our dataset use casual language with various expressions to describe the condition/treatment rather than structured formal language with unified medical terminologies. This would lead to the inconsistency of the entity annotation. Also, the dataset contains considerable long-distance relations, which poses difficulties for annotators to identify the correct governor/dependent entities. Another challenge for annotators is to distinguish between labels such as LOCATION/OBJECT, and PROPERTY/TYPE.

	Count
Total Posts	1,150
Average Word Count	198.53
Entity	9786
Relation	3645

Table 3: Corpus statistics.

Table 3 presents the corpus statistics. We currently have 1,150 annotated posts with 9,786 entities and 3,645 relations. Detailed statistics on specific labels will be provided in Section 5.

5 Experiments

For automatic entity recognition and relation extraction, we adopt the state-of-art joint model for mention detection, coreference resolution, and relation extraction (Xu and Choi, 2022). Focusing on task interactions between mention detection and relation extraction, the model incorporates graph propagation and graph compatibility, which improves decision-making. Since our dataset does not include coreference annotation, the coreference evaluation is not performed in this paper.

Pretraining Though there are existing pre-trained language models for the medical domain (Lee et al., 2019; Alsentzer et al., 2019), they are trained on biomedical literature, clinical notes, and discharge summaries. Due to the novelty of the dataset, these language models may not provide good representations for online posts due to the different language styles. To take advantage of pre-trained language models, we continue to train 3 models, namely BERT (Devlin et al., 2019), RoBERTa (Liu et al., 2019), and SpanBERT (Joshi et al., 2020), on 1,068,330 disease-related social media posts.

Preprocessing The joint model requires input documents to be segmented into sentences. Since our annotated dataset is not pre-segmented, one additional preprocessing step is necessary before experimenting with the model. Initially, we utilize ELIT tokenizer (He et al., 2021) to segment posts, followed by remapping all token index, offset and label index. However, the tokenizer fails to process some posts due to the following problems: (1) inappropriate spacing (e.g., lacking space between two sentences in '...vomiting.She...'); (2) unknown characters between entities (e.g., '�' in entity 'Alzeheimer�s'); (3) period after digits and abbreviations (e.g., 'Type 1.' or 'MS.'). Therefore, we create rules to filter out and segment the problematic posts.

Iterative Stratify Split Table 4 shows that the dataset is imbalanced, especially for the entity labels. For instance, PATIENT CONDITION has 2949 instances in the corpus, while COLOR only has 12 instances. As a result, the model generalizability may be hindered, if we randomly split the dataset. To avoid a skewed train/dev/test split, we employ the iterative stratification algorithm designed for multi-label data (Sechidis et al., 2011). Detailed sampling statistics are provided in Table 4.

	Train	Dev	Test	Total
Post	859	118	173	1150
Avg length	198.15	207.71	194.13	198.53
Entity				
PCON	2,189	344	416	2,949
UCON	1,259	173	244	1,676
MED	924	125	165	1,214
CCON	835	94	170	1,099
LOC	565	87	104	756
PROC	492	50	111	653
TYPE	375	60	73	508
DEG	309	44	63	416
PROP	165	18	30	213
OBJ	91	12	16	119
QUANT	72	11	16	99
SIZE	39	4	8	51
PROF	16	2	3	21
COLOR	9	1	2	12
Relation				
ATTR	1,644	246	316	2,206
TREAT	1,088	135	216	1,439

Table 4: Statistics for iterative stratify split.

Results Table 5 gives the results of the joint model on entity recognition and relation extraction tasks using 6 different pre-trained language models as the encoder. It is apparent from this table that by using language models pre-trained on the one million unlabeled data, most models achieve better performance, with an increase ranging from 3.9% to 7.8% for entity and from 3.5% to 5.1% for relation. It is not surprising that SpanBERT-large-med

	Entity			Relation		
	P	R	F1	P	R	F1
BERT-large	55.9	70.1	62.2 (±0.80)	33.1	54.5	41.1 (±0.52)
BERT-large-med	66.8	73.6	70.0 (±0.22)	40.0	**57.6**	47.2 (±1.00)
RoBERTa-large	65.9	73.1	69.3 (±0.53)	42.0	52.9	46.7 (±1.22)
RoBERTa-large-med	65.0	72.3	68.4 (±0.10)	35.9	48.4	41.2 (±1.37)
SpanBERT-large	63.4	71.1	67.0 (±0.57)	45.4	51.3	48.2 (±0.34)
SpanBERT-large-med	**67.5**	**74.9**	**70.9** (± 0.51)	**48.85**	55.0	**51.7** (± 0.66)
Merged Labels						
Condition	68.1	**78.8**	**73.0** (±0.98)	45.4	51.4	48.2 (±1.96)
Treatment	**70.1**	75.2	72.4 (±1.34)	**47.1**	**54.1**	**50.3** (±0.35)

Table 5: Experiment results comparing different pre-trained language models. All scores are the average scores based on 3-5 rounds of experiments. The *med* suffix indicates the model is trained on the one million unlabeled data. The *Merged Labels* section gives results on tagsets with merged labels (e.g., 'Condition' means entity labels under *Condition* subcategory are merged into one tag).

reaches the highest F1 scores for both entity (70.9) and relation (51.7), since it provides an improved prediction on spans and is proved to be promising on span selection tasks.

Besides experimenting with pre-trained language models, we also trained models with various merged tagsets. As in the *Merged Labels* section in Table 5, most merged tagset settings bring an increase on entity F1 scores. However, none of the merged label settings outperforms the original label setting in terms of the relation F1 score.

Postprocessing For this task, higher precision scores are preferable to higher recall scores since less false positive output would benefit the subsequent medical decision-making processes. Hence, we attempt to improve the precision score through postprocessing. First, we adjust the top span ratio for entity extraction, which controls the pruning rate of candidate entities according to their mention scores. Top span ratios ranging from 0.4 to 0.1 are tested, which leads to an average of 1~2 percent increase in precision. Then, we filter out singleton[2] attribute entities with no relation attached, which gives a precision increase of ~2 percent.

6 Error Analysis

Further analysis is conducted based on the model prediction on the test dataset. Table 6 displays the breakdown of results using pre-trained SpanBERT-large model. The best F1 scores are obtained on MEDICINE, CAREGIVER CONDITION, and PATIENT CONDITION since most entities in these labels are likely to be medical terms. The relatively low F1 score of UNSPECIFIED CONDITION is due to mislabeling it as PATIENT CONDITION or CAREGIVER CONDITION. The primary reason for the low performance on OBJECT is that the model is prone to predict anatomical structures as LOCATION. The majority of the entities in PROPERTY are common words, such as 'short' and 'double', which leads to a high false positive rate.

	Count(%)		Results		
	Cor	Spu	P	R	F1
Entity					
MED	86.1	5.5	78.0	80.7	79.3
CCON	84.1	9.4	73.7	76.9	75.3
PCON	83.2	13.5	72.9	75.8	74.4
LOC	81.7	15.4	70.2	77.3	73.6
UCON	77.9	11.5	66.2	68.6	67.4
PROC	73.9	23.4	60.3	70.1	64.8
DEG	61.9	11.1	60.0	57.4	58.6
QUANT	68.8	18.8	57.9	57.9	57.9
PROF	66.7	0	66.7	50.6	57.1
TYPE	57.5	5.5	56.8	53.2	54.9
SIZE	62.5	37.5	45.5	62.5	52.6
COLOR	50.0	0	50.0	50.0	50.0
OBJ	37.5	25.0	40.0	35.3	37.5
PROP	33.3	20.0	26.3	28.6	27.4
Relation					
ATTR	-	-	55.7	58.7	57.1
TREAT	-	-	44.0	48.0	45.9

Table 6: SpanBERT-large-med result breakdown. **Count** shows the proportion of correctly predicted entity count and spurious entity count for each label.

[2]Singleton refers to the single entity without ingoing or outgoing relations attached to other entities.

One crucial problem that is observed from the predicted results is the spurious problem. In other words, the model predicts entities that do not exist in the gold annotation. There is a total of 178 spurious entities produced in 173 posts. The majority of predicted spurious entity types are condition labels (100 spurious entities detected) and treatment labels (35 spurious entities detected). The reasons for this problem are threefold:

1. During the annotation process, we do not annotate singletons such as 'pain' and 'problem' unless they have modifiers that are labeled as attributes (e.g., 'thyroid problem'). The model fails to rule out this kind of singletons.
2. Certain terms such as 'B12' could be treatment for diseases (labeled as MEDICINE in 'B12 supplement') or non-entity (as in 'B12 level'). The model fails to distinguish between these two scenarios.
3. Since most attribute entities we labeled could be non-entity in most times, the model is likely to produce false positive responses. Taking 'severe' as an example, the model may mistakenly label it as DEGREE in 'severe situation', since the model has seen many instances (e.g., 'severe anxiety disorder').

Subsequently, relation extraction also suffers from the spurious problem. Since the relation is generated based on the detected entities, the model would predict relations on spurious entities. Moreover, long-distance relations pose challenges to the relation extraction task. For instance, when more than one condition are labeled in the post, the model is prone to attach the treatment to the closer condition rather than the corresponding one.

7 Conclusion

To facilitate medical text mining in the social media context, we develop an annotation scheme of disease-related posts for the condition-relation extraction. Following the guidelines, we present a reliable corpus[3] with 9,785 entities and 3,645 relations, which is a valuable addition to the limited corpora in this field. Additionally, we experiment with automatic entity recognition and relation extraction, providing a promising model for mining online medical posts. We also conduct a detailed error analysis that may shed light on future work.

The findings of our work suggest potential directions for further studies in this domain. Possible progress could be made by increasing the corpus size since the current corpus is relatively small. Also, the model structure could be designed to solve the spurious problem.

[3] `https://github.com/emorynlp/REDSM` We distribute the dev and test dataset and part of the training dataset (add up to 50% of the corpus), as requested by the sponsor.

Acknowledgements

We gratefully acknowledge the support of the Real Life Sciences grant. Any opinions, findings, conclusions, or recommendations expressed in this material are those of the authors and do not necessarily reflect the views of Real Life Sciences.

References

Emily Alsentzer, John Murphy, William Boag, Wei-Hung Weng, Di Jindi, Tristan Naumann, and Matthew McDermott. 2019. Publicly available clinical BERT embeddings. In *Proceedings of the 2nd Clinical Natural Language Processing Workshop*, pages 72–78, Minneapolis, Minnesota, USA. Association for Computational Linguistics.

Nestor Alvaro, Yusuke Miyao, and Nigel Collier. 2017. Twimed: Twitter and pubmed comparable corpus of drugs, diseases, symptoms, and their relations. *JMIR Public Health Surveill*, 3(2):e24.

Samina Amin, M. Irfan Uddin, Saima Hassan, Atif Khan, Nidal Nasser, Abdullah Alharbi, and Hashem Alyami. 2020. Recurrent neural networks with tf-idf embedding technique for detection and classification in tweets of dengue disease. *IEEE Access*, 8:131522–131533.

Timothy Baldwin, Paul Cook, Marco Lui, Andrew MacKinlay, and Li Wang. 2013. How noisy social media text, how diffrnt social media sources? In *Proceedings of the Sixth International Joint Conference on Natural Language Processing*, pages 356–364, Nagoya, Japan. Asian Federation of Natural Language Processing.

Marco Basaldella, Fangyu Liu, Ehsan Shareghi, and Nigel Collier. 2020. COMETA: A corpus for medical entity linking in the social media. In *Proceedings of the 2020 Conference on Empirical Methods in Natural Language Processing (EMNLP)*, pages 3122–3137, Online. Association for Computational Linguistics.

Adrian Benton, Margaret Mitchell, and Dirk Hovy. 2017. Multitask learning for mental health conditions with limited social media data. In *Proceedings of the 15th Conference of the European Chapter of the Association for Computational Linguistics: Volume 1, Long Papers*, pages 152–162, Valencia, Spain. Association for Computational Linguistics.

Alex Brandsen, Suzan Verberne, Milco Wansleeben, and Karsten Lambers. 2020. Creating a dataset for named entity recognition in the archaeology domain. In *Proceedings of the Twelfth Language Resources and Evaluation Conference*, pages 4573–4577, Marseille, France. European Language Resources Association.

Rion Brattig Correia, Ian B. Wood, Johan Bollen, and Luis M. Rocha. 2020. Mining social media data for biomedical signals and health-related behavior. *Annual Review of Biomedical Data Science*, 3(1):433–458.

Jacob Devlin, Ming-Wei Chang, Kenton Lee, and Kristina Toutanova. 2019. BERT: Pre-training of deep bidirectional transformers for language understanding. In *Proceedings of the 2019 Conference of the North American Chapter of the Association for Computational Linguistics: Human Language Technologies, Volume 1 (Long and Short Papers)*, pages 4171–4186, Minneapolis, Minnesota. Association for Computational Linguistics.

Felix Drinkall, Stefan Zohren, and Janet Pierrehumbert. 2022. Forecasting COVID-19 caseloads using unsupervised embedding clusters of social media posts. In *Proceedings of the 2022 Conference of the North American Chapter of the Association for Computational Linguistics: Human Language Technologies*, pages 1471–1484, Seattle, United States. Association for Computational Linguistics.

George Gkotsis, Anika Oellrich, Sumithra Velupillai, Maria Liakata, Tim JP Hubbard, Richard JB Dobson, and Rina Dutta. 2017. Characterisation of mental health conditions in social media using informed deep learning. *Scientific reports*, 7(1):1–11.

Han He, Liyan Xu, and Jinho D. Choi. 2021. Elit: Emory language and information toolkit.

George Hripcsak and Adam S Rothschild. 2005. Agreement, the f-measure, and reliability in information retrieval. *Journal of the American medical informatics association*, 12(3):296–298.

Saahil Jain, Ashwin Agrawal, Adriel Saporta, Steven QH Truong, Du Nguyen Duong, Tan Bui, Pierre Chambon, Yuhao Zhang, Matthew P. Lungren, Andrew Y. Ng, Curtis P. Langlotz, and Pranav Rajpurkar. 2021. Radgraph: Extracting clinical entities and relations from radiology reports.

Antonio Jimeno Yepes, Andrew MacKinlay, and Bo Han. 2015. Investigating public health surveillance using Twitter. In *Proceedings of BioNLP 15*, pages 164–170, Beijing, China. Association for Computational Linguistics.

Antonio Jimeno-Yepes, Andrew D. MacKinlay, Bo Han, and Qiang Chen. 2015. Identifying diseases, drugs, and symptoms in twitter. *Studies in health technology and informatics*, 216:643–647.

Mandar Joshi, Danqi Chen, Yinhan Liu, Daniel S. Weld, Luke Zettlemoyer, and Omer Levy. 2020. SpanBERT: Improving Pre-training by Representing and Predicting Spans. *Transactions of the Association for Computational Linguistics*, 8:64–77.

Sarvnaz Karimi, Alejandro Metke-Jimenez, Madonna Kemp, and Chen Wang. 2015. Cadec: A corpus of adverse drug event annotations. *Journal of Biomedical Informatics*, 55:73–81.

Jan-Christoph Klie, Michael Bugert, Beto Boullosa, Richard Eckart de Castilho, and Iryna Gurevych. 2018. The inception platform: Machine-assisted and knowledge-oriented interactive annotation. In *Proceedings of the 27th International Conference on Computational Linguistics: System Demonstrations*, pages 5–9. Association for Computational Linguistics. Event Title: The 27th International Conference on Computational Linguistics (COLING 2018).

Jinhyuk Lee, Wonjin Yoon, Sungdong Kim, Donghyeon Kim, Sunkyu Kim, Chan Ho So, and Jaewoo Kang. 2019. BioBERT: a pre-trained biomedical language representation model for biomedical text mining. *Bioinformatics*, 36(4):1234–1240.

Yinhan Liu, Myle Ott, Naman Goyal, Jingfei Du, Mandar Joshi, Danqi Chen, Omer Levy, Mike Lewis, Luke Zettlemoyer, and Veselin Stoyanov. 2019. Roberta: A robustly optimized bert pretraining approach.

Andrew MacKinlay, Antonio Jimeno Yepes, and Bo Han. 2015. Identification and analysis of medical entity co-occurrences in twitter. In *Proceedings of the ACM Ninth International Workshop on Data and Text Mining in Biomedical Informatics*, DTMBIO '15, page 22, New York, NY, USA. Association for Computing Machinery.

Azadeh Nikfarjam, Abeed Sarker, Karen O'Connor, Rachel Ginn, and Graciela Gonzalez. 2015. Pharmacovigilance from social media: mining adverse drug reaction mentions using sequence labeling with word embedding cluster features. *Journal of the American Medical Informatics Association*, 22(3):671–681.

Karen O'Connor, Pranoti Pimpalkhute, Azadeh Nikfarjam, Rachel Ginn, Karen L Smith, and Graciela Gonzalez. 2014. Pharmacovigilance on twitter? mining tweets for adverse drug reactions. In *AMIA annual symposium proceedings*, volume 2014, page 924. American Medical Informatics Association.

Pinal Patel, Disha Davey, Vishal Panchal, and Parth Pathak. 2018. Annotation of a large clinical entity corpus. In *Proceedings of the 2018 Conference on Empirical Methods in Natural Language Processing*, pages 2033–2042, Brussels, Belgium. Association for Computational Linguistics.

Abeed Sarker, Karen O'connor, Rachel Ginn, Matthew Scotch, Karen Smith, Dan Malone, and Graciela Gonzalez. 2016. Social media mining for toxicovigilance: automatic monitoring of prescription medication abuse from twitter. *Drug safety*, 39(3):231–240.

Sanja Scepanovic, Enrique Martin-Lopez, Daniele Quercia, and Khan Baykaner. 2020. Extracting medical entities from social media. In *Proceedings of the ACM Conference on Health, Inference, and Learning*, CHIL '20, pages 170–181, New York, NY, USA. Association for Computing Machinery.

Sarah Schulz, Jurica Ševa, Samuel Rodriguez, Malte Ostendorff, and Georg Rehm. 2020. Named entities in medical case reports: Corpus and experiments. In *Proceedings of the 12th Language Resources and Evaluation Conference*, pages 4495–4500, Marseille, France. European Language Resources Association.

Konstantinos Sechidis, Grigorios Tsoumakas, and Ioannis Vlahavas. 2011. On the stratification of multi-label data. In *Machine Learning and Knowledge Discovery in Databases*, pages 145–158, Berlin, Heidelberg. Springer Berlin Heidelberg.

Isabel Segura-Bedmar, Paloma Martínez, and María Herrero-Zazo. 2013. SemEval-2013 task 9 : Extraction of drug-drug interactions from biomedical texts (DDIExtraction 2013). In *Second Joint Conference on Lexical and Computational Semantics (*SEM), Volume 2: Proceedings of the Seventh International Workshop on Semantic Evaluation (SemEval 2013)*, pages 341–350, Atlanta, Georgia, USA. Association for Computational Linguistics.

Özlem Uzuner. 2009. Recognizing Obesity and Comorbidities in Sparse Data. *Journal of the American Medical Informatics Association*, 16(4):561–570.

Özlem Uzuner, Ira Goldstein, Yuan Luo, and Isaac Kohane. 2008. Identifying Patient Smoking Status from Medical Discharge Records. *Journal of the American Medical Informatics Association*, 15(1):14–24.

Özlem Uzuner, Imre Solti, and Eithon Cadag. 2010. Extracting medication information from clinical text. *Journal of the American Medical Informatics Association*, 17(5):514–518.

Özlem Uzuner, Brett R South, Shuying Shen, and Scott L DuVall. 2011. 2010 i2b2/VA challenge on concepts, assertions, and relations in clinical text. *Journal of the American Medical Informatics Association*, 18(5):552–556.

Liyan Xu and Jinho Choi. 2022. Modeling task interactions in document-level joint entity and relation extraction. In *Proceedings of the 2022 Conference of the North American Chapter of the Association for Computational Linguistics: Human Language Technologies*, pages 5409–5416, Seattle, United States. Association for Computational Linguistics.

Antonio Jimeno Yepes and Andrew MacKinlay. 2016. NER for medical entities in Twitter using sequence to sequence neural networks. In *Proceedings of the Australasian Language Technology Association Workshop 2016*, pages 138–142, Melbourne, Australia.

A Appendix

A.1 Annotation Output

Raw Sentence: I had very bad de realisation when I was first diagnosed with schizoaffective disorder. The doctor came to the house and immediately knew what to do. I had to have a massive dose of tranquillisers over three days. It worked very weel.

#Text=I had very bad de realisation when I was first diagnosed with schizoaffective disorder .

1-1	0-1	I	_	_	_
1-2	2-5	had	_	_	_
1-3	6-10	very	_	_	_
1-4	11-14	bad	Degree	Attribute	1-5[1_0]
1-5	15-17	de	Patient Condition[1]	_	_
1-6	18-29	realisation	Patient Condition[1]	_	_
……					
1-13	62-77	schizoaffective	Patient Condition[2]	_	_
1-14	78-86	disorder	Patient Condition[2]	_	_
1-15	86-87	.	_	_	_

#Text=The doctor came to the house and immediately knew what to do .

2-1	88-91	The	_	_	_
……					

#Text=I had to have a massive dose of tranquillisers over three days .

……					
3-9	182-196	tranquillisers	Medicine	Treatment	1-13[2_0]
……					

#Text=It worked very well .

4-1	214-216	It	_	_	_

Figure 6: Exported annotation example after segmentation and remapping (See Section 5). Framed text is the raw text collected from social media forums. Tokens of each sentence have 6 properties: token position (e.g., sentence ID - token ID), token offset, token, entity label, relation label, and disambiguation ID (e.g., governor sentence ID [multi-unit entity ID]).

Integration of Heterogeneous Knowledge Sources for Biomedical Text Processing

Parsa Bagherzadeh and Sabine Bergler
CLaC Labs
Concordia Univesity, Montréal, Canada
{p_bagher, bergler}@cse.concordia.ca

Abstract

Recently, research into bringing outside knowledge sources into current neural NLP models has been increasing. Most approaches that leverage external knowledge sources require laborious and non-trivial designs, as well as tailoring the system through intensive ablation of different knowledge sources, an effort that discourages users to use quality ontological resources. In this paper, we show that multiple large heterogeneous KSs can be easily integrated using a decoupled approach, allowing for an automatic ablation of irrelevant KSs, while keeping the overall parameter space tractable. We experiment with BERT and pre-trained graph embeddings, and show that they interoperate well without performance degradation, even when some do not contribute to the task.

1 Introduction

Integration of external knowledge sources (KSs) is seen as a daunting task by the community. Most KSs like ontologies are large and complex. Thus, a majority of the current efforts focus on leveraging a single task relevant KS using hand-tailored architectures (Goodwin and Demner-Fushman, 2020; Peters et al., 2019; Bagherzadeh et al., 2018). During the design process, knowledge sources are often selected through an ablation study, which is laborious and makes the result task-dependent. Thus for every new task the set of relevant knowledge sources has to be identified with a similar study.

It is possible to ignore the tailoring step and use all available KSs, trusting that the training process will properly weigh the heterogeneous KSs given the internal dynamics of the model. This ideal case requires sufficient training data, but most tasks (like many biomedicial tasks) have only small or moderate-sized training data, a common problem for large, monolithic machine learning systems (Glasmachers, 2017). In those systems all KSs are always contributing their expertise, which can result in decreased rather than improved performance. We explore here a way to integrate several large, heterogeneous KSs with partly overlapping, partly divergent, and possibly even contradictory expertise in such a way that they interoperate well without adaptation, with no resulting performance decrease as well as low parameter implications. We use an integration of six KSs as our experiment system and test it over seven different shared task datasets to assess its robustness. We visualize and inspect the contribution of each KS and analyze the parameter space in detail.

The question is how to integrate multiple heterogeneous KSs so that the same system can be used for multiple, unrelated tasks without manual adaptation and without large overhead. We argue that a system with decoupled modules is suitable for this purpose. Decoupled modules can be activated conditioned on the input, allowing the system to ignore an irrelevant KS and thus preventing performance loss with fewer parameter updates at each training step (Shazeer et al., 2017). In this paradigm, instead of hand-picking KSs, an internal and automatic ablation is performed at each step for all KSs, making it easy to use the same system for different tasks, with the least detrimental effects.

Our KSs consist of the pre-trained language model BERT, as well as six structured knowledge repositories designed for human useage: WordNet, DBpedia, ConceptNet, MeSH, GO, and UMLS. For all but ConceptNet we found open source graph embeddings, and we embed ConceptNet using RDF2Vec (see Section 2).

The recently proposed multi-input RIM framework (Bagherzadeh and Bergler, 2021) comes close to our ideas and we use it here for decoupled integration of our KSs. (Bagherzadeh and Bergler, 2021) showed successful decoupled integration of

Proceedings of the 13th International Workshop on Health Text Mining and Information Analysis (LOUHI), pages 229 - 238
December 7, 2022 ©2022 Association for Computational Linguistics

simple KSs like gazetteer lists that were task appropriate but did not report on experiments with large, structured KSs.

We test the same system on 7 different biomedical shared task datasets and show that our heterogeneous KSs interoperate well and achieve synergy, despite their overlap in coverage. Our results improve on two baselines contributed by the knowledge-enhanced models bioBERT (Lee et al., 2020) and KB-BERT (Hao et al., 2020). The system is competitive with state of the art systems (see Table 1).

2 Heterogeneous knowledge sources

Specialized ontological resources contain quality curated information and are often very large and complex. A graph-based knowledge representation is symbolic and discrete, making it hard to use in a machine learning framework, as most machine learning models prefer conducting computations on continuous data. The past few years have seen several techniques to embed graph structures into vector spaces. Inspired by distributional word representations (Mikolov et al., 2013), where each word is embedded in a low dimensional space, graph embedding models embed a graph into a vector space. In graph embedding models, entities (nodes) and relations (edges) are represented by vectors or matrices (Bordes et al., 2013; Ristoski and Paulheim, 2016).

Inspection of graph embedding models shows that they can capture a fair amount of ontological information. For instance (Nayyeri et al., 2021) show that related concepts are often close to each other in the vector space. We use the following ontologicial resources encoded using a pre-trained graph embedding:

WordNet is a lexical databese that defines word senses by their relations to other senses (Miller, 1995). The most important relation in WordNet is synonymy that is used to group synonymous senses into synsets.

DBpedia (Auer et al., 2007) extracts knowledge from Wikipedia info boxes, providing a large number of facts, largely focused on named entities that have Wikipedia articles.

We use the pre-trained RDF2Vec (Ristoski and Paulheim, 2016) embeddings of WordNet and DBpedia, which are available from KGvec2go website.[1]

ConceptNet is a large multi-lingual graph of general knowledge (Speer et al., 2017). ConceptNet uses closed class of 36 relations. To embed ConceptNet a set of graph embeddings is obtained in-house, using RDF2Vec.

MeSH or Medical Subject Headings is a hierarchical vocabulary, produced by the US National Library of Medicine (NLM) (Lipscomb, 2000). It is used for indexing, cataloging, and searching of biomedical and health-related information in PubMed.[2] MeSH is also embedded using a pre-trained graph embedding called MeSH2Vec (Guo et al., 2020).

GO or Gene Ontology (Ashburner et al., 2000) is a controlled vocabulary that describes gene- and protein-related terms. We use the pre-trained GO2Vec embeddings (Zhong et al., 2019) for encoding the Gene Ontology.

UMLS or Unified Medical Language System (Bodenreider, 2004) is a rich and large semantic network of biomedical vocabularies developed by NLM. UMLS comprises 127 semantic types and 54 semantic relations. Currently UMLS encompases 222 biomedical vocabularies including MeSH, GO, DrugBank, etc. For UMLS, we use the embeddings provided by (Maldonado et al., 2019).

KS	Size	Reference
WordNet	300	(Ristoski and Paulheim, 2016)
ConceptNet	200	In-House
GO	100	(Grover and Leskovec, 2016)
MeSH	64	(Guo et al., 2020)
UMLS	50	(Maldonado et al., 2019)
DBpedia	200	(Ristoski and Paulheim, 2016)

Table 1: Summary of pre-trained graph embeddings used in experiments

Table 1 provides a summary of the pre-trained graph embeddings used in the experiments. In this paper, the pre-trained graph embeddings are used off-the-shelf, without any special adjustments. We do not fine-tune the graph embeddings for three reasons. First, ontological resources represent

[1] http://kgvec2go.org

[2] https://pubmed.ncbi.nlm.nih.gov/

facts that should not be biased depending on the task. In a decoupled approach, the modules are responsible for representation learning and any task-specific adaptations are performed by the modules. Second, using graph embeddings as is enhances reproducibility of the model, as all future replications can use the same embeddings. Third, freezing the pre-trained graph embeddings significantly reduces the number of training parameters (see Section 4.5).

3 Tasks

We choose seven biomedically oriented datasets from different shared task competitions that range from simple classification tasks over multi-label classification and relation extraction to sequence labeling tasks. Comparing results for the same system on such a variety of tasks and datasets (including NER on Spanish!) allows us to be confident that the decoupled integration together with sparse activation in the miRIM architecture successfully avoids interference of the KSs and performance degradation.

BB-Rel or Bacteria Biotope which is part of the BioNLP 2019 challenge focuses on the extraction of two types of relations namely `Lives_In` and `Exhibits` (Bossy et al., 2019). `Lives_In` relations link a microorganism entity to its location. `Exhibits` relations on the other hand link a microorganism entity to a phenotype entity. To evaluate the test predictions we use the online tool provided by the organizers.[3]

ChemProt or BioCreative VI track 5 involves detection of relations between mentions of chemicals and genes/proteins in medical journals (Krallinger et al., 2017). The ChemProt task provides a manually annotated corpus, where domain experts have exhaustively labeled all chemical and gene mentions, and all binary interactions between them corresponding to a specific set of biologically relevant relation types, called ChemProt relation classes (CPRs).

DDI or SemEval 2013 task 9.b (Segura-Bedmar et al., 2013) is a relation extraction task for drug-drug interaction mentions in DrugBank (Wishart et al., 2018) and MedLine abstracts.

HoC or Hallmarks of Cancer (Baker et al., 2015) is a multi-label classification task where zero or more labels are assigned to sentences from PubMed abstracts describing cancer hallmarks. Note that the HoC data set is not pre-spitted into train, development, and test sets. We therefore randomly split the data with 60%, 20%, and 20% ratios for train, development, and test respectively.

LitCov or BioCreative VII track 5 (Chen et al., 2021) concerns multi-label classification of abstracts from Covid-related articles into 7 classes, namely: *Treatment, Mechanism, Prevention, Case Report, Diagnosis, Transmission*, and *Epidemic Forecasting*.

LivNER is a sequence labeling task that requires recognition and classification living things into the two categories HUMAN and SPECIES in Spanish clinical reports. Note that since LivNER is a recent challenge, the gold standard labels for the official test set is not disclosed, thus, we used a hold out test set from the training data.

PPI or BioCreative III Article Classification Task (ACT) is a binary task in which biomedical articles describing protein-protein interactions (PPI) must be identified (Krallinger et al., 2011).

Task	Metric	Train/Dev/Test split
BB-Rel	F1	1000/64/500
ChemProt	F1	1682/612/800
DDI	F1	500/214/191
HoC	F1	10.4k/3.5k/3.5k
LitCov	mac-F1	24.9k/6.2k/2.5k
LivingNER	μF1	500/250/250
PPI	Acc	6280/6000

mac-F1: macro-F1, **μF1**: micro-F1, **Acc**: Accuracy

Table 2: Size and evaluation metric for datasets

Table 2 provides a summary of the biomedical tasks. The tasks differ in their complexity, number of training samples, as well as the type of knowledge they require. The diversity of the biomedical tasks allows to evaluate the efficacy of the decoupled integration of heterogeneous KSs.

[3] http://bibliome.jouy.inra.fr/demo/BioNLP-OST-2019-Evaluation/index.html

4 Experiments

4.1 Decoupled framework

As described in (Bagherzadeh and Bergler, 2021), mi-RIM is an architecture of M decoupled recurrent modules $f_1, \ldots, f_M$, where each module f_m operates on a different input, making it possible to integrate different KSs.

In mi-RIM, each KS_m (for instance a pre-trained model) provides its representation x_t^m for a token at position t to the module f_m. Module f_m selects its input using an attention mechanism:

$$\tilde{x}_t^m = Attention(h_{t-1}^m, X_t^m, X_t^m) \quad (1)$$

where $Attention(h_{t-1}^m, X_t^m, X_t^m)$ is the dot-product attention (Vaswani et al., 2017) with h_{t-1}^m as query and X_t^m as both key and value, and $X_t^m = [\mathbf{0}\,; x_t^m]$, where **0** is an all-zero vector and ; denotes row-level concatenation. This attention mechanism allows a module to ignore the input from a KS by attending more to the null input (the all-zero vector).

Once all modules have selected their input, M sets of attention scores are available. Among the modules, a set of top-k modules with the least attention to the null input are selected as active modules, denoted by $\mathcal{F}_t$. As argued by (Goyal et al., 2019), sparse activity leads to competition among modules which leads to developing more specialized expertise for them. We show that this input selection mechanism allows for an automatic ablation of KSs, identifying and blocking irrelevant ones and thus preventing a module to be updated by its corresponding KS.

The active modules are updated using their selected input to obtain temporary hidden representations $\tilde{h}_t^m$ $(m \in F_t)$:

$$\tilde{h}_t^m = f_m(\tilde{x}_t^m, h_{t-1}^m) \quad m \in \mathcal{F}_t \quad (2)$$

where $f_m(\tilde{x}_t^m, h_{t-1}^m)$ denotes a single recurrence of f_m with $\tilde{x}_t^m$ as input and h_{t-1}^m as previous hidden state. For the inactive modules, the temporary hidden representation is copied from the previous position, in other words, $\tilde{h}_t^m = h_{t-1}^m$.

The active modules then interact with each other via another attention mechanism to obtain their actual hidden representations:

$$h_t^m = Attention(\tilde{h}_t^m, \tilde{H}_t, \tilde{H}_t) \quad m \in \mathcal{F}_t \quad (3)$$

where $\tilde{H}_t = [\tilde{h}_t^1; \ldots; \tilde{h}_t^M]$. The actual hidden state for inactive modules is the same as their temporary hidden state ($h_t^m = \tilde{h}_t^m \; m \notin F_t$).

Because the input selection and interaction mechanisms are attention based and attention can take a variable number of argument representation, new KS modules can be added to an existing model without major changes.

(Bagherzadeh and Bergler, 2021) provided a proof of concept for integration of language models and a few gazetteer lists on simple tweet-related biomedical tasks. Here, instead, we test the decoupled mi-RIM framework on complex tasks and on a more diverse set of KSs.

4.2 Preprocessing and implementation details

We use a GATE pipeline (Cunningham et al., 2002) for preprocessing with CoreNLP (Manning et al., 2014) plugin for tokenization and sentence splittting. For LivNER we use the Spanish version of CoreNLP for preprocessing[4]. The integration of KSs requires minimal prepossessing. Tokens are matched against each ontology using a simple case-insensitive exact match approach, by matching for the longest possible span. The exact matching approach is widely used for incorporating external KSs. For instance (Goodwin and Demner-Fushman, 2020) successfully use exact matching to incorporate information from ConceptNet.

We use the PyTorch library (Paszke et al., 2017) for mi-RIM implementation. We use 7 LSTM modules[5] to accommodate the BERT and the graph embeddings. The hidden size of all modules is set to $d_h = 128$. All models are trained using the Adam optimizer (Kingma and Ba, 2015) with a learning rate of $lr = 5 \times 10^{-6}$.

4.3 Numerical results

Table 3 reports the results. The first 2 rows of the table report the performance of BERT and BERT (frozen) as the sole KSs, forming baselines for the experiments. We use the same system with all knowledge sources for all tasks to observe how the system behaves for widely different tasks with different knowledge requirements. For LivNER,

[4]see `https://stanfordnlp.github.io/CoreNLP/human-languages.html`

[5]see `https://pytorch.org/docs/stable/generated/torch.nn.LSTMCell.html`

KSs	M	k	mic-F1 BB-Rel	mic-F1 ChemProt	F1 DDI	mac-F1 LitCov	F1 PPI	F1 HoC	mic-F1 LivNER
BERT (frozen)	1	1	58.3	68.3	85.7	75.5	68.3	79.1	85.3
BERT	1	1	62.9	74.2	87.3	79.2	70.2	83.1	87.9
BERT (frozen), All Graph Emb.	7	7	64.1	70.9	87.2	79.2	72.6	82.4	88.9
		6	64.7	71.4	87.8	79.7	73.1	82.7	89.3
		5	64.9	72.2	88.3	80.6	73.8	83.4	89.5
		4	66.0	73.4	88.6	81.1	**74.2**	83.8	90.4
		3	**66.1**	**74.1**	**88.9**	**81.3**	73.3	**84.3**	**90.6**
		2	64.7	72.6	87.6	79.1	72.6	82.2	89.4
		1	62.9	70.1	86.3	76.7	70.9	81.6	88.8
BERT, All Graph Emb.	7	7	66.3	76.2	89.8	83.2	73.8	85.6	91.2
		6	66.9	76.8	90.1	84.2	74.1	85.9	91.5
		5	67.4	77.0	90.7	84.7	74.6	86.3	91.8
		4	**67.6**	77.4	91.3	85.1	**75.7**	86.5	92.3
		3	66.8	**78.8**	**91.8**	**85.6**	74.9	**86.9**	**92.8**
		2	66.2	77.3	89.0	83.0	73.1	85.2	91.6
		1	64.5	75.5	88.2	81.9	72.2	84.4	89.7
BioBERT			65.3	75.2	89.9	81.7	73.8	84.6	NA
KB-BERT			65.8	76.1	90.3	81.5	72.7	85.1	NA
SOTA			64.8[1]	77.2[2]	92.2[3]	88.7[4]	NA	NA	NA

1. (Zhang et al., 2019) 2. (Gu et al., 2021) 3. (Luo et al., 2020) 4. (Fang and Wang, 2021)

Table 3: Decoupled Integration of KSs using a mi-RIM. The same system is used for all tasks

which is a Spanish task, we use the Spanish version of BERT (Cañete et al., 2020).

The table indicates the number of modules M: for the simple BERT baseline, there is only one module (namely, BERT). The experimental system has 7 modules, 6 graph embeddings and BERT's language model.

Column 3 indicates the degree of enforced sparsity k. When all modules are active ($k = 7$), all KSs contribute their information and update their corresponding modules. In this case, the system corresponds to a monolithic model and shows small improvements (of 2-5%). This shows that to some extent, the monolithic model can ignore irrelevant KSs using its inner dynamics.

Integration of BERT with the extant graph embeddings never shows loss of performance compared to the respective baseline. This is a strong result, considering how heterogeneous the KSs are and how varied and small the datasets and tasks. The strongest results of the decoupled design with sparse activation are for $k = 3$ or $k = 4$. This can be attributed to the competition among KSs, allowing for their contribution only if they are relevant to the task using input selection. This mitigates the inclusion of irrelevant KSs. For instance, LitCov and DDI tasks do not require knowledge on genes, thus GO is an irrelevant KS. Nevertheless, its inclusion does not lead to a performance decrease for the two tasks compared to the BERT and BERT (frozen) baselines. They are, however, the only two tasks for which SOTA outperforms our experimental system for $k = 3$.

Extreme sparsity ($k = 1$ or $k = 2$) shows lower performance than $k = 3$ but never below the BERT baselines. $k = 1$ is generally lower than $k = 7$ but still close to SOTA performances. This shows that although the system is forced to ablate most of the KSs, it can still find a combination that improves overall performance. Note that $k = 1$ is not equivalent to injecting only a single KS into the system since the miRIM architecture makes decisions at the token level and in certain cases the computation graph for $k = 1$ may include all 7 KSs.

As discussed in the next section, sparsity generally leads to a significant reduction in the number of parameters.

Although LitCov (which has the largest training set) benefits the most from the integration of

KSs compared to its BERT baseline, other tasks with smaller sized training data also show sizable improvements, which are more pronounced with sparse activity of the modules. This demonstrates the benefits of an automatic internal ablation mechanism for integration of large heterogeneous KSs.

In general, a decoupled approach also allows to reuse embeddings of KSs. Consider LivNER, which is a Spanish task. We use the same system as for the English tasks and only replace BERT with its Spanish version. Note that a language model trained on Spanish text has significantly different representations compared to its English version, however, as the results suggest, it interoperates well with the other (English) KSs. This recommends the approach also for underresourced languages.

The pre-trained graph embeddings also interoperate well with frozen BERT. The results show that once integrated with frozen BERT (which has no fine tuning on the target task datasets), the lexical information in the knowledge sources effectively compensates for the loss. In most cases, integration of the off-the-shelf KSs with frozen BERT outperforms fine-tuned BERT significantly with almost 100M less parameters. This is very attractive for training on small or moderated-sized data, with less potential for overfitting (Li et al., 2021) or in resource limited situations.

Table 3 also reports on other knowledge enhanced models such as BioBERT (Lee et al., 2020) and KB-BERT (Hao et al., 2020), as well as the state-of-the-art (SOTA). With sparse activity ($k = 4$ or $k = 3$), integration of lexical KSs with BERT always outperforms both BioBERT and KB-BERT, showing that the automatic ablation of discrete KSs is competitive with domain specific pre-training.

Note that the k values for best-preforming settings fall within an arrow interval ($k = 3$ or $k = 4$), suggesting that automatic mechanisms can be used to determine k during training.

4.4 Analysis of results

In precision-oriented applications such as biomedical tasks, users require to understand why and how a prediction is made (Amini and Kosseim, 2019). In a decoupled approach, the activity of each module is often transparent for inspection. Likewise, in mi-RIM, contributions of KS are transparent. Each module selects its input from its corresponding KS using an attention mechanism and if the input is deemed relevant, the module has a high chance of activation. The activation patterns can be traced, providing insight into the functionality of the system. Consider Example 1 (from HoC task):

(1) *Unlike insulin, ghrelin inhibited Akt kinase activity as well as up-regulated gluconeogenesis*

In this example, the term *gluconeogenesis* is matched with UMLS, MeSH, GO, ConceptNet, and DBpedia. Note that BERT also provides a representation for the term. Figure 1 shows the activation patterns of mi-RIM for Example 1. The gray regions indicate activity for a module.

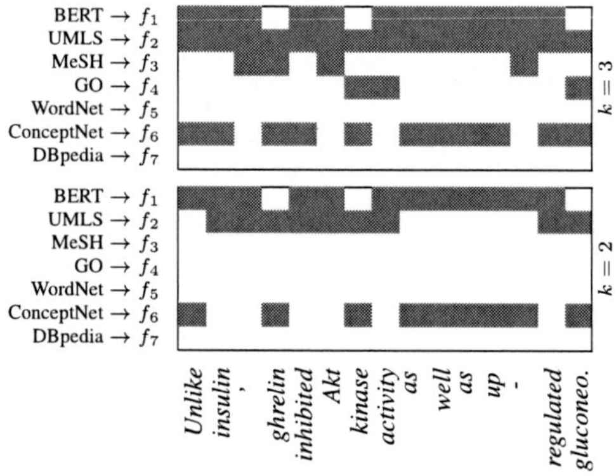

Figure 1: Activation patterns of mi-RIM for Example 1

For the term *gluconeogenesis*, when $k = 3$, modules f_2, f_4, and f_6 (corresponding to UMLS, GO, and ConceptNet respectively) win the competition and are active. Note that the model has selected a very specialized KS for genes (GO), a more comprehensive KS (UMLS), and a general KS (ConceptNet). This suggests that the model is trying to balance the expertise of active KSs. In this light, the activity of ConceptNet versus the inactivity of MeSH is interesting where the general resource ConceptNet is selected over the more specialized MeSH. A similar pattern is also observed when $k = 2$, where ConceptNet is selected over GO, suggesting that it is a more robust resource.

The activation patterns suggest that an automatic and internal ablation is performed by the decoupled model. This suggests that an established system of M KSs can be used for different tasks without pre-ablating relevant KSs because contributions of irrelevant KSs are mitigated by input selection.

4.5 Parameter space and inference time

Let Θ_{mod} denote the set of training parameters implicated by all modules and $|\Theta_{mod}|$ denotes the overall number of parameters. Due to conditional computation in mi-RIM, the number of trained parameters $|\Theta_{mod}|'$ (sample-wise) is linked to the value of k. If $k = M$, all modules are part of the computation graph, i.e. all parameters are trained: $|\Theta_{mod}|' = |\Theta_{mod}|$.

However, when $k < M$ (sparse activity) $\frac{k}{M}|\Theta_{mod}| \leq |\Theta_{mod}|' \leq |\Theta_{mod}|$. The best case ($\frac{k}{M}|\Theta_{mod}|$) occurs when $M-k$ modules are never active and thus not included in the computation graph. The worst case ($|\Theta_{mod}|$) on the other hand occurs, when all modules are active at least for a single position t, forcing all to be included in the computation graph.

Consider the activation patterns of Figure 1 when $k = 3$. Module f_4 (corresponding to GO) is active only at three positions, leading to the inclusion of the module in the computation graph. Moreover, module f_3 (corresponding to MeSH) shows activity for four positions. Although the top-k activity is set to 3, overall, 5 modules demonstrate activity for at least one position. In this case, $|\Theta_{mod}|' = \frac{5}{7}|\Theta_{mod}|$. Note that the best case when $k = 3$, is $|\Theta_{mod}|' = \frac{3}{7}|\Theta_{mod}|$. Although more reduction is expected with smaller values of k, it is possible that all modules demonstrate activity at least for one position even if $k = 1$.

Figure 2 shows a comparison of the fraction of trained parameters $\frac{|\Theta_{mod}|'}{|\Theta_{mod}|}$ for two different tasks. Sparse activity consistently reduces the number of trained parameters. Note that on average, the fraction of trained parameters never approaches its best case ($\frac{k}{M}$). For instance, when $k = 1$, for HoC, $\frac{|\Theta_{mod}|'}{|\Theta_{mod}|} = 0.46$ while the best case is about 0.14. This shows that on average 3.2 modules show activity at least for one position even tough $k = 1$.

The reported experiments showed that most runs demonstrate their best performance when $k = 4$ or 3. As Figure 2 shows, on average, when $k = 4$ and $k = 3$, 67% and 52% of parameters are trained respectively. This shows that while improving performance, sparse activity can significantly reduce the number of trained parameters.

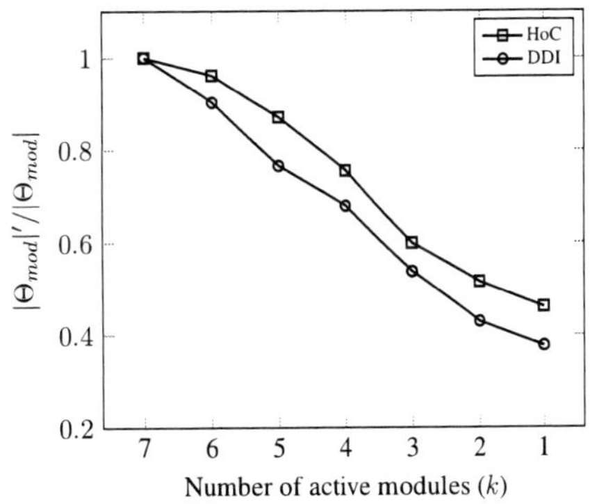

Figure 2: Fraction of trained parameters vs number of active modules

The reduced parameter space allows for training on small or moderate-sized data sets with less potentials for over-fitting (Li et al., 2021).

A brief analysis of the inference time is also provided in Figure 3. We measure the inference time for different values of top-k activity. Note that the reported inference time is the average timing on all tasks, timed on an Intel Corei7 CPU.

As Figure 3 shows, sparse activity significantly reduces the inference time. This is expected since once a KS is not selected, there is no need to update its corresponding module, leading to speed-up in the inference time.

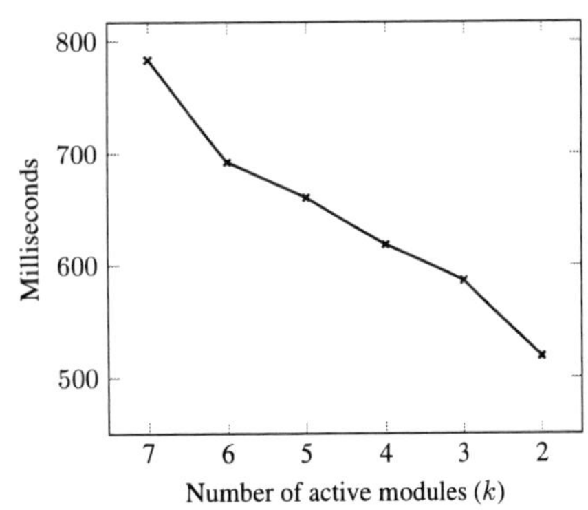

Figure 3: Inference time of mi-RIM with 7 KSs/modules for different k values

5 Conclusion

This paper presents extensive experiments on decoupled integration of heterogeneous KSs such as language models and pre-trained graph embeddings. The same system with all KSs was used for all tasks, without special calibrations, demonstrating reusability of extant knowledge sources.

The tasks differed in terms of complexity as well as their knowledge requirements (specialized or general knowledge). The results show that for the tasks considered here, the KSs interoperate well and they do not confound each other's performances. Moreover, we showed that a system that leverages multiple KS does not necessarily show significant improvement, rather the sparse activity of modules is required to effectively improve performance.

Inspection of activation patterns shows that a decoupled system can ignore irrelevant/redundant KSs, showing an automatic ablation behavior.

We show that in terms of the number of trained parameters, a decoupled approach is efficient. The sparse activity significantly reduces the number of trained parameters. Moreover, since the pre-trained graph embeddings are not fine-tuned, the overall model does not have large parameter implications.

We also stress the ease of reusing and replicating such a decoupled system, since the same pre-trained embeddings will be used by different users. Moreover, the pre-trained embeddings do not have to be stored on the same machine that the model is trained on. KGvec2go[6], for instance, provides an API through which pre-trained embeddings are accessible. This ultimately results in lightweight models.

In conclusion, a decoupled approach allows for robust and efficient integration of heterogeneous KSs, allowing the user to leverage multiple knowledge sources, without any need for special calibration or tailoring.

References

Hessam Amini and Leila Kosseim. 2019. Towards explainability in using deep learning for the detection of anorexia in social media. In *Natural Language Processing and Information Systems*, pages 225–235. Springer International Publishing.

Michael Ashburner, Catherine A Ball, Judith A Blake, David Botstein, Heather Butler, J Michael Cherry, Allan P Davis, Kara Dolinski, Selina S Dwight, Janan T Eppig, et al. 2000. Gene ontology: tool for the unification of biology. *Nature genetics*, 25(1):25–29.

Sören Auer, Christian Bizer, Georgi Kobilarov, Jens Lehmann, Richard Cyganiak, and Zachary Ives. 2007. Dbpedia: A nucleus for a web of open data. In *The semantic web*, pages 722–735. Springer.

Parsa Bagherzadeh and Sabine Bergler. 2021. Multi-input Recurrent Independent Mechanisms for leveraging knowledge sources: Case studies on sentiment analysis and health text mining. In *Proceedings of Deep Learning Inside Out (DeeLIO): The 2nd Workshop on Knowledge Extraction and Integration for Deep Learning Architectures*, pages 108–118.

Parsa Bagherzadeh, Nadia Sheikh, and Sabine Bergler. 2018. CLaC at SMM4H task 1, 2, and 4. In *Proceedings of the 2018 EMNLP Workshop SMM4H: The 3rd Social Media Mining for Health Applications Workshop & Shared Task.*

Simon Baker, Ilona Silins, Yufan Guo, Imran Ali, Johan Högberg, Ulla Stenius, and Anna Korhonen. 2015. Automatic semantic classification of scientific literature according to the hallmarks of cancer. *Bioinformatics*, 32(3):432–440.

Olivier Bodenreider. 2004. The Unified Medical Language System (UMLS): Integrating Biomedical Terminology. *Nucleic acids research*, 32(suppl_1):D267–D270.

Antoine Bordes, Nicolas Usunier, Alberto Garcia-Duran, Jason Weston, and Oksana Yakhnenko. 2013. Translating Embeddings for Modeling Multi-Relational Data. In *Neural Information Processing Systems (NIPS)*, pages 1–9.

Robert Bossy, Louise Deléger, Estelle Chaix, Mouhamadou Ba, and Claire Nédellec. 2019. Bacteria biotope at BioNLP open shared tasks 2019. In *Proceedings of The 5th Workshop on BioNLP Open Shared Tasks*, pages 121–131.

José Cañete, Gabriel Chaperon, Rodrigo Fuentes, Jou-Hui Ho, Hojin Kang, and Jorge Pérez. 2020. Spanish Pre-Trained BERT Model and Evaluation Data. In *PML4DC at ICLR 2020.*

Qingyu Chen, Alexis Allot, and Zhiyong Lu. 2021. Litcovid: an open database of covid-19 literature. *Nucleic acids research*, 49(D1):D1534–D1540.

Hamish Cunningham, Diana Maynard, Kalina Bontcheva, and Valentin Tablan. 2002. GATE: A Framework and Graphical Development Environment for Robust NLP tools and applications. In *Proceedings of the 40th Anniversary Meeting of the Association for Computational Linguistics (ACL'02).*

Li Fang and Kai Wang. 2021. Team bioformer at biocreative vii litcovid track: Multic-label topic classification for covid-19 literature with a compact bert model. In *Proceedings of the seventh BioCreative challenge evaluation workshop.*

Tobias Glasmachers. 2017. Limits of end-to-end learning. In *Asian Conference on Machine Learning*, pages 17–32.

[6]http://kgvec2go.org/

Travis Goodwin and Dina Demner-Fushman. 2020. Enhancing question answering by injecting ontological knowledge through regularization. In *Proceedings of Deep Learning Inside Out (DeeLIO): The First Workshop on Knowledge Extraction and Integration for Deep Learning Architectures*, pages 56–63.

Anirudh Goyal, Alex Lamb, Jordan Hoffmann, Shagun Sodhani, Sergey Levine, Yoshua Bengio, and Bernhard Schölkopf. 2019. Recurrent independent mechanisms. *arXiv preprint arXiv:1909.10893*.

Aditya Grover and Jure Leskovec. 2016. node2vec: Scalable feature learning for networks. In *Proceedings of the 22nd ACM SIGKDD international conference on Knowledge discovery and data mining*, pages 855–864.

Yu Gu, Robert Tinn, Hao Cheng, Michael Lucas, Naoto Usuyama, Xiaodong Liu, Tristan Naumann, Jianfeng Gao, and Hoifung Poon. 2021. Domain-specific language model pretraining for biomedical natural language processing. *ACM Transactions on Computing for Healthcare (HEALTH)*, 3(1):1–23.

Zhen-Hao Guo, Zhu-Hong You, De-Shuang Huang, Hai-Cheng Yi, Kai Zheng, Zhan-Heng Chen, and Yan-Bin Wang. 2020. MeSHHeading2vec: a new method for representing MeSH headings as vectors based on graph embedding algorithm. *Briefings in Bioinformatics*, 22(2):2085–2095.

Boran Hao, Henghui Zhu, and Ioannis Paschalidis. 2020. Enhancing clinical BERT embedding using a biomedical knowledge base. In *Proceedings of the 28th International Conference on Computational Linguistics*, pages 657–661.

Diederik P Kingma and Jimmy Lei Ba. 2015. Adam: A method for stochastic optimization. In *Proceedings of the 3rd International Conference on Learning Representations, ICLR'15*.

Martin Krallinger, Obdulia Rabal, Saber A Akhondi, Martın Pérez Pérez, Jesús Santamaría, GP Rodríguez, et al. 2017. Overview of the biocreative vi chemical-protein interaction track. In *Proceedings of the sixth BioCreative challenge evaluation workshop*, volume 1, pages 141–146.

Martin Krallinger, Miguel Vazquez, Florian Leitner, David Salgado, Andrew Chatr-Aryamontri, Andrew Winter, Livia Perfetto, Leonardo Briganti, Luana Licata, Marta Iannuccelli, et al. 2011. The protein-protein interaction tasks of biocreative iii: classification/ranking of articles and linking bio-ontology concepts to full text. *BMC bioinformatics*, 12(8):1–31.

Jinhyuk Lee, Wonjin Yoon, Sungdong Kim, Donghyeon Kim, Sunkyu Kim, Chan Ho So, and Jaewoo Kang. 2020. Biobert: a pre-trained biomedical language representation model for biomedical text mining. *Bioinformatics*, 36(4):1234–1240.

Zhu Li, Zhi-Hua Zhou, and Arthur Gretton. 2021. Towards an understanding of benign overfitting in neural networks. *arXiv preprint arXiv:2106.03212*.

Carolyn E Lipscomb. 2000. Medical subject headings (MeSH). *Bulletin of the Medical Library Association*, 88(3).

Ling Luo, Zhihao Yang, Mingyu Cao, Lei Wang, Yin Zhang, and Hongfei Lin. 2020. A neural network-based joint learning approach for biomedical entity and relation extraction from biomedical literature. *Journal of biomedical informatics*, 103:103384.

Ramon Maldonado, Meliha Yetisgen, and Sanda M Harabagiu. 2019. Adversarial learning of knowledge embeddings for the unified medical language system. *AMIA Summits on Translational Science Proceedings*, 2019:543.

Christopher D. Manning, Mihai Surdeanu, John Bauer, Jenny Finkel, Steven J. Bethard, and David McClosky. 2014. The Stanford CoreNLP natural language processing toolkit. In *Association for Computational Linguistics (ACL) System Demonstrations*, pages 55–60.

Tomas Mikolov, Wen-tau Yih, and Geoffrey Zweig. 2013. Linguistic regularities in continuous space word representations. In *Proceedings of the 2013 Conference of the North American Chapter of the Association for Computational Linguistics: Human Language Technologies*, pages 746–751.

George A Miller. 1995. Wordnet: a lexical database for english. *Communications of the ACM*, 38(11):39–41.

Mojtaba Nayyeri, Sahar Vahdati, Can Aykul, and Jens Lehmann. 2021. 5* knowledge graph embeddings with projective transformations. In *AAAI 2021*. AAAI Press.

Adam Paszke, Sam Gross, Soumith Chintala, Gregory Chanan, Edward Yang, Zachary DeVito, Zeming Lin, Alban Desmaison, Luca Antiga, and Adam Lerer. 2017. Automatic differentiation in PyTorch. In *Neural Information Processing Systems (NIPS)*.

Matthew E. Peters, Mark Neumann, Robert Logan, Roy Schwartz, Vidur Joshi, Sameer Singh, and Noah A. Smith. 2019. Knowledge enhanced contextual word representations. In *Proceedings of the 2019 Conference on Empirical Methods in Natural Language Processing and the 9th International Joint Conference on Natural Language Processing (EMNLP-IJCNLP)*, pages 43–54.

Petar Ristoski and Heiko Paulheim. 2016. RDF2Vec: RDF graph embeddings for data mining. In *International Semantic Web Conference*, pages 498–514. Springer.

Isabel Segura-Bedmar, Paloma Martínez, and María Herrero-Zazo. 2013. SemEval-2013 task 9 : Extraction of drug-drug interactions from biomedical texts

(DDIExtraction 2013). In *Second Joint Conference on Lexical and Computational Semantics (*SEM), Volume 2: Proceedings of the Seventh International Workshop on Semantic Evaluation (SemEval 2013)*, pages 341–350.

Noam Shazeer, Azalia Mirhoseini, Krzysztof Maziarz, Andy Davis, Quoc Le, Geoffrey Hinton, and Jeff Dean. 2017. Outrageously large neural networks: The sparsely-gated mixture-of-experts layer. *arXiv preprint arXiv:1701.06538*.

Robyn Speer, Joshua Chin, and Catherine Havasi. 2017. Conceptnet 5.5: An open multilingual graph of general knowledge. In *"Proceedings of AAAI 2017 Conference on Artificial Intelligence*, pages 4444–4451.

Ashish Vaswani, Noam Shazeer, Niki Parmar, Jakob Uszkoreit, Llion Jones, Aidan N Gomez, Łukasz Kaiser, and Illia Polosukhin. 2017. Attention is all you need. In *Advances in Neural Information Processing Systems*, pages 5998–6008.

David S Wishart, Yannick D Feunang, An C Guo, Elvis J Lo, Ana Marcu, Jason R Grant, Tanvir Sajed, Daniel Johnson, Carin Li, Zinat Sayeeda, et al. 2018. Drugbank 5.0: a major update to the drugbank database for 2018. *Nucleic acids research*, 46(D1):D1074–D1082.

Qi Zhang, Chao Liu, Ying Chi, Xuansong Xie, and Xiansheng Hua. 2019. A multi-task learning framework for extracting bacteria biotope information. In *Proceedings of The 5th Workshop on BioNLP Open Shared Tasks*.

Xiaoshi Zhong, Rama Kaalia, and Jagath C Rajapakse. 2019. Go2vec: transforming go terms and proteins to vector representations via graph embeddings. *BMC genomics*, 20(9):1–10.

Author Index